Integrative Approach to Fungal Infections and Bioprospecting for Drug Discovery

Research into plant-based antifungals offers a viable alternative to discover potent, eco-friendly, and sustainable treatments. Natural antifungal compounds offer a promising solution, as they tend to have fewer side effects and reduced risk of resistance development compared to synthetic alternatives. This book delves into the application of metagenomics in identifying and harnessing the microbial communities associated with medicinal plants. By integrating bioprospecting techniques with cutting-edge genomic technologies, this work opens up promising new avenues to provide insights into the potential of natural products for developing novel antifungal therapies.

Key Features:

- Combines bioprospecting, microbiology, plant science, and genomic technologies to unlock the therapeutic potential of medicinal plants for antifungal applications.
- Contains comprehensive coverage of metagenomics tools and advanced bioprocessing methods.
- Covers the practical applications for drug development and sustainable solutions for emerging fungal threats.
- Addresses the urgent global health challenge posed by emerging and resistant fungal pathogens, providing new strategies and potential solutions through the bioprospecting of plant-associated microbes.
- Provides an innovative approach to antifungal discovery and focuses on emerging fungal infections.

Medicinal Plants and Natural Products for Human Health

Series Editor: Christophe Wiart

We are at a point in history where the global population is increasingly interested in medicinal plants, natural products, and herbalism. This interest has accelerated with COVID-19 and long COVID symptoms. Rediscovery of the connection between plants and health is responsible for a new generation of botanical therapeutics that include plant-derived pharmaceuticals, multicomponent botanical drugs, dietary supplements, functional foods, and plant-produced recombinant proteins. Many of these products can complement conventional pharmaceuticals in the treatment, prevention, and diagnosis of diseases. This series is a critical reference for anyone involved in the discovery of biopharmaceuticals for the improvement of human health.

Alternative Medicines for Diabetes Management: Advances in Pharmacognosy and Medicinal Chemistry
Edited by: Varma H. Rambaran and Nalini K. Singh

Therapeutic Perspectives of Tea Compounds: Potential Applications against COVID-19
Authored by: Farnoosh Dairpoosh and Kianoosh Dairpoosh

Biologically Active Substances Usable in Food, Pharmaceutical and Agrobiological Fields
Authored by: Zeno Garban and Gheorghe Ilia

Integrative Approach to Fungal Infections and Bioprospecting for Drug Discovery

Volume 1

Edited by
Eti Sharma, Rekha Puria, and Karan Kumar

CRC Press
Taylor & Francis Group
Boca Raton London New York

CRC Press is an imprint of the
Taylor & Francis Group, an **informa** business

First edition published 2026
by CRC Press
2385 NW Executive Center Drive, Suite 320, Boca Raton FL 33431

and by CRC Press
4 Park Square, Milton Park, Abingdon, Oxon, OX14 4RN

CRC Press is an imprint of Taylor & Francis Group, LLC

Library of Congress Cataloging-in-Publication Data
Names: Sharma, Eti editor | Puria, Rekha editor | Kumar, Karan editor
Title: Integrative approach to fungal infections and bioprospecting for
drug discovery / edited by Eti Sharma, Rekha Puria, Karan Kumar.
Description: First edition. | Boca Raton, FL : CRC Press, 2026. |
Series: Medicinal plants and natural products for human health |
Includes bibliographical references and index
Identifiers: LCCN 2025051634 | ISBN 9781041061380 v. 1 hardback |
ISBN 9781041061397 v. 1 paperback | ISBN 9781003633952 v. 1 ebook |
ISBN 9781041269625 v. 2 hardback | ISBN 9781041269571 v. 2 paperback |
ISBN 9781003752196 v. 2 ebook
Subjects: LCSH: Antifungal agents | Antifungal agents—Development |
Mycoses—Treatment | Materia medica, Vegetable | Integrative medicine
Classification: LCC RS431.A62 I58 2026
LC record available at https://lccn.loc.gov/2025051634

ISBN: 9781041061380 (hbk)
ISBN: 9781041061397 (pbk)
ISBN: 9781003633952 (ebk)

DOI: 10.1201/9781003633952

Typeset in Times
by codeMantra

Contents

PART 2 Comprehending Pathobiology of Fungus

Preface

Fungal infections have increasingly emerged as a serious global health concern. The World Health Organization (WHO) has recognized fungal pathogens as a significant worldwide health priority, highlighting the growing threat they pose to human health, agriculture, and the environment. Despite considerable advances in medicine and microbiological research, the incidence of invasive and multidrug-resistant fungal infections continues to rise. This alarming trend calls for greater awareness, surveillance, and coordinated scientific efforts to understand and mitigate the spread of fungal diseases.

The challenge is further compounded by the widespread development of antifungal resistance. Many existing antifungal agents are rapidly losing their efficacy due to misuse, over-prescription, and the remarkable adaptability of fungal species. Consequently, there is an urgent need to explore alternative and sustainable therapeutic strategies. The discovery of novel antifungal compounds from medicinal plants, endophytes, and microbial metabolites, coupled with biotechnological, computational, and metagenomic innovations, represents a promising frontier in the global fight against fungal pathogens.

This book, titled *Integrative Approach to Fungal Infections and Bioprospecting for Drug Discovery*, has been conceptualized to bridge traditional ethnobotanical knowledge with cutting-edge scientific approaches. It provides a comprehensive exploration of fungal infection biology, antifungal mechanisms, and the discovery pipeline of bioactive compounds, integrating perspectives from plant science, microbiology, pharmacognosy, bioinformatics, and nanotechnology.

Volume I introduces the fundamental aspects of fungal infections and antifungal strategies. *Part I: Medicinal Plants, Antifungals, and Metagenomics* discusses the global burden of fungal infections, the antifungal potential of medicinal plants and antimicrobial peptides, the contribution of endophytes, and the role of artificial intelligence in identifying and managing fungal pathogens such as black fungus. *Part II: Comprehending Pathobiology of Fungus* delves into the genetic and molecular basis of antifungal resistance, environmental interactions, biofilm-mediated virulence, and metagenomic insights into host–microbiome relationships.

Volume II extends these concepts into applied and translational domains. *Part I: Medicinal Plants Bioprospecting and Advanced Bioprocessing Techniques* covers downstream processing, fermentation optimization, and computational modeling for antifungal metabolite production. *Part II: Integrative Approaches and Future Applications* highlights recent advancements in biotechnology, nanotechnology, and legal frameworks governing antifungal bioprospecting—emphasizing the pathway from "metagenomics to market."

This book is designed for a broad readership that spans academic, clinical, and industrial domains. It serves as a valuable reference for researchers and academicians in microbiology, mycology, plant biotechnology, pharmacognosy, and pharmaceutical sciences; graduate and postgraduate students seeking a detailed understanding of fungal biology, antifungal mechanisms, and bioprospecting; biotechnologists and industry professionals involved in natural product discovery, bioactive compound formulation, and bioprocess development; healthcare and clinical researchers working on antifungal resistance, infection control, and therapeutic innovation; and policymakers and regulatory experts interested in sustainable bioprospecting and bioethics in antifungal drug development. By integrating ancient medicinal plant knowledge with modern omics technologies, nanotechnology, and artificial intelligence, this book provides a holistic reference to guide researchers and innovators addressing the multifaceted challenges of fungal infections and antifungal drug resistance.

We extend our heartfelt gratitude to all contributing authors, reviewers, and collaborators whose expertise and dedication have enriched this work. Special thanks are also due to the editorial and production teams for their continuous support in bringing this compilation to completion. We hope that this book will serve as both a scientific reference and an inspiration for the next generation of researchers exploring the untapped potential of nature in combating fungal diseases.

Acknowledgment for Reviewers

We sincerely acknowledge the valuable contributions of all reviewers who have generously devoted their time and expertise to evaluate the chapters included in this volume. Their insightful comments and constructive suggestions have significantly enhanced the scientific quality and clarity of the content. We deeply appreciate their support in maintaining the academic integrity of this publication.

List of Reviewers

Dr. Anjana Rustagi
Associate Professor
Department of Botany
Gargi College
University of Delhi
New-Delhi, India

Dr. Shubhashree Mahallik
Assistant Professor
Department of Bioscience and Biotechnology
Fakir Mohan University
Balasore
Orissa, India

Dr. Sharda Pasricha
Professor
Department of Chemistry
Sri Venkateswara College
University of Delhi
New-Delhi, India

Romila Manchanda
Research Assistant II
Michigan State University
United States

Dr. Richa Dubey
Project scientist
CSIR-Indian Institute of Toxicology Research
Lucknow, India

Dr. Manish Kumar Tripathi
Department of Biophysics
All India Institute of Medical Sciences
New Delhi, India

Dr. Dhruv Kumar
Professor & Cluster Head
Allied Health Sciences
School of Health Sciences & Technology (SoHST)
UPES University
Dehradun, India

Dr. Vinay Kumar Bari
Assistant Professor
Department of Biochemistry
Central University of Punjab
Post Office-Ghudda
Bhatinda, India

Dr. Moiz Ashraf Ansari
Visiting Faculty
Department of Veterinary Pathobiology
Texas A&M University
U.S.

Dr. Venkata Saibabu Thokala
Research Associate
THISTI
Faridabad, India

About the Editors

Eti Sharma is currently working as an Assistant Professor at the School of Biotechnology, Gautam Buddha University, Greater Noida, India. She earned her Ph.D. in Molecular Biology and Biotechnology from G.B. Pant University of Agriculture and Technology, Pantnagar, India. She has been awarded several prestigious fellowships from reputed organizations in India. She has established herself in the field of bioprospecting and metabolomics of medicinal plants. She has expertise in medicinal plant research, biofabrication, and plant–microbe interactions. Her current research focuses on exploring the role of plant-associated microbiomes in the production of pharmaceutically significant bioactive compounds. She has completed two edited book proposals, with more currently in progress. Additionally, she has authored numerous research papers in nationally and internationally reputed journals, along with many book chapters.

Rekha Puria is working as an Assistant Professor and the Head of the Department, School of Biotechnology, Gautam Buddha University, Greater Noida, India. She finished her post-doctorate from Duke University Medical Centre, USA, and her doctorate from the Institute of Microbial Technology (IMTECH), India. She has 15 plus years of experience in yeast genetics and genomics. She is proficient in genetic manipulation techniques and genome-scale expression profiling through microarrays and transcriptomics. She extensively uses *Saccharomyces* as a model to understand the concepts of cellular growth and proliferation regulated through both coding and noncoding transcripts. Her research work has added in current understanding of the role of TOR signaling pathway in cell growth under stress conditions. Currently, her interest is in understanding growth, pathogenesis, and drug tolerance of *Candida auris* and the identification of novel antifungals against *Candida*. Her research is sponsored by several grants from the Government of India. Her work has been published in a number of international journals with high impact factor. In addition, she has authored several book chapters.

Karan Kumar is a Postdoctoral Research Associate in The Blank Lab (Industrial Systems Biotechnology, iAMB) at RWTH Aachen University, working on the EIC Pathfinder project HYDROCOW. His current research integrates pangenome-guided genome-scale metabolic models, DBTL experimentation, and machine-learning-assisted design to engineer *Xanthobacter* sp. SoF1 for "food-from-air" protein production from CO_2/H_2, in collaboration with Solar Foods, FGen/Ginkgo Bioworks, and the University of Groningen. He earned his Ph.D. in Energy Science and Engineering from the Indian Institute of Technology (IIT) Guwahati as a Prime Minister's Research Fellow (2018–2023), where he combined clostridial ABE fermentation on lignocellulosic hydrolysates with comparative/pangenomics, GEMs (FBA/FVA), omics integration, DoE-driven bioprocess optimization, and sono-enzymatic upgrading to n-butyl levulinate. His dissertation was recognized with the institute's

Best PhD Thesis Award (2024). He has authored over 13 peer-reviewed journal articles and 10 book chapters in leading international publications. His publication record spans leading venues, including two first-author articles in *Process Safety and Environmental Protection*, and additional publications in *Biotechnology & Bioengineering*, *Biomass Conversion & Biorefinery*, and *Process Biochemistry*. He also co-authored papers in *Algal Research* and *Energy Conversion and Management* and contributed invited chapters to multiple volumes published in 2024. Collectively, his contributions cover the domains of biofuel process intensification, metabolic modeling, and computational bioprocess engineering. Since joining RWTH, he has co-authored an invited review in press in *Trends in Biotechnology* on gas precision fermentation with hydrogen-oxidizing bacteria, as well as a 2025 review in *Renewable and Sustainable Energy Reviews* on fourth-generation biodiesel, reflecting his broadened systems biotechnology agenda in Aachen. He led Winter/Virtual Internships in bioinformatics and mathematical biology that trained 30+ students and yielded a preprint, journal papers, book chapters, and student awards. He has been proactive in disseminating his research. He delivered oral presentations at Clostridia XVI (INSA Toulouse, 2022), VII-SEEC (2022), VI-SEEC (2021), and AOSS-4 (2019); presented an e-poster at ISMB 2022; and gave a 3-Minute Thesis talk at RIC 2022. His talks received multiple accolades, including the Best Paper Award (VII-SEEC 2022), Springer Best Paper Award (NERC-2022), and Best Oral Presentation (BREEECH-2021). He also secured competitive travel grants from PMRF, DBT-CTEP, and SERB-ITS (Young Scientist).

List of Contributors

Yakshit Arora
Amity Institute of Biotechnology
Amity University
Noida, India

Shumaiza Asif
Department of Infectious Disease Biology
Regional Centre for Biotechnology – Faridabad
Faridabad, India

Tushar Bandhu
Regional Centre for Biotechnology – Faridabad
Faridabad, India

Himanshi Bihaniya
Regional Centre for Biotechnology – Faridabad
Faridabad, India

Ananya Chamola
Department of Education (CIE)
University of Delhi
New Delhi, India

Navaneet Chaturvedi
Amity Institute of Biotechnology
Amity University
Noida, India

Nidhee Chaudhary
Centre of Biotechnology and Biochemical Engineering
Amity Institute of Biotechnology
Amity University
Noida, India

Ashik Francis
Department of Infectious Disease Biology
Regional Centre for Biotechnology – Faridabad
Faridabad, India

Garima Gangwar
Gautam Budha University
Greater Noida, India

Chaitanya Ullhas Gatkal
Regional Centre for Biotechnology – Faridabad
Faridabad, India

Chaitali Ghosh
Department of Zoology
Gargi College
University of Delhi
New Delhi, India

Emmanuel Ijabani
Department of Infectious Disease Biology,
Regional Centre for Biotechnology – Faridabad
Faridabad, India

Nishtha Kaushik
School of Biotechnology
Gautam Buddha University
Greater Noida, India

Bablu Kumar
Department of Oncology and Hematology-Oncology
Università degli Studi di Milano
Milan, Italy
and
Department of Biosciences
Biotechnology and Environment University of Bari A. Moro
Bari, Italy

Manoj Kumar
Faculty of Chemistry and Biochemistry
Ruhr-University Bochum
Bochum, Germany

Patel Manish Chandrashekhar Kumar
Regional Centre for Biotechnology – Faridabad
Faridabad, India

Harsh Lalwani
Amity Institute of Biotechnology
Amity University
Noida, India

Pesaru Sai Laya
Department of Biotechnology
National Institute of Technology
Warangal, India

Ashwinipriyadarshini Megur
Department of Biological Models
Institute of Biochemistry
Life Sciences Center
Vilnius University
Vilnius, Lithuania

Samrath Singh Narula
Amity Institute of Biotechnology
Amity University
Noida, India

R. Nischitha
National Fungal Culture Collection of India
Biodiversity and Palaeobiology–Fungi, MACS
Agharkar Research Institute
Pune, India

Ritesh Pandey
Department of Applied Science
Shambhunath Institute of Engineering and Technology
Prayagraj, India

Pisal Ketali Prakash
Department of Biotechnology
National Institute of Technology
Warangal, India

Priya Prasad
Department of Biotechnology
National Institute of Technology
Warangal, India

Seema Raj
School of Basic and Applied Sciences
K. R. Mangalam University
Gurugram, India

Jitendra Singh Rathore
Department of Biological Sciences and Engineering
Netaji Subhash University of Technology
Delhi, India

Kamal Rawal
Amity Institute of Biotechnology
Amity University
Noida, India

Garima Singh
School of Biotechnology
Gautam Buddha University
Greater Noida, India

Shweta Singh
Department of Immunology and Microbiology
School of Medicine
University of Texas Rio Grande Valley
Edinburg, Texas

Anil Thakur
Department of Infectious Disease Biology
Regional Centre for Biotechnology – Faridabad
Faridabad, India

Part 1

Medicinal Plants, Antifungals, and Metagenomics

1 Fungal Infections
An Emerging Global Threat to Human Health, Current Trends, and Challenges

Garima Singh, Ananya Chamola, Jitendra Singh Rathore, and Chaitali Ghosh

1.1 INTRODUCTION TO FUNGAL PATHOGENS

The fungal kingdom, with up to 6 million distinct species, exhibits a significant global impact on human health. Over 600 fungal species have been recognized to be associated with human illness, participating either as commensals within the human microbiome or as pathogens causing severe, life-threatening infectious diseases. Individuals with compromised immunity are highly susceptible, whereas immunocompetent individuals can also be infected, particularly under conditions involving high fungal loads or exposure to virulent or emerging strains. The rising global prevalence of invasive fungal infections, combined with the increasing incidence of resistance to all currently available antifungal drugs, underscores the emerging threat these organisms represent to human health.

A huge number of COVID-19 patients have also been documented to have fungal co-infections, such as mucormycosis, candidiasis, and aspergillosis. These fungal infections were the primary cause of morbidity and mortality in certain COVID-19 patients (Bhatt et al., 2021; Mina et al., 2022). Additionally, the global proliferation of fungal diseases of growing significance has been recognized. The World Health Organization (WHO) published the first list of 19 fungal priority infections on October 25, 2022, which were determined to pose the biggest risk to public health. The first worldwide effort to identify and rank fungal diseases, their effects on public health worldwide, along with unfulfilled needs for studying and developing are WHO Fungal Priority Diseases List. Priority fungal pathogens have been divided by the WHO into three categories: critical, high, and medium priority, as listed in Table 1.1. The WHO's fungal priority pathogen classification identifies those that require immediate attention because of their high prevalence, severity of disease, and limited therapeutic options (*WHO Fungal Priority Pathogens List to Guide Research, Development and Public Health Action*, 2022). Three different categories were established in accordance with the threat level and severity about the fungal pathogen. The critical priority group contains those pathogens that have the highest public health threat in regard to high mortality rates, multidrug resistance,

DOI: 10.1201/9781003633952-2

TABLE 1.1
Fungal Pathogen Priority List, as per WHO (WHO Fungal Priority Pathogens List to Guide Research, Development and Public Health Action, 2022)

Critical Priority Group	High Priority Group	Medium Priority Group
Candida auris	*Nakaseomyces glabrata* (*Candida glabrata*)	*Lomentospora prolificans*
Candida albicans	*Mucorales*	*Scedosporium* spp.
Cryptococcus neoformans	*Candida tropicalis*	*Pichia kudriavzevii* (*Candida krusei*)
Aspergillus fumigatus	*Histoplasma* spp.	*Coccidioides* spp.
	Eumycetoma causative agents	*Talaromyces marneffei*
	Fusarium spp.	*Cryptococcus gattii*
	Candida parapsilosis	*Paracoccidioides* spp.
		Pneumocystis jirovecii

and limited treatment options. The high priority group involves fungi that are increasingly linked to invasive infections and emerging resistance. The medium priority group demonstrates that these pathogens are less prevalent and less impactful but have the potential to be very impactful and cause outbreaks or could evolve into resistant pathogens eventually more strongly impacting immunocompromised individuals. This tiered approach is intended to inform future research and interventions from a public health standpoint, as well as future development of diagnostics and antifungal treatments.

This increasing threat has led international health authorities to enhance their surveillance and give high priority to research on the most important fungal pathogens (Garvey et al., 2022). Most troublesome among these are a limited set of species with demonstrated increasing levels of drug resistance, healthcare facility adaptability, and high levels of mortality, particularly among the vulnerable population. These pathogens are not only challenging to treat but also pose significant diagnostic and management problems in clinical practice (Pagano & Fernández, 2025). The subsequent section discusses the most critical of these fungal pathogens—*Candida albicans, Candida auris, Cryptococcus neoformans, Aspergillus fumigatus,* and *Candida glabrata*—each of which is responsible for playing a critical role in the prevalence of invasive fungal infections worldwide (Fang et al., 2023).

1.1.1 *Candida auris*

Candida auris is an emerging species posing a global fungal threat to human health. This yeast can cause severe infections, including bloodstream infections (candidemia), as well as affecting the heart, brain, eyes, bones, and other internal organs (De Gaetano et al., 2024). It's particularly dangerous for people who are critically ill or have weakened immune systems, such as those undergoing cancer treatments, bone marrow transplants, or organ transplants. Other factors that increase the risk of infection include kidney problems, long hospital stays (over 10–15 days), use of mechanical ventilation, having central venous catheters, receiving total parenteral nutrition, and being in a state of sepsis. Additionally, previous use of antifungal drugs, especially triazoles, is linked to a higher risk of developing a *C. auris* infection.

C. auris has been implicated in nosocomial outbreaks, highlighting urgent requirements for strong infection prevention strategies to mitigate its transmission within healthcare settings (Eix & Nett, 2025). However, optimal prevention strategies are yet to be determined. The overall mortality rate associated with invasive candidiasis caused by *C. auris* ranges from 29% to 53% (Alvarez-Moreno et al., 2023). Patients with *C. auris* candidemia typically face longer hospital stays or more time in the ICU compared to those infected with other types of Candida. On average, adults with *C. auris* candidemia stay in the hospital for 46–68 days, while pediatric patients with more severe cases can be hospitalized for anywhere from 70 to 140 days (Kim et al., 2024). Over the past 10 years, outbreaks in numerous countries have contributed to an increase in the incidence of *C. auris* infections. Also, the COVID-19 pandemic reported an increase in the number of *C. auris* infection cases globally (Lankarani et al., 2022).

Invasive *C. auris* candidiasis can be moderately prevented, while having no existing vaccine. Therefore, surveillance of at-risk individuals and control of colonization are extremely critical to monitor those vulnerable to *Candida* infections. There is moderate accessibility to conventional diagnostics, while it is hard to identify *C. auris* (Keighley et al., 2021). However, the affordability and availability of evidence-based treatment are also very low. Invasive candidiasis is typically best treated with echinocandins. However, after confirming in vitro susceptibility, other antifungals can also be employed. WHO included echinocandins in the essential medicine list (EML) in 2021 (WHO Model List of Essential Medicines-22nd List, 2021).

The highest worldwide resistance rates of *C. auris* are observed for fluconazole (87%–100%). Whereas, the amphotericin B and echinocandins show global resistance rates of 8%–35% and 0%–8%, respectively. Approximately 4% isolates of *C. auris* have been reported to be resistant to all three classes of antifungal medications in South America, India, and the United States. Therefore, *C. auris* is the first classified multidrug-resistant *Candida* species (Keighley et al., 2021). Synergies between antifungal medications (in vitro and in vivo) must be analyzed to upgrade present treatment regimens against *C. auris*, narrowing the knowledge gap.

1.1.2 *CANDIDA ALBICANS*

Candida albicans is a common fungus that is part of normal human microbiota, residing in areas like gut, vagina, throat, mouth, and skin without causing harm in healthy individuals (Talapko et al., 2021). However, when it overgrows or spreads to other tissues, it can lead to infections such as vulvovaginal candidiasis, esophageal candidiasis, skin infections, and oral thrush. In immunocompromised or critically ill patients (Lewis & Williams, 2017; Mohamed et al., 2019; Rodríguez-Cerdeira et al., 2020; Schlager et al., 2018), *C. albicans* can also cause more serious, invasive infections affecting the central nervous system, eyes, bones, heart, blood (candidemia), and other internal organs (Talapko et al., 2021).

Therefore, *C. albicans* can also be the reason for life-threatening disease with high motility and classified as an opportunistic fungus. Even with the availability of effective antifungal treatment against invasive candidiasis, its approximate mortality ranges from 20% to 50%. However, 4% of invasive candidates may develop secondary growth (Pfaller & Diekema, 2007). The rising concern regarding *C. albicans* is its resistance to azoles (Whaley et al., 2017).

Newborns may contract vaginal candidiasis during delivery (Messina et al., 2024). The obese may also develop *C. albicans* in deep skin folds. Proper monitoring and prevention of colonization for patients at risk for *C. albicans* infection is important, as there is no available vaccine against it. There is very low preventability of *C. albicans*-caused invasive candidiasis. Abdominal candidiasis is one of the few *C. albicans*-caused illnesses that is even challenging to diagnose. However, the use of echinocandins is the preferred treatment for invasive candidiasis following the step-down process with azoles, and they are still inaccessible in several countries.

1.1.3 *Aspergillus fumigatus*

Aspergillus fumigatus is a widely distributed environmental mold. *A. fumigatus* is highly pathogenic to humans, causing aspergillosis. The term aspergillosis can be used for a wide range of infections, such as semi-invasive disease to acute invasive aspergillosis, colonization, and allergic reactions (Machado et al., 2024). Critically ill patients, especially those suffering from chronic lung disease or immunocompromised patients (cancer or transplant, hematological malignancies), are at higher risk for serious invasive infections (invasive aspergillosis).

The hospital stays for patients related to *A. fumigatus* infection range widely (21–532 days) (Kang et al., 2025). *A. fumigatus* can cause life-threatening disease as azole-resistant invasive aspergillosis, with a high mortality rate of 47%–88% (Kulirankal et al., 2024). The prevalence of invasive aspergillosis through *A. fumigatus* is variable, geographically. However, it has high preventability with no available vaccine.

Invasive aspergillosis can be avoided by prophylactic antifungal treatment for high-risk individuals (Baddley & Apewokin, 2008). Even azole-naive patients must be screened for azole resistance, other than obviously high-risk patients dealing with cancer, cystic fibrosis, or intensive care unit patients. The affordability and availability of evidence-based treatments, as well as access to traditional diagnostics, are generally poor. However, the use of azoles is the mainstay. Other efficient medications, such as liposomal amphotericin B, are easily accessible in wealthy nations but scarce in low- or middle-income countries (LMICs) (Lee et al., 2024). An urgent issue is the increase in resistance to antifungals. The increasing prevalence of resistant aspergillosis in humans is an outcome of the extensive application of azole fungicides to prevent crop losses (Verweij et al., 2020).

1.1.4 *Cryptococcus neoformans*

Cryptococcus neoformans, a pathogenic yeast found throughout the world, thrives in environments such as decaying wood and soil. *C. neoformans* can infect humans once they inhale fungal cells from the surroundings (Coelho et al., 2013). Despite starting in the lungs, cryptococcosis can spread to the blood and central nervous system, where it can result in cryptococcal meningitis and cryptococcaemia, respectively. There is no human-human transfer. HIV infection is the main risk factor, and most patients have weakened immune systems. However, infection can also be found in people who appear healthy, and organ transplant recipients and other people on drugs that impair immunity are also at risk. HIV infection, autoimmune disorders, decompensated liver cirrhosis, and iatrogenic immunosuppression are risk factors for invasive cryptococcal illness (Rajasingham et al., 2017).

With a death rate of 41%–61%, *C. neoformans* cryptococcosis is an extremely dangerous illness, particularly for those who are HIV-positive. For patients infected with *C. neoformans*, the typical length of hospital stay is 18–39 days; this is mostly seen for patients with HIV. Blindness, increased intracranial pressure requiring shunts, and severe renal impairment are a few side effects of the *C. neoformans* infection and its treatment (Bongomin et al., 2017). Although there is no vaccine to prevent invasive cryptococcosis, prophylactic and preventive treatment greatly lowers the incidence of cryptococcal meningitis in the high-risk populations.

Global access to diagnostics might be quickly increased with an efficient, quick, affordable, and simple way to conduct a lateral flow test if one is available. Fluconazole can be used to treat localized cryptococcosis, while amphotericin B combined with flucytosine and then step-down to fluconazole is used to treat severe and widespread cases (WHO, 2018). Many nations still do not have access to the treatments, even though they are listed on the WHO Essential Medicines List (EML). Only amphotericin B has a clinical breakpoint, and little is known about antifungal resistance. Furthermore, fluconazole susceptibility is decreased. Clinical studies aiming at lowering morbidity and mortality are required to fill in knowledge gaps. Molecular typing of *C. neoformans* would provide a more precise evaluation of antifungal resistance rates for various genotypes in conjunction with additional information on antifungal susceptibility. Synergy studies between in vivo and in vitro would allow for the extension and enhancement of the existing *C. neoformans* treatment alternatives. To fully understand the impact of *C. neoformans* and its spread across different patient groups and regions, future studies should focus on long-term complications, risk factors, and clinical outcomes. Prospective cohort studies combined with global surveillance data will provide a clearer picture of the disease's burden and its molecular epidemiology.

1.1.5 *Nakaseomyces glabrata* (*Candida glabrata*)

The commensal yeast *Nakaseomyces glabrata*, also known as *Candida glabrata*, is found all over the world and can cause disease. With an incidence often second only to *C. albicans*, it is a major cause of candidiasis. It may lead to invasive candidiasis, which can impact internal organs, the heart, blood, bones, eyes, and central nervous system (Pfaller & Diekema, 2007).

With an all-cause 20%–50% mortality rate at 30 days, invasive candidiasis caused by *C. glabrata* is an extremely dangerous illness. Those with affected host immunity are at risk factors for infection. The consequences and aftereffects of infection have not received much attention. Few data are available to quantify the influence on hospital stay duration, albeit this is probably comparable to other *Candida* species, where descriptions of durations ranging from two to eight weeks are common (Beardsley et al., 2024). It's uncertain to report if staying is directly related to infection.

The prevalence of *C. glabrata* as a percentage of all invasive *Candida* infections has been rising over the past ten years, according to trends. Similar to information on other *Candida* species, *C. glabrata* causes invasive candidiasis that can be prevented. It is unknown whether evidence-based treatments are available and affordable, or whether conventional diagnostics are accessible. Echinocandins are typically used to treat invasive candidiasis; however, if in vitro susceptibility has been confirmed, other antifungals, such as azoles, may be employed. This species exhibits high azole minimum inhibitory concentrations (MICs), and echinocandin resistance appears to be increasing recently (Alexander et al., 2013; Castanheira et al., 2022). Clinical trials aimed at enhancing results and avoiding infection are required to close the knowledge gap. Data is severely lacking, from middle-class and particularly low-income countries, regarding every facet of the infection. There is a special need for cohort studies that fill in the gaps regarding attributed complications and sequelae, mortality, and time of hospital stay.

1.2 THE SURGE IN ANTIFUNGAL RESISTANCE

Fungal infections are arising to become a significant public health concern and a generally overlooked aspect of the global developing challenge of antibiotic resistance. Antifungal resistance is the ability of a fungus to grow even at antifungal concentrations that kill or stop the growth of fungi in individual cultures.

Human-infecting pathogenic fungi are rapidly developing resistance to the currently existing approved systemic antifungal medications during a time of significant worldwide environmental change and growing at-risk populations. When a patient no longer responds to a medication given at the prescribed dosage, it is known as therapeutic failure, which can be caused by a variety of host, medication, and microbiological factors. Individuals with a weakened immune system are less likely to react to treatment because the antifungal medication cannot combat the infection with the support of a strong immune response (Brown & Netea, 2012). Drug-impermeable biofilms frequently colonize the surfaces of implanted devices such as mechanical heart valves and indwelling catheters, which can lead to refractory infections. The delivery of medication within the host is recognized to contribute to therapeutic failure; for example, an intra-abdominal abscess might produce drug resistance because fungi are exposed to suboptimal drug concentrations (Shields et al., 2014). In patients with persistent illnesses, inadequate adherence to medication schedules may result in less than the required drug exposure. Resistance is linked to drug exposure through long-term therapy, repetitive administration, or preventative use.

Similarly, environmental reservoirs of resistant organisms have been cultivated by exposure to agricultural fungicides that have molecular targets that are equivalent to those of systemic antifungals. There are two types of microbiological antibiotic resistance: primary (intrinsic), which may be developed in some fungi that have never been exposed to drugs before spontaneously or due to ineffective binding to drug targets, and secondary (acquired), which occurs when resistance mechanisms are acquired, leading to growth of fungal cells even at higher antifungal drug concentrations than the other members of the naturally occurring population (Shor & Perlin, 2015).

In the literature, the phrases "antifungal tolerance" and "antifungal resistance" frequently appear interchangeably, albeit incorrectly. It is believed that tolerance develops as a result of physiological, genetic, or epigenetic adaptation to the drug; the likelihood of developing tolerance is influenced by one's genetic background. Antifungal tolerance is also further discussed in later parts of this chapter. Antibacterial tolerance and persistence, in which almost all cells or few cells, respectively, survive bactericidal drug treatment through transitory metabolic quiescence of varying durations (Balaban et al., 2019), which is starkly different from antifungal tolerance.

Point mutations (~10^{-6} to 10^{-8} per cell per generation), gene duplications, and transposon insertions (~10^{-3} to 10^{-4} per cell per generation) are among the classes of changes that are common to bacteria and fungi and can result in drug resistance. There are a few similarities in antibacterial and antifungal resistance, which are further discussed. Resistance to antifungal medications, such as azoles, can arise due to subtherapeutic concentrations or insufficient exposure to them; therapeutic drug monitoring is necessary to maximize pharmacokinetics and pharmacodynamics as a crucial component for supervising the antifungal resistance, which is also comparable to bacterial resistance as seen in β-lactams.

In agriculture, fungal resistance, some of which is clinically relevant, is rapidly increasing due to the widespread use of fungicides (such as azole-resistant *A. fumigatus*), while bacterial antimicrobial resistance is rapidly increasing due to the widespread use of antibiotics in agriculture and livestock (such as livestock-associated methicillin-resistant *Staphylococcus aureus*). Resistance in both bacteria and fungi is also observed through trade and travel due to the clonal growth of bacteria containing antibiotic resistance genes (ARGs) (e.g., *S. aureus* clone EMRSA-15 (ST22) SCCmec), active worldwide dissemination of microbial resistance through travel and trade, local passive spread in the air (e.g., in concentrated animal feed operations), and fungal long-range passive spread in the air (e.g., triazole-resistant *A. fumigatus*), active spread by hospital-acquired transmission and travel (e.g., fluconazole-resistant *C. auris* clades I–IV), respectively. Nosocomial healthcare outbreaks and a worldwide spread of antimicrobial resistance are also observed in both of these microbes due to the commensal carriage in the gut, mucosa, and epithelium (e.g., *Candida* spp.) (Figure 1.1).

The constant necessity to protect intensively cultivated animals and genetically homogeneous crop cultures from fungal infections, as well as to protect materials from fungal saprotrophic decay, dictates the use of fungicides. Resistance to all of the main families of fungicides, including benzimidazoles, anilinopyrimidines, strobilurins, succinate dehydrogenase inhibitors, and sterol demethylation inhibitors (DMIs), including azoles (Fisher et al., 2018), has evolved as a result of environmental pressure. In addition to requiring the creation of resistance management plans and the breeding of more disease-resistant crops, environmental resistance is closely related to the emergence of antifungal-resistant IFDs in humans as a result of the use of sterol 14α-DMIs in the environment and clinical settings (Verweij et al., 2020; Fisher et al., 2018).

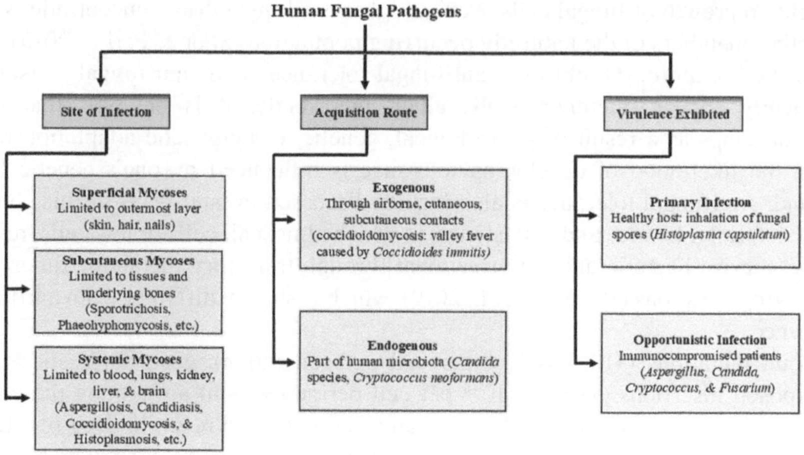

FIGURE 1.1 Schematic representation of the site of infection, acquisition route, and virulence exhibited by human fungal pathogens.

In contrast to bacteria, fungi frequently have many chromosomes and are multinucleate and/or multicellular. This type of genomic architecture offers more chances for genetic alterations that support adaptations and the development of resistance. For instance, many kinds of mutations, such as whole chromosomal and segmental aneuploidies, can generate clinically meaningful resistance and/or tolerance to azoles (Selmecki et al., 2006; Suwunnakorn et al., 2016; Kwon-Chung & Chang, 2012; Ksiezopolska et al., 2021). Drug stress can select for distinct loss of heterozygosity events, which can lead to an increase in resistance or tolerance in diploid organisms (Forche et al., 2011). Hypermutator fungal lineages in *C. glabrata* (Healey et al., 2016) and *Cryptococcus* spp. (Billmyre et al., 2017) may also be the cause of antifungal resistance occurrences, although the extent to which these particular changes are in charge for increased mutation rates is still unknown (Singh et al., 2018). MSH2 mutations, which arise either naturally among pathogenic fungal lineages or as a result of rapid in-host adaptation to drug exposure (Rhodes et al., 2017), are the main mechanism by which the known molecular drivers of fungal hypermutator status converge upon DNA mismatch repair processes (Billmyre et al., 2017; Singh et al., 2018). Fungal hypermutator lineages have relatively low fitness costs, in contrast to bacterial hypermutator lineages, which frequently experience severe fitness deficits (Healey et al., 2016; Rhodes et al., 2017; Boyce et al., 2017).

About 4.0% of *C. auris* showed resistance to all three medication classes, azoles, echinocandins, and polyenes, while multiple drug resistance of at least two antifungal classes is seen in ≥40.0% of cases (Fasciana et al., 2020). The results are also supported by Chow and colleagues, who examined three hundred *C. auris* isolates and discovered that 80.0% of them were resistant to fluconazole, 23.0% to amphotericin B, and 7.0% to micafungin, while 24.0% were resistant to at least two antifungal classes and 1.0% to all three classes (Ruiz & Lorenz, 2020; Chow et al., 2018; Zamith-Miranda et al., 2019). By blocking the synthesis of ergosterol, specifically lanosterol 14-α-demethylase, which is encoded by the ERG11 gene, fluconazole inhibits cell proliferation (Ruiz & Lorenz, 2020). Another significant mechanism of antifungal resistance is efflux pumps, particularly in the early phases of biofilm formation. The biofilm matrix itself can prevent medication diffusion, which increases the biofilm's resistance to antifungals as it ages (Zamith-Miranda et al., 2019). Therefore, the pathophysiology of *Candida* species and its resistance to antifungals depend heavily on biofilm formation (Finkel & Mitchell, 2010). In addition to expressing a stronger capacity to form biofilms, *C. auris* possesses 686 biofilm-related proteins, including ribosomal proteins, transporters, several enzymes, and transcription factors (Ciurea et al., 2021).

1.3 SPECIES WIDE IMPACTS OF RESISTANCE IN CHANGING ENVIRONMENT

Environmental changes have a major impact on antifungal resistance, a serious worldwide health concern. Whereas, host susceptibility is also enhanced by human factors such as compromised immunity, aging population, and excessive use of immunosuppressive treatments. Climate change has a major impact on the basis of antifungal resistance. Anthropogenic climate change, primarily due to greenhouse gas emissions, produces conditions that affect 58% of all known infectious diseases in humans, including fungal infections (Mora et al., 2022). In particular, fungal growth, pathogenicity, and spread are being impacted by climatic extremes, increasing temperatures, and altered precipitation patterns (Buckley & Roughgarden, 2004; Casadevall, 2020; Fisher et al., 2020). Pathogenic fungus can be dispersed geographically due to these conditions, subjecting populations to novel, potentially more virulent, or drug-resistant strains. Environmental variations due to climate change facilitate the appearance, adaptation, and transmission of drug-resistant fungal pathogens, a major public health issue (George et al., 2025).

1.3.1 Fungal Thermotolerance

It is predicted that, due to heat extremes and rising global temperatures, fungi would adjust to the normal human body temperature, perhaps creating novel fungal illnesses. Thermotolerant fungi can pose severe new threats because they can capitalize on this ecological niche (Smith & Casadevall, 2023)

The pathogen *C. auris* has spread around the world and is well-known for its adaptability. It is more thermally and halotolerant than other *Candida*, which enables it to thrive in harsh environments with high temperatures and salinity (Arora et al., 2021; Casadevall et al., 2021; Escandón, 2022; Welsh et al., 2017). Its discovery in aquatic ecosystems and animal hosts is a subject of fear regarding zoonotic transmission and temperature adaptation in the long run. Ever since its discovery in 2009, *C. auris* has been documented in more than 50 countries across 6 continents in association with various healthcare-associated epidemics (Kim et al., 2024). With almost a 200% increase in clinical cases and a 275% increase in positive screening cultures between 2019 and 2021, the prevalence of *C. auris* in the United States has dramatically increased over the years (Lyman et al., 2023).

Thermostasis is one of the most notable pathogenicity characteristics of the fungus *A. fumigatus*. The heat shock (HS) response, which is regulated by heat shock transcription factors (HSFs), is an adaptation mechanism used by opportunistic fungal pathogens in response to high temperatures (Fabri et al., 2021).

Candida orthopsilosis, one of the *Candida parapsilosis* complex species derived from temperate marine environments, highlights the shift in fungal ecology due to increasing ocean temperatures (del Olmo et al., 2023). Thermotolerant strains of *C. orthopsilosis* have likely been favored and their dissemination in marine environments facilitated by global warming. As a result, nosocomial outbreaks and *C. orthopsilosis* isolation rates have surged worldwide, with Brazil and Italy generating the most prominent reports (Luo et al., 2023; Tavanti et al., 2007).

Climate change has caused changes in the geographic distribution of the *Cryptococcus gattii* species complex, which may be more pathogenic because of increased thermotolerance and melanin synthesis (Cogliati, 2021; Fernandes et al., 2016; Granados & Castañeda, 2006; Kidd et al., 2007). Sub-Saharan Brazil, Africa, Colombia, Australia, Canada, and the United States (including Oregon, California, Washington, and other states) have all documented this species complex (Bruner et al., 2018; Poplin et al., 2024). The frequency of *C. gattii* species complex infections in people with weakened immune systems is frequently increasing.

1.3.2 Enlarged Geographical Ranges

The movement and growth of fungal pathogens are facilitated by changing precipitation patterns and rising global temperatures, which allow them to colonize previously inhabitable areas. Furthermore, severe weather conditions accelerate the spread of fungus, raising the possibility of endemic and newly emerging fungal illnesses (Anderson et al., 2006; Taylor et al., 2022).

Once restricted to particular geographic areas, dimorphic fungi are especially vulnerable to environmental changes brought on by climate change. Human exposure to these fungi is rising as a result of climate change-related disruptions that enable them to establish themselves in new ecological niches (Benedict et al., 2015; Brown et al., 2018; Mazi et al., 2023). Even in historically endemic areas, the incidence of these fungi could be extremely variable, ranging from low-incidence areas to zones with hyperendemic outbursts. Spatial and temporal changes in fungal occurrence present strong challenges to effective prevention and disease surveillance measures. This problem is made more difficult by climate change, which can create new endemic zones and disrupt environmental conditions (Wang et al., 2023).

Several endemic fungal pathogens are changing their geographic ranges, mostly due to climate change. This includes the *Blastomyces dermatitidis* complex, which was traditionally confined to North America's river valleys and Great Lakes basin but now is occurring in regions of Canada, Africa, and India due to extreme weather conditions and drought (Ashraf et al., 2020; Brown et al., 2018; Pfister et al., 2011; Williams et al., 2020). *Histoplasma capsulatum*, which is previously primarily found in the Americas, is spreading to temperate climates such as Italy, Argentina, and Turkey as rising temperatures allow more hardy strains (Antinori et al., 2006; Campbell & Berliner, 1973; Medoff et al., 1986). Likewise, *Coccidioides* species, previously limited to dry regions of the southwestern United States and Latin America, have spread to Missouri, Washington State, and Portugal, facilitated by changing precipitation and dust storms (Carvalho et al., 2010; Gade et al., 2021; Gorris et al., 2019; Laws et al., 2021). *Paracoccidioides* species, which are native to some regions of South and Central America, are becoming more common in North America, Europe, Africa, and Asia due to factors such as deforestation and El Niño (Barrozo et al., 2009, 2010; Restrepo et al., 2001). *Talaromyces marneffei*, which has hitherto been reported in Southeast Asia, can be seen as expanding northwards into China as a consequence of increased humidity and more intense monsoons (Bulterys et al., 2013; Zheng et al., 2015). Finally, *Emergomyces* species have come to be known for worldwide distribution, although certain climate-related effects and range changes are less well characterized (Jiang et al., 2018; Vinayagamoorthy et al., 2023).

1.3.3 Natural Environmental Stressors and Displacement of Population

According to the United Nations High Commissioner for Refugees (UNHCR), approximately 24 million people are displaced each year as a result of disasters, of which the majority are weather-related (*Climate Crisis Fuels Flooding and Deepens Displacement | UNHCR India*, 2024). Displaced populations tend to settle in climate-exposed zones with poor resources and infrastructure, making them more susceptible to health hazards such as fungal infections. Severe weather, compounded by global warming, damages ecosystems and infrastructure, setting the stage for fungal development. Such events kill human health directly through injury and exposure to infected environments, as well as indirectly through the clogging of healthcare systems. Aerosolized spores of fungi and infected water are the primary vectors, especially for those with compromised immune systems or compromised skin (Van Rhijn & Bromley, 2021; Zeng et al., 2019). The combined effect of physical trauma, psychological stress, immunosuppression, and compromised healthcare infrastructure dramatically increases the risk of fungal infections among survivors of disasters. Increased fungal infections have been linked to a variety of natural disasters, including dust storms, earthquakes, tornadoes, wildfires, tsunamis, and volcanic eruptions.

Several dust disasters have also been linked to fungal infections. In 2011, after a tornado in Joplin, Missouri, USA, 13 patients with severe trauma developed mucormycosis related to *Apophysomyces trapeziformis,* due to probable environmental exposure to water and organic material. Penetrating trauma and multiple wounds were the major risk factors (Neblett Fanfair et al., 2012). In another case, a dust storm in Arvin, Kern County, California in 1977 caused pulmonary and disseminated infection by *Coccidioides immitis*, with disease reported as far away as 500 km from where the storm occurred. Similarly, after the 1994 Northridge earthquake in Simi Valley and Ventura County, California, there was an accelerated number of cases of acute coccidioidomycosis within two weeks, through the dispersal of fungal spores by high winds (Flynn et al., 1979). In 2017, California wildfires also added to the increased risk of coccidioidomycosis several months later, further exacerbated by soil disturbance from fighting the fires (Mulliken et al., 2023). Furthermore, the 1985 volcanic eruption in Armero, Colombia, was associated with eight cases of mucormycosis in patients with necrotizing skin lesions, which reflected fungal infection after the natural disaster.

Also, Flooding is particularly conducive to the growth and dissemination of fungi. Mold develops on water-damaged structures. while contact with fungal infections is heightened by dirty water and wreckage (Kawakami et al., 2012; Kontoyiannis et al., 2019). These risks are especially problematic in regions susceptible to tsunamis and hurricanes, where large numbers of people could be affected simultaneously.

1.4 RESISTANCE DETECTION, TRANSMISSION, AND SURVEILLANCE

Molecular diagnostic techniques can be used to detect fungus species that are inherently resistant to antifungals as well as to uncover genetic markers that may be linked to antifungal resistance, though they are critically underutilized (Garcia-Effron, 2021). Because of their sensitivity, clinical specimens can be directly applied, eliminating the requirement for culture and speeding up turnaround times. Real-time PCR can identify species in the *Aspergillus fumigati* complex, such as *Aspergillus lentulus* and *Aspergillus felis*, which are challenging to distinguish using traditional techniques and may have higher MIC values to azole antifungals (Chong et al., 2016b). The most effective method for detecting the changes linked to azole resistance is still DNA sequencing, which restricts clinical use, especially direct sample testing (Perlin & Wiederhold, 2017).

Fluconazole resistance in *Candida* species is uncommon, except for inherently resistant species like *Candida krusei*. Before the appearance of *C. auris*, the pathogen *C. glabrata* had the highest frequency of acquired resistance, with a resistance rate roughly 10%–12% globally (Lamoth et al., 2017). This was altered by the appearance of *C. auris*, for which the majority of clinical isolates show acquired resistance to fluconazole (Chowdhary et al., 2017; Arendrup et al., 2017; Lockhart et al., 2016). Isolates of clades I and III are mostly always resistant to fluconazole, isolates of clade II are nearly always susceptible to it, and isolates of clade IV exhibit varying resistance according to the circulating clone (Chow et al., 2020). This suggests that resistance is clade-specific. PCR can identify and distinguish resistant *Candida* species, including *C. auris, C. glabrata, and C. krusei*, which may help with infection prevention and patient care (Leach et al., 2019). For the detection of multidrug-resistant tuberculosis, the use of fully automated molecular platforms such as the T2 Biosystems or Becton Dickinson Max offers quick testing procedures that require little specialized training. It is necessary to broaden the scope of this prospective near-patient test to discover resistance-related changes in fungus species that are typically vulnerable. The development of real-time PCR techniques is constrained because azole resistance in *Candida* spp. is linked to a variety of pathways and ensuing mutations. Rapid molecular tests have been made possible by the identification of specific mutations and associated hotspots in *C. auris* strains that are resistant to azoles and echinocandins through the sequencing of the ERG11 and FKS1 genes (Hou et al., 2018). Most echinocandin resistance in *Candida* species is linked to a limited number of mutations in the FKS1 gene, and PCR tests have been created (Pham et al., 2013). There are currently no commercial PCR assays available to identify mutations linked to yeast antifungal resistance; hence, improved commercial development and regulatory body assistance are necessary to improve diagnosis.

Recent studies have connected the development of azole-resistant infections in patients to the use of azole fungicides, which share chemical similarities with azole antifungal drugs used in medicine (Rhodes et al., 2022; Gonzalez-Jimenez et al., 2021; Kang et al., 2021). According to research conducted on one cohort based in Europe, the 90-day death rate for patients with azole-resistant invasive aspergillosis was around 25% more than that of patients with azole-susceptible infections. Additionally, patients with voriconazole-resistant aspergillosis who were initiated on incorrect antifungal medication (such as voriconazole monotherapy) had a greater 42-day mortality rate than those who received proper therapy, according to the same study (Lestrade et al., 2018). In another investigation, patients with azole resistance-associated mutations in the Cyp51A gene had a 6-week death rate that was 2.7 times higher than that of patients with hematologic malignancies (Chong et al., 2016a). According to these investigations, treatment failure and resistant infections in azole-naive individuals are two notable characteristics of infections caused by azole-resistant *A. fumigatus*. Patients with chronic pulmonary aspergillosis can also develop azole-resistant infections, in addition to those with invasive aspergillosis. To find possible resistance-associated mutations, direct sequencing of genes encoding drug target proteins (such as CYP51A in A. fumigatus or ERG11 in *Candida* spp.) has been employed (Garcia-Effron, 2021). Commercial real-time PCR assays were launched based on the high frequency of common mutations (e.g., TR34/L98H and TR46/Y121F/T289A in *A. fumigatus*), and their diagnostic use is growing because of the high sensitivity and specificity of PCR-based methods (Chong et al., 2016a; Montesinos et al., 2016).

1.5 THERAPEUTIC APPROACHES FOR TACKLING ANTIFUNGAL RESISTANCE

Modifications that either directly or indirectly impact the drug–target interaction are usually the cause of antifungal resistance. Resistance can be caused by genetic alterations to the target binding site (e.g., mutation of the genes encoding β-glucan synthase for echinocandins or lanosterol demethylase for azoles, Robbins et al., 2017), overexpression of the availability of the target, and/or changes to the appropriate drug concentration (e.g., increased drug efflux activity for intracellular drugs like azoles, Robbins et al., 2017, or inhibition of prodrug activation for flucytosine, Edlind & Katiyar, 2010). We have earlier discussed in "The Surge in antifungal resistance", the generalized fungal resistance mechanisms. Antifungal tolerance, as opposed to antifungal resistance, involves a variety of general stress responses and epigenetic pathways that describe the capacity of drug-susceptible cells to proliferate at drug concentrations higher than the minimum inhibitory concentration (MIC) (Berman & Krysan, 2020). Fluconazole-treated *C. albicans* isolates have been the subject of the most rigorous measurement and characterization of tolerance, which is most noticeable with fungistatic medications. Though the clinical significance of the same is still up for debate.

The development of resistance to antifungal drugs is essentially an evolutionary reaction to the drug's selective pressure. The size of the population subjected to the selective pressure, the rate with which cells double, the number of distinct pathways (causes of genetic changes and physiological mechanisms) that confer resistance, and the fitness costs associated with each of them all influence the likelihood that resistance will emerge as a result of genetic changes. Crucially, the environment or the host may be the source of antifungal medication resistance. A variety of pathogenic fungi, including molds (Ballard et al., 2018) and yeasts (Shields et al., 2012), experience treatment failure due to in vivo resistance, which further evolves into de novo resistance in people undergoing antifungal therapy. Antifungal drug resistance can develop in treated persons as a result of drug exposure. Echinocandin resistance, for instance, is more prevalent in people who have already received echinocandin treatment (Alexander et al., 2013), while azole-resistant genotypes of *A. fumigatus* (Ballard et al., 2018) and *C. neoformans* (Rhodes et al., 2017) emerge over lengthy treatment regimens. *A. fumigatus* strains resistant to antifungal medications used in clinical medicine to treat aspergillosis can be selected for by the application of azole fungicides such as tebuconazole, propiconazole, epoxiconazole, and difenoconazole (Verweij et al., 2015).

Antifungal medications must get to the infection site to work. The pharmacokinetic characteristics (absorption, distribution, metabolism, and excretion) of each antifungal medication vary greatly, and the variations in tissue-specific penetration are much more noticeable. Drug concentrations that are either temporarily high or persistently low may hasten the development of resistance. However, there is a danger of toxicity associated with utilizing too high dosages of medications. Regular therapeutic medication monitoring is therefore necessary to avoid the development of resistance while minimizing unpleasant responses, as well as to optimize the dosage to maximize therapeutic potential. Although physiologically based modeling approaches have started to provide some insight into tissue-specific pharmacokinetics, they remain largely unknown (Hens et al., 2017; Li et al., 2019; Gerhart et al., 2019; Campoli et al., 2013).

To investigate pharmacokinetics across clinical cohorts, real-world studies are increasingly employing therapeutic drug monitoring (Figure 1.2). For instance, monitoring of patients with cystic fibrosis has shown a high prevalence of subtherapeutic levels of azoles along with a high probability (>20%) of developing a resistant infection after two years (Di Paolo et al., 2021). For these reasons, in sensitive patient cohorts, therapeutic medication monitoring through antifungal stewardship programs needs to be implemented more effectively. In addition, drug resistance may be avoided through the prudent use of medication combinations. Micafungin, for instance, blocks several fungal and human efflux pumps; as a result, it may increase the intracellular retention and effectiveness of medications like azoles.

Unlike opportunistic fungal diseases, *C. auris* can spread and remain in intensive care units and other healthcare settings, causing nosocomial outbreaks that are severe and uncontrollable. A frequent treatment for *C. auris*, which is typically resistant to fluconazole, is echinocandin monotherapy. This is widely exhibited in several *Candida* yeasts that are the main contributors to nosocomial bloodstream infections, leading to widespread antifungal resistance (Steinmann et al., 2015;

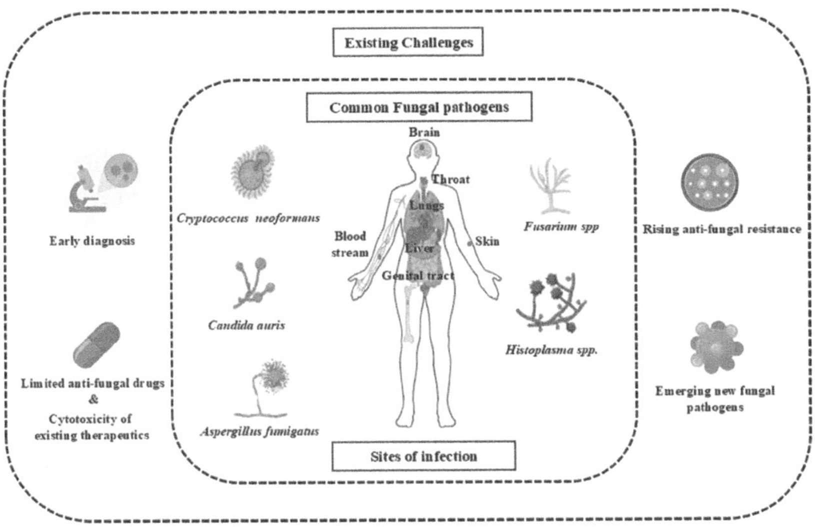

FIGURE 1.2 Schematic representation of existing challenges of fungal infection in human beings.

Pristov & Ghannoum, 2019). For example, it is well-known that *C. albicans* developed azole resistance after receiving long-term fluconazole treatment for oral candidiasis in HIV-positive patients (Johnson et al., 1995). Since a progressive loss of echinocandin activity has also been documented with extended caspofungin therapy for *C. albicans* oesophagitis (Laverdière et al., 2006), this issue is not limited to azole antifungals. However, if human pathogenic fungi have previously been exposed to fungicides in the environment, environmental resistance may develop (Verweij et al., 2020). Azoles, echinocandins, and polyenes are the three primary types of antifungal medications utilized in both clinical and therapeutic treatment settings for *C. auris* (Fasciana et al., 2020; Chowdhary et al., 2017).

It will be necessary for future research to determine the probability of resistance and tolerance mechanisms developing further. Pharmacometric techniques enable the simulation of model predictions (Hope et al., 2016); the hollow fiber model, for instance, simulates the human pharmacokinetics of antimicrobials by utilizing the pharmacokinetic data that is currently accessible (Tängdén et al., 2017). Furthermore, widespread necrosis at the infection site makes medicine distribution difficult and leads to unfavorable results. Improved pharmacodynamic models are required for diseases with low medication penetration at the infection site to maximize dosage schedules and avoid treatment failure. To find circumstances that reduce the evolution of antifungal resistance in situ, dual pharmacokinetic/pharmacodynamic techniques may enable integrative, dynamic investigations of the interaction between (unbound) drug concentrations, pathogen growth, and kill kinetics (Chen et al., 2015). Even though a lot of new drugs are coming to the market in recent times, with various advantages, the development process remains expensive, and unique side effects can be observed via their use, leading to noble drawbacks.

1.6 CURRENT AND FUTURE INTERVENTIONS

Our capacity to treat invasive fungal diseases will be safeguarded and enhanced by increasing surveillance and antifungal stewardship programs, which both call for better diagnosis of invasive fungal diseases and antifungal resistance; reducing environmental–clinical dual usage of antifungals; and optimizing resilient combination therapies using currently licensed medications, which are some of the strategies available to limit the evolution of human fungal pathogens to chemical control. Treating people who are at risk of developing resistance with innovative antifungal drugs that are patented exclusively for clinical use will be a major component of future initiatives to mitigate the effects of antifungal resistance. Using immunotherapies and/or vaccinations against invasive fungal diseases should address the compromised immunity that predisposes people to these diseases and lower the chance of acquiring invasive fungal diseases.

The widespread preventive and experimental administration of antifungals to treat suspected invasive fungal diseases in those who are persistently at risk and/or are critically ill, and patients with hemato-oncology, remains a problem when viewed from resistance prevention. To maximize antifungal use and maintain the small antifungal arsenal, effective antifungal management is necessary (Logan et al., 2020; Michallet et al., 2020). This is particularly important for highly contagious fungal diseases like Candida species and skin-infecting Trichophyton species (Kano et al., 2020).

Antifungal stewardship programs have continuously shown an improvement in measures like timely and appropriate antifungal prescriptions, use of diagnostics and drug monitoring, and a reduction in antifungal consumption, thereby reducing selective pressures and the development of resistance in largely single-center, systematic observational studies (Bienvenu et al., 2018; Hart et al., 2019; Rautemaa-Richardson et al., 2018; Talento et al., 2021). Antifungal stewardship showed a lack of negative effect on invasive fungal diseases, and though these studies weren't intended to show better clinical results, there was no effect on duration of stay at the hospital and in-hospital deaths (Whitney et al., 2019).

Combining the azole and echinocandin classes for invasive aspergillosis appears to boost fungal mortality and improve survival, according to consistent evidence from both in vitro and animal models (Kirkpatrick et al., 2002; Petraitis et al., 2003, 2017). In a randomized clinical trial, the death rate for patients receiving this combination was 19%, while the mortality rate for patients receiving azole monotherapy was 28% (Marr et al., 2015). Despite the study's small sample size, which meant the survival advantage fell short of traditional statistical significance, the strategy outlined appears promising. Before proceeding to adequately powered phase III trials, more work is required to systematically explore combinations of known and novel antifungal agents in experimental models and phase II clinical studies. Animal models indicate a role for combination therapy in azole-resistant invasive aspergillosis (Seyedmousavi et al., 2013).

Utilizing host-directed strategies to control antifungal resistance is another way to safeguard currently available antifungals. These consist of antibodies to fungal targets (Ambati et al., 2021), fungal vaccines (Oliveira et al., 2021), and immunotherapy (Armstrong-James et al., 2017). Host-directed immunotherapies, including recombinant cytokines, monoclonal antibodies, and fungus-specific modified

T cells, have been developed since invasive fungal diseases are most prevalent in immunocompromised hosts (Armstrong-James et al., 2017). These developments demonstrate how host-directed strategies may reduce the demand for antifungal medications. Furthermore, although they must be evaluated in clinical trials, cell-based treatments such as chimeric antigen receptor (CAR) T cell therapy and dendritic cell transfer have demonstrated encouraging outcomes in vitro (Fisher et al., 2022).

REFERENCES

Alexander, B. D., Johnson, M. D., Pfeiffer, C. D., Jiménez-Ortigosa, C., Catania, J., Booker, R., Castanheira, M., Messer, S. A., Perlin, D. S., & Pfaller, M. A. (2013). Increasing echinocandin resistance in *Candida glabrata*: Clinical failure correlates with presence of FKS mutations and elevated minimum inhibitory concentrations. *Clinical Infectious Diseases*, 56(12), 1724–1732. https://doi.org/10.1093/cid/cit136

Alvarez-Moreno, C. A., Morales-López, S., Rodriguez, G. J., Rodriguez, J. Y., Robert, E., Picot, C., Ceballos-Garzon, A., Parra-Giraldo, C. M., & Le Pape, P. (2023). The mortality attributable to candidemia in *C. auris* is higher than that in other candida species: Myth or reality? *Journal of Fungi*, 9(4), 430. https://doi.org/10.3390/JOF9040430

Ambati, S., Yang, L., Cai, H., Mullick, A. E., Sodhi, N., & Maybruck, B. T. (2021). Antifungal liposomes directed by dectin-2 offer a promising therapeutic option for pulmonary aspergillosis. *mBio*, 12(3), e00030-21. https://doi.org/10.1128/mBio.00030-21

Anderson, H., Honish, L., Taylor, G., Johnson, M., Tovstiuk, C., Fanning, A., Tyrrell, G., Rennie, R., Jaipaul, J., Sand, C., & Probert, S. (2006). Histoplasmosis cluster, golf course, Canada. *Emerging Infectious Diseases*, 12(1), 163–165. https://doi.org/10.3201/EID1201.051083

Antinori, S., Magni, C., Nebuloni, M., Parravicini, C., Corbellino, M., Sollima, S., Galimberti, L., Ridolfo, A. L., & Wheat, L. J. (2006). Histoplasmosis among human immunodeficiency virus-infected people in Europe: Report of 4 cases and review of the literature. *Medicine*, 85(1), 22–36. https://doi.org/10.1097/01.MD.0000199934.38120.D4

Arendrup, M. C., Prakash, A., Meletiadis, J., Sharma, C., & Chowdhary, A. (2017). Comparison of EUCAST and CLSI reference microdilution MICs of eight antifungal compounds for *Candida auris* and associated tentative epidemiological cutoff values. *Antimicrobial Agents and Chemotherapy*, *61*(6). https://doi.org/10.1128/aac.00485-17

Armstrong-James, D., Meintjes, G., & Brown, G. D. (2017). Immunotherapeutic approaches to treatment of fungal diseases. *The Lancet Infectious Diseases*, 17(12), e393–e402. https://doi.org/10.1016/S1473-3099(17)30442-5

Arora, P., Singh, P., Wang, Y., Yadav, A., Pawar, K., Singh, A., Padmavati, G., Chowdhary, A., & Xu, J. (2021). Environmental isolation of *Candida auris* from the coastal wetlands of Andaman islands, India. mBio, 12(2), 1–9. https://doi.org/10.1128/MBIO.03181-20

Ashraf, N., Kubat, R. C., Poplin, V., Adenis, A. A., Denning, D. W., Wright, L., McCotter, O., Schwartz, I. S., Jackson, B. R., Chiller, T., & Bahr, N. C. (2020). Re-drawing the maps for endemic mycoses. *Mycopathologia*, 185(5), 843–865. https://doi.org/10.1007/S11046-020-00431-2

Baddley, J. W., & Apewokin, S. (2008). Prevention of invasive aspergillosis in high-risk patients: Universal versus preemptive, targeted treatment. *Current Fungal Infection Reports*, 2(2), 61. https://doi.org/10.1007/S12281-008-0010-6

Balaban, N. Q., Helaine, S., Lewis, K., Ackermann, M., Aldridge, B., Andersson, D. I., Brynildsen, M. P., Bumann, D., Camilli, A., Collins, J. J., Dehio, C., Fortune, S., Ghigo, J., Hardt, W., Harms, A., Heinemann, M., Hung, D. T., Jenal, U., Levin, B. R.,... Zinkernagel, A. (2019). Definitions and guidelines for research on antibiotic persistence. *Nature Reviews Microbiology*, 17(7), 441–448. https://doi.org/10.1038/s41579-019-0196-3

Ballard, E., Melchers, W. J., Zoll, J., Brown, A. J., Verweij, P. E., & Warris, A. (2018). In-host microevolution of *Aspergillus fumigatus*: A phenotypic and genotypic analysis. *Fungal Genetics and Biology*, 113, 1–13. https://doi.org/10.1016/j.fgb.2018.02.003

Barrozo, L. V., Benard, G., Silva, M. E. S., Bagagli, E., Marques, S. A., & Mendes, R. P. (2010). First description of a cluster of acute/subacute paracoccidioidomycosis cases and its association with a climatic anomaly. *PLoS Neglected Tropical Diseases*, *4*(3). https://doi.org/10.1371/JOURNAL.PNTD.0000643

Barrozo, L. V., Mendes, R. P., Marques, S. A., Benard, G., Siqueira Silva, M. E., & Bagagli, E. (2009). Climate and acute/subacute paracoccidioidomycosis in a hyper-endemic area in Brazil. *International Journal of Epidemiology*, 38(6), 1642–1649. https://doi.org/10.1093/ije/dyp207

Beardsley, J., Kim, H. Y., Dao, A., Kidd, S., Alastruey-Izquierdo, A., Sorrell, T. C., Tacconelli, E., Chakrabarti, A., Harrison, T. S., Bongomin, F., Gigante, V., Galas, M., Siswanto, S., Dagne, D. A., Roitberg, F., Sati, H., Morrissey, C. O., & Alffenaar, J. W. (2024). *Candida glabrata* (*Nakaseomyces glabrata*): A systematic review of clinical and microbiological data from 2011 to 2021 to inform the World Health Organization Fungal Priority Pathogens List. *Medical Mycology*, 62(6), myae041. https://doi.org/10.1093/MMY/MYAE041

Benedict, K., Thompson, G. R., Deresinski, S., & Chiller, T. (2015). Mycotic infections acquired outside areas of known endemicity, United States. *Emerging Infectious Diseases*, 21(11), 1935–1941. https://doi.org/10.3201/EID2111.141950

Berman, J., & Krysan, D. J. (2020). Drug resistance and tolerance in fungi. *Nature Reviews Microbiology*, 18(6), 319–331. https://doi.org/10.1038/s41579-019-0322-2

Bhatt, K., Agolli, A., H. Patel, M., Garimella, R., Devi, M., Garcia, E., Amin, H., Domingue, C., Del Castillo, R. G., & Sanchez-Gonzalez, M. (2021). High mortality co-infections of COVID-19 patients: Mucormycosis and other fungal infections. *Discoveries*, *9*(1), e126. https://doi.org/10.15190/D.2021.5

Bienvenu, A. L., Desbois, A., Louberton, M., Picot, S., & Garraud, O. (2018). A systematic review of interventions and performance measures for antifungal stewardship programmes. *Journal of Antimicrobial Chemotherapy*, 73(2), 297–305. https://doi.org/10.1093/jac/dkx383

Billmyre, R. B., Clancey, S. A., & Heitman, J. (2017). Natural mismatch repair mutations mediate phenotypic diversity and drug resistance in *Cryptococcus deuterogattii*. *eLife*, *6*. https://doi.org/10.7554/elife.28802

Bongomin, F., Gago, S., Oladele, R. O., & Denning, D. W. (2017). Global and multi-national prevalence of fungal diseases—Estimate precision. *Journal of Fungi*, 3(4), 57. https://doi.org/10.3390/JOF3040057

Boyce, K. J., Wang, Y., Verma, S., Shakya, V. P. S., Xue, C., & Idnurm, A. (2017). Mismatch repair of DNA replication errors contributes to microevolution in the pathogenic fungus *Cryptococcus neoformans*. *mBio*, *8*(3). https://doi.org/10.1128/mbio.00595-17

Brown, E. M., McTaggart, L. R., Dunn, D., Pszczolko, E., Tsui, K. G., Morris, S. K., Stephens, D., Kus, J. V., & Richardson, S. E. (2018). Epidemiology and geographic distribution of blastomycosis, histoplasmosis, and coccidioidomycosis, Ontario, Canada, 1990–2015. *Emerging Infectious Diseases*, *24*(7), 1257–1266. https://doi.org/10.3201/EID2407.172063

Brown, G. D., & Netea, M. G. (2012). Exciting developments in the immunology of fungal infections. *Cell Host & Microbe*, 11(5), 422–424. https://doi.org/10.1016/j.chom.2012.04.010

Bruner, K. T., Franco-Paredes, C., Henao-Martínez, A. F., Steele, G. M., & Chastain, D. B. (2018). *Cryptococcus gattii* complex infections in HIV-infected patients, southeastern United States. *Emerging Infectious Diseases*, 24(11), 1998–2002. https://doi.org/10.3201/EID2411.180787

Buckley, L. B., & Roughgarden, J. (2004). Biodiversity conservation: Effects of changes in climate and land use. *Nature*, *430*(6995). https://doi.org/10.1038/NATURE02717

Bulterys, P. L., Le, T., Quang, V. M., Nelson, K. E., & Lloyd-Smith, J. O. (2013). Environmental predictors and incubation period of AIDS-associated *Penicillium marneffei* infection in Ho Chi Minh City, Vietnam. *Clinical Infectious Diseases*, 56(9), 1273–1279. https://doi.org/10.1093/CID/CIT058

Campbell, C. C., & Berliner, M. D. (1973). Virulence differences in mice of type A and B *Histoplasma capsulatum* yeasts grown in continuous light and total darkness. *Infection and Immunity*, 8(4), 677–678. https://doi.org/10.1128/IAI.8.4.677-678.1973

Campoli, P., Perlin, D. S., Kristof, A. S., White, T. C., Filler, S. G., & Sheppard, D. C. (2013). Pharmacokinetics of posaconazole within epithelial cells and fungi: Insights into potential mechanisms of action during treatment and prophylaxis. *The Journal of Infectious Diseases*, 208(10), 1717–1728. https://doi.org/10.1093/infdis/jit358

Carvalho, C., Ferreira, I., Gaião, S., Guimarães, S., Costa, R., Santos, J., Sampaio, S., Bustorff, M., Oliveira, G., & Pestana, M. (2010). Cerebral coccidioidomycosis after renal transplantation in a non-endemic area: Case report. *Transplant Infectious Disease*, 12(2), 151–154. https://doi.org/10.1111/J.1399-3062.2009.00456.X

Casadevall, A. (2020). Climate change brings the specter of new infectious diseases. *Journal of Clinical Investigation*, 130(2), 553–555. https://doi.org/10.1172/JCI135003

Casadevall, A., Kontoyiannis, D. P., & Robert, V. (2021). Environmental *Candida auris* and the global warming emergence hypothesis. *mBio*, 12(2), 1–3. https://doi.org/10.1128/mBio.00360-21

Castanheira, M., Deshpande, L. M., Davis, A. P., Carvalhaes, C. G., & Pfaller, M. A. (2022). Azole resistance in *Candida glabrata* clinical isolates from global surveillance is associated with efflux overexpression. *Journal of Global Antimicrobial Resistance*, 29, 371–377. https://doi.org/10.1016/J.JGAR.2022.05.004

Chen, G., Mulla, W. A., Kucharavy, A., Tsai, H., Rubinstein, B., Conkright, J., McCroskey, S., Bradford, W. D., Weems, L., Haug, J. S., Seidel, C. W., Berman, J., & Li, R. (2015). Targeting the adaptability of heterogeneous aneuploids. *Cell*, 160(4), 771–784. https://doi.org/10.1016/j.cell.2015.01.026

Chong, G. M., Van Der Beek, M. T., Von Dem Borne, P. A., Boelens, J., Steel, E., Kampinga, G. A., Span, L. F. R., Lagrou, K., Maertens, J. A., Dingemans, G. J. H., Gaajetaan, G. R., Van Tegelen, D. W. E., Cornelissen, J. J., Vonk, A. G., & Rijnders, B. J. A. (2016a). PCR-based detection of *Aspergillus fumigatus* Cyp51A mutations on bronchoalveolar lavage: A multicentre validation of the AsperGenius assay ® in 201 patients with haematological disease suspected for invasive aspergillosis. *Journal of Antimicrobial Chemotherapy*, 71(12), 3528–3535. https://doi.org/10.1093/jac/dkw323

Chong, G., Vonk, A., Meis, J., Dingemans, G., Houbraken, J., Hagen, F., Gaajetaan, G., Van Tegelen, D., Simons, G., & Rijnders, B. (2016b). Interspecies discrimination of *A. fumigatus* and siblings *A. lentulus* and *A. felis* of the Aspergillus section Fumigati using the AsperGenius® assay. *Diagnostic Microbiology and Infectious Disease*, 87(3), 247–252. https://doi.org/10.1016/j.diagmicrobio.2016.11.020

Chow, N. A., Gade, L., Tsay, S. V., Forsberg, K., Greenko, J. A., Southwick, K. L., Barrett, P. M., Kerins, J. L., Lockhart, S. R., Chiller, T. M., Litvintseva, A. P., Adams, E., Barton, K., Beer, K. D., Bentz, M. L., Berkow, E. L., Black, S., Bradley, K. K., Brooks, R., … Zhu, Y. (2018). Multiple introductions and subsequent transmission of multidrug-resistant *Candida auris* in the USA: A molecular epidemiological survey. *The Lancet Infectious Diseases*, 18(12), 1377–1384. https://doi.org/10.1016/s1473-3099(18)30597-8

Chow, N. A., Muñoz, J. F., Gade, L., Berkow, E. L., Li, X., Welsh, R. M., Forsberg, K., Lockhart, S. R., Adam, R., Alanio, A., Alastruey-Izquierdo, A., Althawadi, S., Araúz, A. B., Ben-Ami, R., Bharat, A., Calvo, B., Desnos-Ollivier, M., Escandón, P.,

Gardam, D., … Cuomo, C. A. (2020). Tracing the evolutionary history and global expansion of *Candida auris* using population genomic analyses. *mBio*, *11*(2). https://doi.org/10.1128/mbio.03364-19

Chowdhary, A., Prakash, A., Sharma, C., Kordalewska, M., Kumar, A., Sarma, S., Tarai, B., Singh, A., Upadhyaya, G., Upadhyay, S., Yadav, P., Singh, P. K., Khillan, V., Sachdeva, N., Perlin, D. S., & Meis, J. F. (2017). A multicentre study of antifungal susceptibility patterns among 350 *Candida auris* isolates (2009–17) in India: Role of the ERG11 and FKS1 genes in azole and echinocandin resistance. *Journal of Antimicrobial Chemotherapy*, 73(4), 891–899. https://doi.org/10.1093/jac/dkx480

Ciurea, C. N., Mare, A. D., Kosovski, I., Toma, F., Vintilă, C., & Man, A. (2021). *Candida auris* and other phylogenetically related species – A mini-review of the literature. *GERMS*, 11(3), 441–448. https://doi.org/10.18683/germs.2021.1281

Climate crisis fuels flooding and deepens displacement | UNHCR India. (2024). Retrieved April 28, 2025, from https://www.unhcr.org/in/news/stories/climate-crisis-fuels-flooding-and-deepens-displacement

Coelho, C., Bocca, A. L., & Casadevall, A. (2013). The intracellular life of *Cryptococcus neoformans*. *Annual Review of Pathology*, 9, 219. https://doi.org/10.1146/ANNUREV-PATHOL-012513-104653

Cogliati, M. (2021). Global warming impact on the expansion of fundamental niche of *Cryptococcus gattii* VGI in Europe. *Environmental Microbiology Reports*, 13(3), 375–383. https://doi.org/10.1111/1758-2229.12945

De Gaetano, S., Midiri, A., Mancuso, G., Avola, M. G., & Biondo, C. (2024). *Candida auris* outbreaks: Current status and future perspectives. *Microorganisms*, 12(5), 927. https://doi.org/10.3390/MICROORGANISMS12050927

Di Paolo, M., Hewitt, L., Nwankwo, E., Ni, M., Vidal-Diaz, A., Fisher, M. C., Armstrong-James, D., & Shah, A. (2021). A retrospective 'real-world' cohort study of azole therapeutic drug monitoring and evolution of antifungal resistance in cystic fibrosis. *JAC-Antimicrobial Resistance*, *3*(1). https://doi.org/10.1093/jacamr/dlab026

Edlind, T. D., & Katiyar, S. K. (2010). Mutational analysis of flucytosine resistance in *Candida glabrata*. *Antimicrobial Agents and Chemotherapy*, 54(11), 4733–4738. https://doi.org/10.1128/aac.00605-10

Eix, E. F., & Nett, J. E. (2025). *Candida auris*: Epidemiology and antifungal strategy. *Annual Review of Medicine*, 76(1), 57. https://doi.org/10.1146/ANNUREV-MED-061523-021233

Escandón, P. (2022). Novel environmental niches for *Candida auris*: Isolation from a Coastal Habitat in Colombia. *Journal of Fungi*, 8(7). https://doi.org/10.3390/JOF8070748

Fabri, J. H. T. M., Rocha, M. C., Fernandes, C. M., Persinoti, G. F., Ries, L. N. A., da Cunha, A. F., Goldman, G. H., Del Poeta, M., & Malavazi, I. (2021). The heat shock transcription factor HsfA is essential for thermotolerance and regulates cell wall integrity in *Aspergillus fumigatus*. *Frontiers in Microbiology*, 12, 656548. https://doi.org/10.3389/fmicb.2021.656548

Fang, W., Wu, J., Cheng, M., Zhu, X., Du, M., Chen, C., Liao, W., Zhi, K., & Pan, W. (2023). Diagnosis of invasive fungal infections: Challenges and recent developments. *Journal of Biomedical Science*, 30(1). https://doi.org/10.1186/S12929-023-00926-2

Fasciana, T., Cortegiani, A., Ippolito, M., Giarratano, A., Di Quattro, O., Lipari, D., Graceffa, D., & Giammanco, A. (2020). *Candida auris*: An overview of how to screen, detect, test and control this emerging pathogen. *Antibiotics*, 9(11), 778. https://doi.org/10.3390/antibiotics9110778

Fernandes, K. E., Dwyer, C., Campbell, L. T., & Carter, D. A. (2016). Species in the *Cryptococcus gattii* complex differ in capsule and cell size following growth under capsule-inducing conditions. *mSphere*, *1*(6). https://doi.org/10.1128/MSPHERE.00350-16

Finkel, J. S., & Mitchell, A. P. (2010). Genetic control of *Candida albicans* biofilm development. *Nature Reviews Microbiology*, 9(2), 109–118. https://doi.org/10.1038/nrmicro2475

Fisher, M. C., Alastruey-Izquierdo, A., Berman, J., Bicanic, T., Bignell, E. M., Bowyer, P., Bromley, M., Brüggemann, R., Garber, G., Cornely, O. A., Gurr, S. J., Harrison, T. S., Kuijper, E., Rhodes, J., Sheppard, D. C., Warris, A., White, P. L., Xu, J., Zwaan, B., & Verweij, P. E. (2022). Tackling the emerging threat of antifungal resistance to human health. *Nature Reviews Microbiology*, 20(9), 557–571. https://doi.org/10.1038/s41579-022-00720-1

Fisher, M. C., Gurr, S. J., Cuomo, C. A., Blehert, D. S., Jin, H., Stukenbrock, E. H., Stajich, J. E., Kahmann, R., Boone, C., Denning, D. W., Gow, N. A. R., Klein, B. S., Kronstad, J. W., Sheppard, D. C., Taylor, J. W., Wright, G. D., Heitman, J., Casadevall, A., & Cowen, L. E. (2020). Threats posed by the fungal kingdom to humans, wildlife, and agriculture. *mBio*, *11*(3). https://doi.org/10.1128/MBIO.00449-20

Fisher, M. C., Hawkins, N. J., Sanglard, D., & Gurr, S. J. (2018). Worldwide emergence of resistance to antifungal drugs challenges human health and food security. *Science*, 360(6390), 739–742. https://doi.org/10.1126/science.aap7999

Flynn, N. M., Hoeprich, P. D., Kawachi, M. M., Lee, K. K., Lawrence, R. M., Goldstein, E., Jordan, G. W., Kundargi, R. S., & Wong, G. A. (1979). An unusual outbreak of windborne coccidioidomycosis. *New England Journal of Medicine*, 301(7), 358–361. https://doi.org/10.1056/NEJM197908163010705

Forche, A., Abbey, D., Pisithkul, T., Weinzierl, M. A., Ringstrom, T., Bruck, D., Petersen, K., & Berman, J. (2011). Stress alters rates and types of loss of heterozygosity in *Candida albicans*. *mBio*, *2*(4). https://doi.org/10.1128/mbio.00129-11

Gade, L., McCotter, O. Z., Bowers, J. R., Waddell, V., Brady, S., Carvajal, J. A., Sunenshine, R., Komatsu, K. K., Engelthaler, D. M., Chiller, T., & Litvintseva, A. P. (2021). The detection of coccidioides from ambient air in Phoenix, Arizona: Evidence of uneven distribution and seasonality. *Medical Mycology*, 58(4), 552–559. https://doi.org/10.1093/MMY/MYZ093

Garcia-Effron, G. (2021). Molecular markers of antifungal resistance: Potential uses in routine practice and future perspectives. *Journal of Fungi*, 7(3), 197. https://doi.org/10.3390/jof7030197

Garvey, M., Meade, E., & Rowan, N. J. (2022). Effectiveness of front line and emerging fungal disease prevention and control interventions and opportunities to address appropriate eco-sustainable solutions. *Science of the Total Environment*, 851, 158284. https://doi.org/10.1016/J.SCITOTENV.2022.158284

George, M. E., Gaitor, T. T., Cluck, D. B., Henao-Martínez, A. F., Sells, N. R., & Chastain, D. B. (2025). The impact of climate change on the epidemiology of fungal infections: implications for diagnosis, treatment, and public health strategies. *Therapeutic Advances in Infectious Disease*, *12*, 20499361251313840. https://doi.org/10.1177/20499361251313841

Gerhart, J. G., Watt, K. M., Edginton, A., Wade, K. C., Salerno, S. N., Benjamin, D. K., Smith, P. B., Hornik, C. P., Cohen-Wolkowiez, M., Duara, S., Ross, A., Shattuck, K., Stewart, D. L., Neu, N., Gonzalez, D., Furda, G., Benjamin, D., Capparelli, E., Kearns, G. L., … Wade, K. (2019). Physiologically-based pharmacokinetic modeling of fluconazole using plasma and cerebrospinal fluid samples from preterm and term infants. *CPT Pharmacometrics & Systems Pharmacology*, 8(7), 500–510. https://doi.org/10.1002/psp4.12414

Gonzalez-Jimenez, I., Garcia-Rubio, R., Monzon, S., Lucio, J., Cuesta, I., & Mellado, E. (2021). Multiresistance to nonazole fungicides in *Aspergillus fumigatus* TR 34 /L98H azole-resistant isolates. *Antimicrobial Agents and Chemotherapy*, 65(9). https://doi.org/10.1128/aac.00642-21

Gorris, M. E., Treseder, K. K., Zender, C. S., & Randerson, J. T. (2019). Expansion of coccidioidomycosis endemic regions in the United States in response to climate change. *GeoHealth*, 3(10), 308–327. https://doi.org/10.1029/2019GH000209

Granados, D. P., & Castañeda, E. (2006). Influence of climatic conditions on the isolation of members of the *Cryptococcus neoformans* species complex from trees in Colombia from 1992–2004. *FEMS Yeast Research*, 6(4), 636–644. https://doi.org/10.1111/J.1567-1364.2006.00090.X

Hart, E., Nguyen, M., Allen, M., Clark, C. M., & Jacobs, D. M. (2019). A systematic review of the impact of antifungal stewardship interventions in the United States. *Annals of Clinical Microbiology and Antimicrobials*, 18, 24. https://doi.org/10.1186/s12941-019-0323-z

Healey, K. R., Zhao, Y., Perez, W. B., Lockhart, S. R., Sobel, J. D., Farmakiotis, D., Kontoyiannis, D. P., Sanglard, D., Taj-Aldeen, S. J., Alexander, B. D., Jimenez-Ortigosa, C., Shor, E., & Perlin, D. S. (2016). Prevalent mutator genotype identified in fungal pathogen *Candida glabrata* promotes multi-drug resistance. *Nature Communications*, 7(1). https://doi.org/10.1038/ncomms11128

Hens, B., Pathak, S. M., Mitra, A., Patel, N., Liu, B., Patel, S., Jamei, M., Brouwers, J., Augustijns, P., & Turner, D. B. (2017). In silico modeling approach for the evaluation of gastrointestinal dissolution, supersaturation, and precipitation of posaconazole. *Molecular Pharmaceutics*, 14(12), 4321–4333. https://doi.org/10.1021/acs.molpharmaceut.7b00396

Hope, W., Drusano, G. L., & Rex, J. H. (2016). Pharmacodynamics for antifungal drug development: An approach for acceleration, risk minimization and demonstration of causality. *Journal of Antimicrobial Chemotherapy*, 71(11), 3008–3019. https://doi.org/10.1093/jac/dkw298

Hou, X., Lee, A., Jiménez-Ortigosa, C., Kordalewska, M., Perlin, D. S., & Zhao, Y. (2018). Rapid detection of ERG11-associated azole resistance and FKS-associated echinocandin resistance in *Candida auris*. *Antimicrobial Agents and Chemotherapy*, *63*(1). https://doi.org/10.1128/aac.01811-18

Jiang, Y., Dukik, K., Muñoz, J. F., Sigler, L., Schwartz, I. S., Govender, N. P., Kenyon, C., Feng, P., van den Ende, B. G., Stielow, J. B., Stchigel, A. M., Lu, H., & de Hoog, S. (2018). Phylogeny, ecology and taxonomy of systemic pathogens and their relatives in Ajellomycetaceae (Onygenales): Blastomyces, Emergomyces, Emmonsia, Emmonsiellopsis. *Fungal Diversity*, 90(1), 245–291. https://doi.org/10.1128/JCM.01325-20

Johnson, E. M., Warnock, D. W., Luker, J., Porter, S. R., & Scully, C. (1995). Emergence of azole drug resistance in Candida species from HIV-infected patients receiving prolonged fluconazole therapy for oral candidosis. *Journal of Antimicrobial Chemotherapy*, 35(1), 103–114. https://doi.org/10.1093/jac/35.1.103

Kang, S. E., Sumabat, L. G., Melie, T., Mangum, B., Momany, M., & Brewer, M. T. (2021). Evidence for the agricultural origin of resistance to multiple antimicrobials in *Aspergillus fumigatus*, a fungal pathogen of humans. *G3 Genes Genomes Genetics*, *12*(2). https://doi.org/10.1093/g3journal/jkab427

Kang, Y., Li, Q., Ma, W., Xu, C., Jia, W., & Wang, P. (2025). Epidemiological characteristics of patients with invasive pulmonary aspergillosis infected with *Aspergillus fumigatus* from a tertiary hospital in Ningxia, China. *Scientific Reports*, 15(1), 13036. https://doi.org/10.1038/S41598-025-97587-Y

Kano, R., Kimura, U., Kakurai, M., Hiruma, J., Kamata, H., Suga, Y., & Harada, K. (2020). Trichophyton indotineae sp. nov.: A new highly terbinafine-resistant anthropophilic dermatophyte species. *Mycopathologia*, 185(6), 947–958. https://doi.org/10.1007/s11046-020-00455-8

Kawakami, Y., Tagami, T., Kusakabe, T., Kido, N., Kawaguchi, T., Omura, M., & Tosa, R. (2012). Disseminated associated with tsunami lung. *Respiratory Care*, 57(10), 1674–1678. https://doi.org/10.4187/RESPCARE.01701

Keighley, C., Garnham, K., Harch, S. A. J., Robertson, M., Chaw, K., Teng, J. C., & Chen, S. C. A. (2021). Candida auris: Diagnostic challenges and emerging opportunities for the clinical microbiology laboratory. *Current Fungal Infection Reports*, 15(3), 116. https://doi.org/10.1007/S12281-021-00420-Y

Kidd, S. E., Bach, P. J., Hingston, A. O., Mak, S., Chow, Y., MacDougall, L., Kronstad, J. W., & Bartlett, K. H. (2007). *Cryptococcus gattii* dispersal mechanisms, British Columbia, Canada. *Emerging Infectious Diseases*, 13(1), 51–57. https://doi.org/10.3201/EID1301.060823

Kim, H. Y., Nguyen, T. A., Kidd, S., Chambers, J., Alastruey-Izquierdo, A., Shin, J. H., Dao, A., Forastiero, A., Wahyuningsih, R., Chakrabarti, A., Beyer, P., Gigante, V., Beardsley, J., Sati, H., Morrissey, C. O., & Alffenaar, J. W. (2024). *Candida auris*-A systematic review to inform the world health organization fungal priority pathogens list. *Medical Mycology*, *62*(6). https://doi.org/10.1093/MMY/MYAE042

Kirkpatrick, W. R., Perea, S., Coco, B. J., & Patterson, T. F. (2002). Efficacy of caspofungin alone and in combination with voriconazole in a guinea pig model of invasive aspergillosis. *Antimicrobial Agents and Chemotherapy*, 46(8), 2564–2568. https://doi.org/10.1128/AAC.46.8.2564-2568.2002

Kontoyiannis, D. P., Shah, E. C., Wurster, S., Aitken, S. L., Graviss, L., Raad, I. I., & Chemaly, R. F. (2019). Culture-documented invasive mold infections at Md Anderson Cancer Center in Houston, Texas, pre- and post-hurricane harvey. *Open Forum Infectious Diseases*, *6*(4). https://doi.org/10.1093/OFID/OFZ138,

Ksiezopolska, E., Schikora-Tamarit, M. À., Beyer, R., Nunez-Rodriguez, J. C., Schüller, C., & Gabaldón, T. (2021). Narrow mutational signatures drive acquisition of multidrug resistance in the fungal pathogen *Candida glabrata*. *Current Biology*, 31(23), 5314–5326. e10. https://doi.org/10.1016/j.cub.2021.09.084

Kulirankal, K. G., Mary, A., Moni, M., Pillai, G. S., & Sathyapalan, D. T. (2024). Long-term survival following medical management of *Aspergillus endocarditis* with dissemination as a consequence of steroid therapy in severe COVID-19 pneumonia. *Medical Mycology Case Reports*, 43, 100638. https://doi.org/10.1016/J.MMCR.2024.100638

Kwon-Chung, K. J., & Chang, Y. C. (2012). Aneuploidy and drug resistance in pathogenic fungi. *PLoS Pathogens*, 8(11), e1003022. https://doi.org/10.1371/journal.ppat.1003022

Lamoth, F., Lockhart, S. R., Berkow, E. L., & Calandra, T. (2017). Changes in the epidemiological landscape of invasive candidiasis. *Journal of Antimicrobial Chemotherapy*, 73(suppl_1), i4–i13. https://doi.org/10.1093/jac/dkx444

Lankarani, K. B., Akbari, M., Tabrizi, R., Vali, M., Sekhavati, E., Heydari, S. T., Khodadadi, H., & Ahmadizar, F. (2022). *Candida auris*: Outbreak fungal pathogen in COVID-19 pandemic: A systematic review and meta-analysis. *Iranian Journal of Microbiology*, 14(3), 276. https://doi.org/10.18502/IJM.V14I3.9753

Laverdière, M., Lalonde, R. G., Baril, J., Sheppard, D. C., Park, S., & Perlin, D. S. (2006). Progressive loss of echinocandin activity following prolonged use for treatment of *Candida albicans* oesophagitis. *Journal of Antimicrobial Chemotherapy*, 57(4), 705–708. https://doi.org/10.1093/jac/dkl022

Laws, R. L., Jain, S., Cooksey, G. S., Mohle-Boetani, J., McNary, J., Wilken, J., Harrison, R., Leistikow, B., Vugia, D. J., Windham, G. C., & Materna, B. L. (2021). Coccidioidomycosis outbreak among inmate wildland firefighters: California, 2017. *American Journal of Industrial Medicine*, 64(4), 266–273. https://doi.org/10.1002/AJIM.23218

Leach, L., Russell, A., Zhu, Y., Chaturvedi, S., & Chaturvedi, V. (2019). A rapid and automated sample-to-result *Candida auris* real-time PCR assay for high-throughput testing of surveillance samples with the BD Max open system. *Journal of Clinical Microbiology*, 57(10). https://doi.org/10.1128/jcm.00630-19

Lee, J. S. F., Cohen, R. M., Khan, R. A., Burry, J., Casas, E. C., Chung, H. Y., Costa, L. H., Ford, N., Galvao, D. L. N., Giron, N., Jarvis, J. N., Mondal, M., Odionyi, J. J., Casas, C. P., Rangaraj, A., Rode, J., Ruffell, C., Sued, O., & Ribeiro, I. (2024). Paving the way for affordable and equitable liposomal amphotericin B access worldwide. *The Lancet. Global Health*, 12(9), e1552. https://doi.org/10.1016/S2214-109X(24)00225-0

Lestrade, P. P., Bentvelsen, R. G., Schauwvlieghe, A. F. A. D., Schalekamp, S., Van Der Velden, W. J. F. M., Kuiper, E. J., Van Paassen, J., Van Der Hoven, B., Van Der Lee, H. A., Melchers, W. J. G., De Haan, A. F., Van Der Hoeven, H. L., Rijnders, B. J. A., Van Der Beek, M. T., & Verweij, P. E. (2018). Voriconazole resistance and mortality in invasive aspergillosis: A multicenter retrospective cohort study. *Clinical Infectious Diseases*, 68(9), 1463–1471. https://doi.org/10.1093/cid/ciy859

Lewis, M. A. O., & Williams, D. W. (2017). Diagnosis and management of oral candidosis. *British Dental Journal*, 223(9), 675–681. https://doi.org/10.1038/SJ.BDJ.2017.886

Li, X., Frechen, S., Moj, D., Lehr, T., Taubert, M., Hsin, C., Mikus, G., Neuvonen, P. J., Olkkola, K. T., Saari, T. I., & Fuhr, U. (2019). A physiologically based pharmacokinetic model of voriconazole integrating time-dependent inhibition of CYP3A4, genetic polymorphisms of CYP2C19 and predictions of drug–drug interactions. *Clinical Pharmacokinetics*, 59(6), 781–808. https://doi.org/10.1007/s40262-019-00856-z

Lockhart, S. R., Etienne, K. A., Vallabhaneni, S., Farooqi, J., Chowdhary, A., Govender, N. P., Colombo, A. L., Calvo, B., Cuomo, C. A., Desjardins, C. A., Berkow, E. L., Castanheira, M., Magobo, R. E., Jabeen, K., Asghar, R. J., Meis, J. F., Jackson, B., Chiller, T., & Litvintseva, A. P. (2016). Simultaneous emergence of multidrug-resistant *Candida auris* on 3 continents confirmed by whole-genome sequencing and epidemiological analyses. *Clinical Infectious Diseases*, 64(2), 134–140. https://doi.org/10.1093/cid/ciw691

Logan, C., Martin-Loeches, I., & Bicanic, T. (2020). Invasive candidiasis in critical care: Challenges and future directions. *Intensive Care Medicine*, 46(11), 2001–2014. https://doi.org/10.1007/s00134-020-06240-x

Luo, Z., Ning, Y., Yu, S., Xiao, M., Dai, R., Chen, X., Wang, Y., Kang, W., Jiang, Y., Yu, H., Liang, H., Xu, Y., Sun, T., & Zhang, L. (2023). The first established microsatellite markers to distinguish *Candida orthopsilosis* isolates and detection of a nosocomial outbreak in China. *Journal of Clinical Microbiology*, *61*(11). https://doi.org/10.1128/JCM.00806-23

Lyman, M., Forsberg, K., Sexton, D. J., Chow, N. A., Lockhart, S. R., Jackson, B. R., & Chiller, T. (2023). Worsening spread of *Candida auris* in the United States, 2019 to 2021. *Annals of Internal Medicine*, 176(4), 489–495. https://doi.org/10.7326/M22-3469

Machado, M., Fortún, J., & Muñoz, P. (2024). Invasive aspergillosis: A comprehensive review. *Medicina Clínica*, 163(4), 189–198. https://doi.org/10.1016/J.MEDCLI.2024.01.045

Marr, K. A., Schlamm, H. T., Herbrecht, R., Rottinghaus, S. T., Bow, E. J., Cornely, O. A., … Maertens, J. (2015). Combination antifungal therapy for invasive aspergillosis: A randomized trial. *Annals of Internal Medicine*, 162(2), 81–89. https://doi.org/10.7326/M13-2508

Mazi, P. B., Sahrmann, J. M., Olsen, M. A., Coler-Reilly, A., Rauseo, A. M., Pullen, M., Zuniga-Moya, J. C., Powderly, W. G., & Spec, A. (2023). The geographic distribution of dimorphic mycoses in the United States for the modern era. *Clinical Infectious Diseases*, 76(7), 1295–1301. https://doi.org/10.1093/CID/CIAC882

Medoff, G., Maresca, B., Lambowitz, A. M., Kobayashi, G., Painter, A., Sacco, M., & Carratu, L. (1986). Correlation between pathogenicity and temperature sensitivity in different strains of *Histoplasma capsulatum*. *The Journal of Clinical Investigation*, 78(6), 1638–1647. https://doi.org/10.1172/JCI112757

Messina, A., Mariani, A., Brandolisio, R., Tavella, E., Germano, C., Lipari, G., Leo, L., Masturzo, B., & Manzoni, P. (2024). Candidiasis in pregnancy: Relevant aspects of the pathology for the mother and the fetus and therapeutic strategies. *Tropical Medicine and Infectious Disease*, 9(5), 114. https://doi.org/10.3390/TROPICALMED9050114

Michallet, M., Sobh, M., Deray, G., Gangneux, J., Pigneux, A., Larrey, D., Ribaud, P., Mira, J., Nivoix, Y., Yakoub-Agha, I., Timsit, J., Alfandari, S., & Herbrecht, R. (2020). Antifungal stewardship in hematology: Reflection of a multidisciplinary group of experts. *Clinical Lymphoma Myeloma & Leukemia*, 21(1), 35–45. https://doi.org/10.1016/j.clml.2020.08.010

Mina, S., Yaakoub, H., Annweiler, C., Dubée, V., & Papon, N. (2022). COVID-19 and fungal infections: A double debacle. *Microbes and Infection*, *24*(8). https://doi.org/10.1016/j.micinf.2022.105039

Mohamed, A. A., Lu, X. L., & Mounmin, F. A. (2019). Diagnosis and treatment of esophageal candidiasis: Current updates. *Canadian Journal of Gastroenterology & Hepatology*, *2019*. https://doi.org/10.1155/2019/3585136

Montesinos, I., Delforge, M., Ajjaham, F., Brancart, F., Hites, M., Jacobs, F., & Denis, O. (2016). Evaluation of a new commercial real-time PCR assay for diagnosis of *Pneumocystis jirovecii* pneumonia and identification of dihydropteroate synthase (DHPS) mutations. *Diagnostic Microbiology and Infectious Disease*, 87(1), 32–36. https://doi.org/10.1016/j.diagmicrobio.2016.10.005

Mora, C., McKenzie, T., Gaw, I. M., Dean, J. M., von Hammerstein, H., Knudson, T. A., Setter, R. O., Smith, C. Z., Webster, K. M., Patz, J. A., & Franklin, E. C. (2022). Over half of known human pathogenic diseases can be aggravated by climate change. *Nature Climate Change*, 12(9), 869–875. https://doi.org/10.1038/S41558-022-01426-1

Mulliken, J. S., Hampshire, K. N., Rappold, A. G., Fung, M., Babik, J. M., & Doernberg, S. B. (2023). Risk of systemic fungal infections after exposure to wildfires: A population-based, retrospective study in California. *The Lancet Planetary Health*, 7(5), e381–e386. https://doi.org/10.1016/S2542-5196(23)00046-3

Neblett Fanfair, R., Benedict, K., Bos, J., Bennett, S. D., Lo, Y.-C., Adebanjo, T., Etienne, K., Deak, E., Derado, G., Shieh, W.-J., Drew, C., Zaki, S., Sugerman, D., Gade, L., Thompson, E. H., Sutton, D. A., Engelthaler, D. M., Schupp, J. M., Brandt, M. E., … Park, B. J. (2012). Necrotizing cutaneous mucormycosis after a Tornado in Joplin, Missouri, in 2011. *New England Journal of Medicine*, 367(23), 2214–2225. https://doi.org/10.1056/NEJMOA1204781

Oliveira, L. V. N., Wang, R., Specht, C. A., & Levitz, S. M. (2021). Vaccines for human fungal diseases: Close but still a long way to go. *NPJ Vaccines*, 6(1), 33. https://doi.org/10.1038/s41541-021-00294-8

del Olmo, V., Mixão, V., Fotedar, R., Saus, E., Al Malki, A., Księżopolska, E., Nunez-Rodriguez, J. C., Boekhout, T., & Gabaldón, T. (2023). Origin of fungal hybrids with pathogenic potential from warm seawater environments. *Nature Communications*, 14(1), 1–14. https://doi.org/10.1038/s41467-023-42679-4

Pagano, L., & Fernández, O. M. (2025). Clinical aspects and recent advances in fungal diseases impacting human health. *Journal of Antimicrobial Chemotherapy*, 80(Suppl 1), i2. https://doi.org/10.1093/JAC/DKAF004

Perlin, D. S., & Wiederhold, N. P. (2017). Culture-Independent molecular methods for detection of antifungal resistance mechanisms and fungal identification. *The Journal of Infectious Diseases*, 216(suppl_3), S458–S465. https://doi.org/10.1093/infdis/jix121

Petraitis, V., Petraitiene, R., Groll, A. H., Hermanns, R. C., & Walsh, T. J. (2003). Combination therapy in treatment of experimental pulmonary aspergillosis: Synergistic interaction between an antifungal triazole and an echinocandin. *Journal of Infectious Diseases*, 187(12), 1834–1843. https://doi.org/10.1086/375351

Petraitis, V., Petraitiene, R., & Walsh, T. J. (2017). Combination therapy with isavuconazole and micafungin for treatment of experimental invasive pulmonary aspergillosis. *Antimicrobial Agents and Chemotherapy*, 61(10), e00305-17. https://doi.org/10.1128/AAC.00305-17

Pfaller, M. A., & Diekema, D. J. (2007). Epidemiology of invasive candidiasis: A persistent public health problem. *Clinical Microbiology Reviews*, 20(1), 133. https://doi.org/10.1128/CMR.00029-06

Pfister, J. R., Archer, J. R., Hersil, S., Boers, T., Reed, K. D., Meece, J. K., Anderson, J. L., Burgess, J. W., Sullivan, T. D., Klein, B. S., Wheat, L. J., & Davis, J. P. (2011). Non-rural point source blastomycosis outbreak near a yard waste collection site. *Clinical Medicine and Research*, 9(2), 57–65. https://doi.org/10.3121/CMR.2010.958

Pham, C. D., Bolden, C. B., Kuykendall, R. J., & Lockhart, S. R. (2013). Development of a Luminex-based multiplex assay for detection of mutations conferring resistance to echinocandins in *Candida glabrata. Journal of Clinical Microbiology*, 52(3), 790–795. https://doi.org/10.1128/jcm.03378-13

Poplin, V., Smith, C., Caceres, D. H., Herkert, P. F., Jegede, O., Thompson, G. R., Baddley, J. W., Schwartz, I. S., Kubat, R., Deka, M. A., Toda, M., Lockhart, S. R., Chiller, T., Hagen, F., & Bahr, N. C. (2024). Geographical distribution of the *Cryptococcus gattii* species complex: A systematic review. *The Lancet Microbe*, *5*(12). https://doi.org/10.1016/S2666-5247(24)00161-7

Pristov, K., & Ghannoum, M. (2019). Resistance of Candida to azoles and echinocandins worldwide. *Clinical Microbiology and Infection*, 25(7), 792–798. https://doi.org/10.1016/j.cmi.2019.03.028

Rajasingham, R., Smith, R. M., Park, B. J., Jarvis, J. N., Govender, N. P., Chiller, T. M., Denning, D. W., Loyse, A., & Boulware, D. R. (2017). Global burden of disease of HIV-associated cryptococcal meningitis: An updated analysis. *The Lancet. Infectious Diseases*, 17(8), 873. https://doi.org/10.1016/S1473-3099(17)30243-8

Rautemaa-Richardson, R., Rautemaa, V., Al-Wathiqi, F., Moore, C. B., & Richardson, M. D. (2018). Impact of a diagnostics-driven antifungal stewardship programme in a UK tertiary referral teaching hospital. *Journal of Antimicrobial Chemotherapy*, 73(12), 3488–3495. https://doi.org/10.1093/jac/dky354

Restrepo, A., McEwen, J. G., & Castañeda, E. (2001). The habitat of *Paracoccidioides brasiliensis*: How far from solving the riddle? *Medical Mycology*, 39(3), 233–241. https://doi.org/10.1080/714031028

Rhodes, J., Abdolrasouli, A., Dunne, K., Sewell, T. R., Zhang, Y., Ballard, E., Brackin, A. P., Van Rhijn, N., Chown, H., Tsitsopoulou, A., Posso, R. B., Chotirmall, S. H., McElvaney, N. G., Murphy, P. G., Talento, A. F., Renwick, J., Dyer, P. S., Szekely, A., Bowyer, P., … Fisher, M. C. (2022). Population genomics confirms acquisition of drug-resistant *Aspergillus fumigatus* infection by humans from the environment. *Nature Microbiology*, 7(5), 663–674. https://doi.org/10.1038/s41564-022-01091-2

Rhodes, J., Beale, M. A., Vanhove, M., Jarvis, J. N., Kannambath, S., Simpson, J. A., Ryan, A., Meintjes, G., Harrison, T. S., Fisher, M. C., & Bicanic, T. (2017). A population genomics approach to assessing the genetic basis of Within-Host microevolution underlying recurrent cryptococcal meningitis infection. *G3 Genes Genomes Genetics*, *7*(4), 1165–1176. https://doi.org/10.1534/g3.116.037499

Robbins, N., Caplan, T., & Cowen, L. E. (2017). Molecular evolution of antifungal drug resistance. *Annual Review of Microbiology*, 71(1), 753–775. https://doi.org/10.1146/annurev-micro-030117-020345

Rodríguez-Cerdeira, C., Martínez-Herrera, E., Carnero-Gregorio, M., López-Barcenas, A., Fabbrocini, G., Fida, M., El-Samahy, M., & González-Cespón, J. L. (2020). Pathogenesis and clinical relevance of candida biofilms in vulvovaginal candidiasis. *Frontiers in Microbiology*, *11*. https://doi.org/10.3389/FMICB.2020.544480

Ruiz, G. B., & Lorenz, A. (2020). What do we know about the biology of the emerging fungal pathogen of humans, *Candida auris*? *Microbiological Research*, 242, 126621. https://doi.org/10.1016/j.micres.2020.126621

Schlager, A., Ashack, E., Khachemoune, K., Schlager, E., Ashack, K., Khachemoune, A., Facms, F., & Author, C. (2018). Erosio interdigitalis blastomycetica: A review of interdigital candidiasis. *Dermatology Online Journal*, 24(8). https://doi.org/10.5070/D3248041124

Selmecki, A., Forche, A., & Berman, J. (2006). Aneuploidy and isochromosome formation in drug-resistant *Candida albicans*. *Science*, 313(5785), 367–370. https://doi.org/10.1126/science.1128242

Seyedmousavi, S., Brüggemann, R. J., Melchers, W. J. G., Verweij, P. E., & Mouton, J. W. (2013). Efficacy and pharmacodynamics of voriconazole combined with anidulafungin in azole-resistant invasive aspergillosis. *Journal of Antimicrobial Chemotherapy*, 68(2), 385–393. https://doi.org/10.1093/jac/dks418

Shields, R. K., Nguyen, M. H., Press, E. G., & Clancy, C. J. (2014). Abdominal candidiasis is a hidden reservoir of echinocandin resistance. *Antimicrobial Agents and Chemotherapy*, 58(12), 7601–7605. https://doi.org/10.1128/aac.04134-14

Shields, R. K., Nguyen, M. H., Press, E. G., Kwa, A. L., Cheng, S., Du, C., & Clancy, C. J. (2012). The presence of an FKS mutation rather than MIC is an independent risk factor for failure of echinocandin therapy among patients with invasive candidiasis due to *Candida glabrata*. *Antimicrobial Agents and Chemotherapy*, 56(9), 4862–4869. https://doi.org/10.1128/aac.00027-12

Shor, E., & Perlin, D. S. (2015). Coping with stress and the emergence of multidrug resistance in fungi. *PLoS Pathogens*, 11(3), e1004668. https://doi.org/10.1371/journal.ppat.1004668

Singh, A., Healey, K. R., Yadav, P., Upadhyaya, G., Sachdeva, N., Sarma, S., Kumar, A., Tarai, B., Perlin, D. S., & Chowdhary, A. (2018). Absence of azole or echinocandin resistance in *Candida glabrata* isolates in India despite background prevalence of strains with defects in the DNA mismatch repair pathway. *Antimicrobial Agents and Chemotherapy*, *62*(6). https://doi.org/10.1128/aac.00195-18

Smith, D. F. Q., & Casadevall, A. (2023). Disaster mycology. *Biomedica*, 43(Sp. 1), 267–277. https://doi.org/10.7705/BIOMEDICA.6943

Steinmann, J., Hamprecht, A., Vehreschild, M. J. G. T., Cornely, O. A., Buchheidt, D., Spiess, B., Koldehoff, M., Buer, J., Meis, J. F., & Rath, P. (2015). Emergence of azole-resistant invasive aspergillosis in HSCT recipients in Germany. *Journal of Antimicrobial Chemotherapy*, 70(5), 1522–1526. https://doi.org/10.1093/jac/dku566

Suwunnakorn, S., Wakabayashi, H., & Rustchenko, E. (2016). Chromosome 5 of human pathogen *Candida albicans* carries multiple genes for negative control of caspofungin and anidulafungin susceptibility. *Antimicrobial Agents and Chemotherapy*, 60(12), 7457–7467. https://doi.org/10.1128/aac.01888-16

Talapko, J., Juzbašić, M., Matijević, T., Pustijanac, E., Bekić, S., Kotris, I., & Škrlec, I. (2021). *Candida albicans*—The virulence factors and clinical manifestations of infection. *Journal of Fungi*, 7(2), 79. https://doi.org/10.3390/JOF7020079

Talento, A. F., Qualie, M., Cottom, L., Backx, M., & White, P. L. (2021). Lessons from an educational invasive fungal disease conference on hospital antifungal stewardship practices across the UK and Ireland. *Journal of Fungi*, 7(10), 801. https://doi.org/10.3390/jof7100801

Tängdén, T., Martín, V. R., Felton, T. W., Nielsen, E. I., Marchand, S., Brüggemann, R. J., Bulitta, J. B., Bassetti, M., Theuretzbacher, U., Tsuji, B. T., Wareham, D. W., Friberg, L. E., De Waele, J. J., Tam, V. H., & Roberts, J. A. (2017). The role of infection models and PK/PD modelling for optimising care of critically ill patients with severe infections. *Intensive Care Medicine*, 43(7), 1021–1032. https://doi.org/10.1007/s00134-017-4780-6

Tavanti, A., Hensgens, L. A. M., Ghelardi, E., Campa, M., & Senesi, S. (2007). Genotyping of *Candida orthopsilosis* clinical isolates by amplification fragment length polymorphism reveals genetic diversity among independent isolates and strain maintenance within patients. *Journal of Clinical Microbiology*, 45(5), 1455–1462. https://doi.org/10.1128/JCM.00243-07

Taylor, M. L., del Rocío Reyes-Montes, M., Estrada-Bárcenas, D. A., Zancopé-Oliveira, R. M., Rodríguez-Arellanes, G., & Ramírez, J. A. (2022). Considerations about the geographic distribution of histoplasma species. *Applied and Environmental Microbiology*, *88*(7). https://doi.org/10.1128/AEM.02010-21

Van Rhijn, N., & Bromley, M. (2021). The consequences of our changing environment on life threatening and debilitating fungal diseases in humans. *Journal of Fungi*, 7(5), 367. https://doi.org/10.3390/JOF7050367

Verweij, P. E., Chowdhary, A., Melchers, W. J. G., & Meis, J. F. (2015). Azole resistance in *Aspergillus fumigatus*: Can we retain the clinical use of mold-active antifungal azoles? *Clinical Infectious Diseases*, 62(3), 362–368. https://doi.org/10.1093/cid/civ885

Verweij, P. E., Lucas, J. A., Arendrup, M. C., Bowyer, P., Brinkmann, A. J. F., Denning, D. W., Dyer, P. S., Fisher, M. C., Geenen, P. L., Gisi, U., Hermann, D., Hoogendijk, A., Kiers, E., Lagrou, K., Melchers, W. J. G., Rhodes, J., Rietveld, A. G., Schoustra, S. E., Stenzel, K., … Fraaije, B. A. (2020). The one health problem of azole resistance in *Aspergillus fumigatus*: Current insights and future research agenda. *Fungal Biology Reviews*, 34(4), 202–214. https://doi.org/10.1016/J.FBR.2020.10.003

Vinayagamoorthy, K., Gangavaram, D. R., Skiada, A., & Prakash, H. (2023). Emergomycosis, an emerging thermally dimorphic fungal infection: A systematic review. *Journal of Fungi*, 9(10). https://doi.org/10.3390/JOF9101039

Wang, F., Han, R. H., & Chen, S. (2023). An overlooked and underrated endemic mycosis—Talaromycosis and the pathogenic fungus *Talaromyces marneffei*. *Clinical Microbiology Reviews*, 36(1). https://doi.org/10.1128/CMR.00051-22

Welsh, R. M., Bentz, M. L., Shams, A., Houston, H., Lyons, A., Rose, L. J., & Litvintseva, A. P. (2017). Survival, persistence, and isolation of the emerging multidrug-resistant pathogenic yeast *Candida auris* on a plastic health care surface. *Journal of Clinical Microbiology*, 55(10), 2996–3005. https://doi.org/10.1128/JCM.00921-17

Whaley, S. G., Berkow, E. L., Rybak, J. M., Nishimoto, A. T., Barker, K. S., & Rogers, P. D. (2017). Azole antifungal resistance in *Candida albicans* and emerging non-albicans Candida species. *Frontiers in Microbiology*, 7(JAN), 2173. https://doi.org/10.3389/FMICB.2016.02173

Whitney, L., Ashley, E. D., & Barlam, T. F. (2019). Effectiveness of an antifungal stewardship programme at a London teaching hospital 2010–16. *Journal of Antimicrobial Chemotherapy*, 74(1), 234–241. https://doi.org/10.1093/jac/dky404

WHO. (2018). *Guidelines for The Diagnosis, Prevention and Management of Cryptococcal Disease in HIV-Infected Adults, Adolescents and Children: Supplement to the 2016 Consolidated Guidelines on the Use of Antiretroviral Drugs for Treating and Preventing HIV Infection*. https://www.ncbi.nlm.nih.gov/books/NBK531449/

WHO Fungal Priority Pathogens List to Guide Research, Development and Public Health Action. (2022). Retrieved April 24, 2025, from https://www.who.int/publications/i/item/9789240060241

WHO Model List of Essential Medicines -22nd List, 2021. (2021). Retrieved April 24, 2025, from https://www.who.int/publications/i/item/WHO-MHP-HPS-EML-2021.02

Williams, A. P., Cook, E. R., Smerdon, J. E., Cook, B. I., Abatzoglou, J. T., Bolles, K., Baek, S. H., Badger, A. M., & Livneh, B. (2020). Large contribution from anthropogenic warming to an emerging North American megadrought. *Science*, 368(6488), 314–318. https://doi.org/10.1126/science.aaz9600

Zamith-Miranda, D., Heyman, H. M., Cleare, L. G., Couvillion, S. P., Clair, G. C., Bredeweg, E. L., Gacser, A., Nimrichter, L., Nakayasu, E. S., & Nosanchuk, J. D. (2019). Multi-omics signature of *Candida auris*, an emerging and multidrug-resistant pathogen. *mSystems*, *4*(4). https://doi.org/10.1128/msystems.00257-19

Zeng, Z., Ziegler, A. D., Searchinger, T., Yang, L., Chen, A., Ju, K., Piao, S., Li, L. Z. X., Ciais, P., Chen, D., Liu, J., Azorin-Molina, C., Chappell, A., Medvigy, D., & Wood, E. F. (2019). A reversal in global terrestrial stilling and its implications for wind energy production. *Nature Climate Change*, 9(12), 979–985. https://doi.org/10.1038/s41558-019-0622-6?urlappend=%3Futm_source%3Dresearchgate.net%26utm_medium%3Darticle

Zheng, J., Gui, X., Cao, Q., Yang, R., Yan, Y., Deng, L., & Lio, J. (2015). A clinical study of acquired immunodeficiency syndrome associated penicillium marneffei infection from a non-endemic area in China. *PLoS ONE*, *10*(6). https://doi.org/10.1371/journal.pone.0130376

2 Medicinal Plants and Their Antifungal Potential

Ancient Knowledge and Modern Science

Seema Raj, Manoj Kumar, and Nidhee Chaudhary

2.1 INTRODUCTION

Throughout human history, medicinal plants have been revered as nature's pharmacy, deeply integrated into the healthcare practices of civilizations around the world. From ancient Egypt to the Indian subcontinent, and from Chinese dynasties to indigenous communities in Africa and the Americas, plants were the cornerstone of traditional healing systems. Ancient texts like the *Charaka Samhita* (Ayurveda), *Shennong Bencao Jing* (Traditional Chinese Medicine), and *De Materia Medica* (Greek and Roman herbal knowledge) document the therapeutic uses of thousands of plant species, many of which are still in use today.

One of the most significant applications of these plants was the treatment of fungal infections conditions, which were prevalent in humid, tropical environments and were often poorly understood. Traditional healers, without the benefit of microscopes or pharmacological testing, relied on generations of empirical knowledge, observing outcomes, noting synergistic effects in polyherbal formulations, and refining remedies based on community wisdom and ecological availability. Plants such as *Azadirachta indica* (Neem), *Curcuma longa* (Turmeric), *Allium sativum* (Garlic), and *Ocimum sanctum* (Tulsi) became mainstays of folk medicine for their broad-spectrum antimicrobial effects, including antifungal properties. We used to listen "Dadi maa ke nuskhe"; these remedies are impactful in daily life to treat or prevent any disease.

In contrast, modern science approaches medicinal plants through a reductionist and evidence-based lens. Technological advances in analytical chemistry, phytochemistry, and molecular biology have made it possible to isolate specific bioactive constituents, elucidate their structures, and determine their mechanisms of action. For example, compounds like allicin (from garlic), curcumin (from turmeric), and eugenol (from holy basil) have been shown to disrupt fungal cell membranes, inhibit ergosterol synthesis, or modulate immune responses, thereby offering a scientific explanation for the therapeutic efficacy observed in traditional usage.

What distinguishes modern science from ancient practice is quantitation, repeatability, and verification. In vitro and in vivo assays permit controlled determination

 DOI: 10.1201/9781003633952-3

of antifungal activity, and clinical trials define safety profiles and optimal dosages. Molecular docking, gene expression profiling, and high-throughput screening now make it possible to identify potential antifungal agents more quickly and with greater specificity than ever before.

But increasingly clear is that ancient and contemporary knowledge systems are complementary, not competitive.

Ethnobotanical data is a rich source of scientific leads, frequently pointing researchers to species of considerable therapeutic potential. Conversely, modern testing strengthens the validity and incorporation of traditional medicine into modern medicine. Additionally, with the globe struggling to cope with the increasing threat of antifungal resistance, there is an increasing interest in plant-derived alternatives that are not only effective but are less prone to evoke resistance and with fewer side effects. Plants have been the origin of ancient systems of medicine, providing a synergistic, holistic intervention in the form of multi-compound formulations, which current science can now investigate using systems biology and omics tools at its command.

This book functions, therefore, as a bridge between the ancient knowledge that is inherent in indigenous healing systems and the reality of modern biomedical science. This book attempts to review the antifungal prospects of medicinal plants systematically by pointing out relevant species, describing their traditional applications, detailing their phytochemical profiles, and summarizing pharmacological validations. In the process, it underscores the need for a transdisciplinary approach that integrates ethnobotany, phytochemistry, pharmacology, and clinical science.

Finally, bringing together the old and new paradigms promises not only increased knowledge of medicinal plants but also pathways to new, sustainable, and still universally available antifungal medicine. As science works to deconstruct the intricate bioactive matrices of the plants, the knowledge of the old healers is reborn and reaffirmed in the 21st century medicine world.

Although the ethnomedical tradition of Ayurveda, Traditional Chinese Medicine, and Greco-Roman pharmacopoeias offers a rich tradition, this chapter focuses specifically on their antifungal aspects. This historical precedent highlights that the preeminent position of plants in antifungal therapy is not just cultural but also scientifically pertinent in responding to today's growing challenge of antifungal resistance.

2.2 ETHNOBOTANICAL PERSPECTIVES ON ANTIFUNGAL MEDICINAL PLANTS

Ethnobotany refers to the scientific study of human complex reliance on plants, including indigenous communities' knowledge of plant use, particularly in medicine. Medicinal plants from across the globe's diverse communities have been used traditionally to successfully cure numerous diseases, including fungal diseases. Compilations of ethnobotanical knowledge passed from generation to generation provide valuable information concerning the antifungal potential of the majority of plant species (Das et al., 2025).

2.2.1 Traditional Knowledge and Antifungal Applications

Ethnobotanical studies always point to the role of local traditions in the identification of fungicidal plants against fungal diseases. In Morocco's Settat Province, for example, thorough ethnobotanical investigation isolated around 51 medicinal and aromatic plant species that are most commonly used to control various infectious diseases, including fungal infection. The *Lamiaceae*, *Apiaceae*, and *Fabaceae* families were prominent because they possess known therapeutic values (El-Assri et al., 2021).

Also, in the Pader district of Uganda, the traditional communities employ species like *Warburgia ugandensis* and *Eucalyptus globulus* for their antifungal property, particularly against candidiasis. These uses have been corroborated in recent laboratory tests, showing potent antifungal activity against disease-causing species like *Candida albicans*, thus linking traditional ethnobotanical claims with current scientific verification (Akwongo et al., 2025).

2.2.2 Mechanisms of Antifungal Action

The antifungal property of medicinal plants is attributed to the presence of varied phytochemical components like alkaloids, flavonoids, essential oils, phenolic compounds, and terpenoids. The latter ones are antifungal with different mechanisms of action, such as fungal cell membrane disruption, inhibition of ergosterol biosynthesis—a critical fungal cell membrane component—and interference with crucial enzymatic functions.

The most recent extensive review that was conducted on Asian medicinal plants described bioactive phytochemicals of different plant species with high antifungal activity. These compounds exhibited inhibitory action against mycelial growth, inhibit ergosterol biosynthesis pathways, and drastically lowered fungal viability, further supplementing the ethnobotanical data from conventional healers (Alam et al., 2025).

2.2.3 Regional Ethnobotanical Insights

Ethnobotanical studies from different regions offer unique insights into how various cultures approach fungal infections using indigenous plant resources:

Asia: Plants are typically employed by traditional practitioners in Tamil Nadu, India, such as *Acalypha indica*, *Allium sativum*, and *Azadirachta indica* (Neem) for antifungal treatment infections. They have been proven scientifically to possess good antifungal properties, which are compounds whose traditional uses are justified (Kiruthik et al., 2025).

Africa: South Africa's Limpopo Province boasts a rich treasury of traditional medicinal uses, such as antifungal uses. *Artemisia afra* and *Solanum incanum* are used traditionally to fight oral candidiasis, and recent studies have verified their effectiveness in treating *C. albicans* infections (Tlaamela et al., 2024).

Ethiopia: In Ethiopia's Metema district, ethnobotanical surveys have yielded dozens of significant medicinal herbs used by local practitioners. Appropriate among them are *Ziziphus spina-christi* and *Ficus sycomorus*, plants traditionally valued for their multifaceted therapeutic roles, including antifungal applications (Tadesse et al., 2025).

2.2.4 Conservation and Sustainable Use

As global demand for plant antifungals increases, there is a need for sustainable medicinal plant management. Excessive collection threatens plant diversity and devalues traditional use. Ethnobotanical research demands sustainable collection through the active engagement of local communities in conservation in order to protect plant resources and preserve rich indigenous knowledge systems (Dutta et al., 2022).

Ethnobotanical research plays an important role in the development of the knowledge and application of the antifungal properties of medicinal plants. The combination of traditional knowledge and recent scientific confirmation facilitates the development of new antifungal drugs. At the same time, promotion of sustainable use preserves precious medicinal plants and cultural values inherent in traditional knowledge.

2.3 KEY MEDICINAL PLANTS WITH ANTIFUNGAL PROPERTIES

To enhance coherence, medicinal plants discussed in this section can be broadly categorized according to their dominant phytochemical groups:

- Organosulfur-rich species (e.g., *A. sativum*),
- Polyphenol- and tannin-rich plants (e.g., *Punica granatum*, *Camellia sinensis*),
- Terpenoid-dominant essential oils (e.g., *O. sanctum*, *Thymus vulgaris*, *Mentha piperita*, Eucalyptus),
- Alkaloid-rich species (e.g., *Berberis aristata*, *Glycyrrhiza glabra*), etc.

Medicinal plants have been the backbone of traditional medicine across the globe for centuries, and they are an excellent source of medicinal value bioactive compounds. Among their multiple uses, some of them exhibit diverse bioactivities including antifungal, antibacterial, and radioprotective effects (Raj et al., 2022). Various studies are available, which shows the potential of medicinal plants regarding antifungal activity (Alam et al., 2025). Among all the benefits, this section highlights some significant medicinal plants with antifungal activity, and the evidence comes from the latest scientific studies (Table 2.1).

2.3.1 Gymnema sylvestre (Gurmar)

Gymnema sylvestre, traditionally recognized for its antidiabetic properties, has recently attracted attention for its antifungal potential. A groundbreaking study published in 2025 demonstrated that saponins from *G. sylvestre* exhibit significant inhibitory activity against pathogenic fungi including *C. albicans* and *Aspergillus niger*. Both in vitro and in silico analyses revealed that gymnemic acids interact with fungal membrane components, particularly by inhibiting ergosterol synthesis—a vital process for maintaining fungal cell membrane integrity. These findings suggest that *G. sylvestre* may represent a novel source of antifungal agents with potential therapeutic applications (Neel et al., 2025).

TABLE 2.1
Studies on Different Medicinal Plants

Medicinal Plant	Finding	Active Phytochemicals	Fungal Targets	Mode of Action	Citation in APA Style
Gymnema sylvestre (Gurmar)	Demonstrated inhibitory effects	Gymnemic acids (triterpenoid saponins)	*Candida albicans* and *Aspergillus niger*	Disrupts membrane integrity by binding ergosterol, increases permeability, and inhibits cell wall biosynthesis	Neel et al. (2025)
Allium sativum (Garlic)	Antifungal activity	Allicin, Ajoene	*Candida albicans* and *Trichophyton rubrum*	Disrupts thiol enzymes, inhibits ergosterol biosynthesis, and damages membranes	Feknous et al. (2024)
Azadirachta indica (Neem)	Antifungal activity	Azadirachtin, Nimbidin, Triterpenoids	*Candida albicans* and *Aspergillus niger*	Disrupts membrane and inhibits fungal growth	Mohanasundaram et al. (2023)
Punica granatum (Pomegranate)	Antifungal activity	Ellagic acid, Punicalagins, Polyphenols	*Candida albicans*	Inhibits adhesion and mycelial growth	Gowda et al. (2023)
Eucalyptus globulus (Eucalyptus)	Antifungal activity	Eucalyptol, Terpenes	*Candida albicans*	Disrupts membrane and inhibits respiratory enzymes	Akwongo et al. (2025)
Curcuma longa (Turmeric)	Antifungal Activity	Curcumin	*Candida albicans* and *Aspergillus niger*	Disrupts membranes and inhibits hyphal growth	Alajlani (2023)
Ocimum sanctum (Holy Basil)	Antifungal activity	Eugenol, Flavonoids	*Candida* spp. and Dermatophytes	Disrupts ergosterol and inhibits hyphal growth	Alam et al. (2025)
Cuminum cyminum (Cumin)	Antifungal activity	Cuminaldehyde	*Aspergillus flavus*	Inhibits spore germination and suppresses growth	Kedia et al. (2014)
Illicium verum (Star Anise)	Antifungal activity and antiviral	Shikimic acid, Anethole	*Candida albicans*	Interferes with nucleic acid synthesis	Verywell Health (2024)
Zingiber officinale (Ginger)	Antifungal activity	Gingerol, Shogaol, Paradol	*Candida albicans* and *Aspergillus* spp.	Disrupts cell membrane and inhibits mycelial growth	Eldesouky et al. (2023)

(*Continued*)

TABLE 2.1 (*Continued*)
Studies on Different Medicinal Plants

Medicinal Plant	Finding	Active Phytochemicals	Fungal Targets	Mode of Action	Citation in APA Style
Mentha piperita (Peppermint)	Antifungal activity	Menthol, Menthone	*Candida* spp., *Fusarium* spp., and *Aspergillus* spp.	Disrupts membrane and inhibits enzyme	Vakili-Ghartavol et al. (2024)
Nigella sativa (Black Seed)	Antifungal activity	Thymoquinone	*Trichophyton mentagrophytes, Microsporum canis*, and *Microsporum gypseum*	Inhibits hyphal growth and induces oxidative stress	Mahmoudvand et al. (2014)
Cinnamomum zeylanicum (Cinnamon)	Antifungal activity	Cinnamaldehyde, Eugenol	*Rhizopus nigricans, Aspergillus flavus*, and *Penicillium expansum*	Disrupts cell wall integrity and inhibits the hyphal growth	Xing et al. (2010)
Berberis aristata (Indian Barberry)	Antifungal activity	Berberine (alkaloid)	*Candida albicans* and Dermatophytes	Inhibits DNA synthesis and induces apoptosis	Huang et al. (2022)
Calendula officinalis (Marigold)	Antifungal activity	Flavonoids, Saponins, Triterpenoids	*Candida albicans*	Disrupts Membrane and inhibits spore formation	Ahmed et al. (2023)
Camellia sinensis (Green Tea)	Antifungal activity	EGCG (polyphenol)	*Candida* spp.	Inhibits adhesion and forms biofilm	do Monte et al. (2024)
Lawsonia inermis (Henna)	Antifungal activity	Lawsone (quinone)	*Candida albicans*	Generates ROS and disrupts membrane	Hassan et al. (2022)
Piper nigrum (Black Pepper)	Antifungal activity	Piperine, Essential oils	*Aspergillus flavus*	Inhibits the growth	Zhang et al. (2021)
Terminalia chebula (Haritaki)	Antifungal activity	Tannins, Phenolics	Dermatophytes and *Candida* spp.	Precipitates protein and inhibits enzyme	Regar et al. (2025)

2.3.2 *Allium sativum* (Garlic)

Garlic has been utilized for centuries for therapeutic purposes, and its antifungal activity is primarily attributed to organosulfur compounds such as allicin and ajoene. Ajoene has been shown to be active against a range of fungal pathogens such as *C. albicans* and *Trichophyton rubrum*. Clinical trials have proven that topically applied ajoene is as effective as conventional antifungal therapy for tinea pedis (athlete's foot) (Feknous et al., 2024).

2.3.3 *Azadirachta indica* (Neem)

The broad-spectrum antibacterial properties of neem are well known in traditional medicine. Its antifungal action is mainly due to substances like azadirachtin and nimbidin, which destroy fungal cell membranes and inhibit growth. Neem has been shown in experiments to be effective against a variety of fungal species, including *Aspergillus* and *Candida* species (Alajlani, 2023).

2.3.4 *Punica granatum* (Pomegranate)

Extracts from pomegranates have strong antifungal properties against oral infections. The fruit's antifungal action is due to its high polyphenol content, which includes ellagic acid and punicalagins. In vitro, pomegranate extracts have been demonstrated to suppress *C. albicans* development, indicating that they may find use in oral healthcare (Kiruthik et al., 2025).

2.3.5 *Eucalyptus globulus* (Eucalyptus)

Eucalyptus globulus oil contains terpenes and eucalyptol, which are antifungal in nature. Such antifungal activity has been identified in *Candida* species. Eucalyptus oil is also useful in the treatment of respiratory and skin infections (Akwongo et al., 2024).

2.3.6 *Curcuma longa* (Turmeric)

Turmeric is rich in curcumin, which has anti-inflammatory and antimicrobial properties. It has been studied that curcumin has antifungal properties against *C. albicans* and *A. niger*. It has a property to destruct cell membranes of fungal cells, which is why curcumin is used in antifungal medicines (Alajlani, 2023).

2.3.7 *Ocimum sanctum* (Holy Basil/Tulsi)

In Hinduism, Tulsi is a sacred plant, which has enormous medicinal properties. The presence of eugenol and other constituents makes Tulsi oil as antifungal medicine. The extract of *O. sanctum* (Tulsi) inhibits growth of the *Candida* species and is also useful in skin infections as well (Alam et al., 2025).

2.3.8 *Thymus vulgaris* (Thyme)

Thyme has been proven to include strong essential oils including carvacrol and thymol, both of which have shown strong antifungal effects. Thyme oil has been suggested as a naturally obtained antifungal cure because it can stop the growth of several fungus, including *C. albicans* and *Aspergillus* spp. (Alajlani, 2023).

2.3.9 Glycyrrhiza glabra (Licorice)

Licorice (*Glycyrrhiza glabra*) carries flavonoids and glycyrrhizin, which give licorice root its antifungal properties. Research indicates that licorice extracts have the capacity to suppress the development of *C. albicans*, therefore implying that they may be a promising treatment for fungal infections—particularly those occurring orally (Kiruthik et al., 2025).

2.3.10 Cuminum cyminum (Cumin)

Cuminaldehyde is one of the substances and essential oils found in abundance in cumin seeds. Cumin extracts have been proven to inhibit the development of several fungal species, including *C. albicans*, and may therefore be used in antifungal therapy (Kedia et al., 2014).

2.3.11 Illicium verum (Star Anise)

Star anise has antifungal and antiviral properties. The presence of Shikimic Acid represents antimicrobial activity (Verywell Health, 2024).

In conclusion, there are various medicinal plants that have potential as antimicrobial drugs. Nature has all self-healing responses, only to utilize the exploration done by the researchers.

2.3.12 Zingiber officinale (Ginger)

Ginger, commonly used in culinary and medicinal practices, is rich in bioactive compounds such as gingerol, shogaol, and paradol. It exhibits notable antifungal activity, especially against *C. albicans*, by disrupting cell membrane integrity and inhibiting mycelial growth. Its essential oil has also shown efficacy against dermatophytes and *Aspergillus* species (Eldesouky et al., 2023).

2.3.13 Mentha piperita (Peppermint)

Peppermint essential oil, composed mainly of menthol and menthone, has been shown to inhibit fungal growth through membrane disruption and interference with enzyme activity. It is effective against *Candida*, *Fusarium*, and *Aspergillus* species (Vakili-Ghartavol et al., 2024).

2.3.14 Nigella sativa (Black Seed)

Black seed oil contains thymoquinone, an active compound known for antifungal, anti-inflammatory, and antioxidant properties. Studies indicate strong inhibition against *Candida* spp. and some molds (Mahmoudvand et al., 2014).

2.3.15 Cinnamomum zeylanicum (Cinnamon)

Cinnamon essential oil, rich in cinnamaldehyde and eugenol, has demonstrated potent antifungal activity. It disrupts fungal cell wall integrity and inhibits hyphal development. It is effective against *Rhizopus nigricans, Aspergillus flavus*, and *Penicillium expansum* (Xing et al., 2010).

2.3.16 Berberis aristata (Indian Barberry)

The active compound berberine possesses strong antifungal and antimicrobial effects. It disrupts fungal metabolism and DNA synthesis. It is effective particularly against *C. albicans* and dermatophytes (Huang et al., 2022).

2.3.17 Calendula officinalis (Marigold)

Calendula contains flavonoids, saponins, and triterpenoids, which exhibit antifungal activity. It is traditionally used to treat wounds and fungal skin infections and has shown in vitro inhibition against *C. albicans* (Ahmed et al., 2023).

2.3.18 Camellia sinensis (Green Tea)

Green tea polyphenols, especially epigallocatechin gallate (EGCG), have antifungal properties. They inhibit fungal adhesion and biofilm formation, particularly in *Candida* species (do Monte et al., 2024).

2.3.19 Lawsonia inermis (Henna)

Henna leaves, containing lawsone (a naphthoquinone derivative), are known for antifungal and antibacterial activity. It is used traditionally for skin infections and fungal lesions (Hassan et al., 2022)

2.3.20 Piper nigrum (Black Pepper)

The piperine-rich essential oil of black pepper has shown antifungal activity against various fungi, particularly *Aspergillus flavus.* It inhibits fungal cell wall synthesis and growth (Zhang et al., 2021).

2.3.21 Terminalia chebula (Haritaki)

Haritaki has demonstrated antifungal properties through its high tannin and phenolic content. It is effective in treating fungal infections in traditional medicine, particularly skin mycoses (Regar et al., 2025).

2.4 BIOACTIVE PHYTOCHEMICALS WITH ANTIFUNGAL ACTIVITY

Herbal plants synthesize a range of bioactive secondary metabolites that are important for their microbial resistance to pathogens like fungi. They have diverse antifungal activities—encompassing inhibition of fungal cell membranes to interference with ergosterol biosynthesis inhibition, spore germination inhibition, and oxidative stress induction. Recent research continues to unveil the therapeutic value of these medications, which are more being studied as alternatives or adjuvants to synthetic antifungal drugs. A few examples of phytochemicals are represented in Figure 2.1.

2.4.1 Alkaloids

Alkaloids such as berberine, sanguinarine, and chelerythrine are nitrogenous bases that occur widely found in medicinal plants. Berberine, for instance, is antifungal by inhibiting mitochondrial function and cell membrane stability in *C. albicans* and *Trichophyton* spp. (Huang et al., 2022; Shah et al., 2025). The molecules also induce ROS production, leading to fungal apoptosis. Alkaloids like berberine block fungal mitochondrial respiration through interference with electron transport and hence ATP synthesis. This bioenergetic failure is frequently combined with interference with nucleic acid synthesis and the induction of programmed cell death.

2.4.2 Flavonoids

Flavonoids like quercetin, kaempferol, and apigenin have strong antifungal activities, modulate fungal enzyme systems, inhibit hyphal elongation, and synergize with antifungal drugs (Al Aboody & Mickymaray, 2020; Aliaghazadeh et al., 2025). Interestingly, quercetin suppresses biofilm development in *Candida* spp., enhancing the efficacy of fluconazole (do Monte et al., 2024). Flavonoids not only inhibit hyphal elongation but also modulate critical enzymes like chitin synthase and β-glucan synthase, which play a central role in fungal cell wall biosynthesis. Their dual interference renders them effective synergistic compounds with the traditional antifungals.

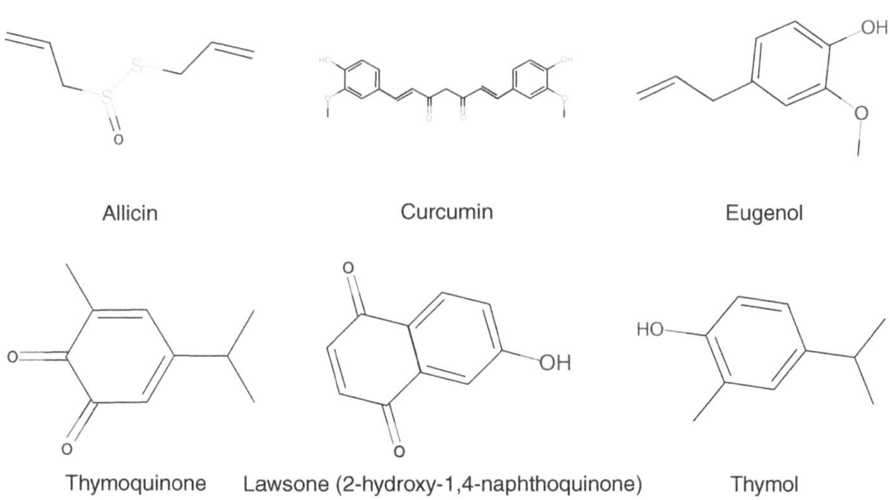

FIGURE 2.1 Some selected phytochemicals.

2.4.3 Terpenoids and Essential Oils

Terpenoids, especially thymol, carvacrol, menthol, and eugenol, are most common in essential oils of plant origin like *T. vulgaris, O. sanctum, and M. piperita*. These act on ergosterol from the fungal membranes, leading to leakage of cytoplasmic contents (Vakili-Ghartavol et al., 2024; Memar et al., 2017; Mezzomo et al., 2025; Sarathambal et al., 2021; Yuan et al., 2024). These drugs disrupt ergosterol in the fungal plasma membrane, resulting in cytoplasmic leakage and loss of viability. They can also make resistance development less likely by targeting several cellular sites at once.

2.4.4 Phenolics and Tannins

Phenolic compounds such as gallic acid, caffeic acid, and ellagic acid possess antifungal properties through protein precipitation and membrane destabilization (Kiruthik et al., 2025). Tannins interact with fungal enzymes, suppressing mycelial growth and spore germination, especially in *Aspergillus fumigatus and Fusarium solani* (Ríos-López et al., 2024).

2.4.5 Saponins

Saponins are glycosides with amphipathic structures capable of forming pores in fungal membranes. Saponins, found in plants like *G. glabra* and *Calendula officinalis*, disrupt membrane sterols and reduce fungal viability (Ahmed et al., 2023). They have also demonstrated potential in enhancing permeability for co-administered antifungal agents.

2.4.6 Coumarins and Quinones

Coumarins (e.g., scopoletin) and quinones (e.g., lawsone and juglone) generate ROS within fungal cells, causing oxidative damage to nucleic acids and proteins. Lawsone from *Lawsonia inermis* has shown efficacy against *C. albicans* and *T. rubrum* (Ali & Khan, 2018; Dar et al., 2024). These compounds also target mitochondrial respiration and ATP synthesis. Along with ROS production, coumarins and quinones disrupt mitochondrial membrane potential, resulting in malfunction in ATP synthesis. The cascade of oxidative stress and bioenergetic failure culminates in apoptosis of the fungus.

2.4.7 Organosulfur Compounds

Organosulfur compounds like allicin and ajoene, primarily found in *A. sativum*, disrupt thiol-containing enzymes, leading to inhibition of key metabolic activities in fungi (Feknous et al., 2024; Khounganian et al., 2023). They are effective against dermatophytes, yeasts, and molds.

2.4.8 Plant Defensins

Plant defensins are small cysteine-rich peptides with high specificity for fungal pathogens. They bind to fungal glucosylceramides and permeabilize membranes. Defensins isolated from *Raphanus sativus* and *Medicago truncatula* have exhibited significant antifungal activity against *Candida* spp., *Aspergillus fumigatus*, and *Fusarium* spp. (Terras et al., 1992; Barimani et al., 2020).

The vast diversity of bioactive phytochemicals in medicinal plants contributes to their broad antifungal efficacy. These compounds not only act via multiple mechanisms but also have potential to overcome resistance limitations posed by synthetic drugs. Ongoing bioassay-guided fractionation, omics-based approaches, and pharmacokinetic studies will be instrumental in developing these phytochemicals into clinically relevant antifungal agents.

2.5 SCIENTIFIC VALIDATION AND PHARMACOLOGICAL RESEARCH

The transition from traditional ethnobotanical knowledge to modern pharmacological application necessitates rigorous scientific validation. Although medicinal plants have been historically employed for antifungal purposes, the empirical evaluation of their safety, efficacy, active compounds, and mechanisms of action is essential for integrating them into formal healthcare systems. Over the past two decades, numerous pharmacological studies have substantiated the antifungal potential of medicinal plants through in vitro, in vivo, and clinical investigations, employing contemporary analytical techniques such as chromatography, spectroscopy, and molecular docking.

2.5.1 Mechanistic Insights and In Vitro Studies

In vitro studies continue to be the initial step toward the validation of antifungal activities of medicinal plants. The studies are generally performed using agar diffusion assays, broth microdilution techniques, and scanning electron microscopy for the observation of morphological alterations in fungal cells. For example, *A. sativum* showed potential inhibition against *C. albicans, A. niger, and Cryptococcus neoformans* (Khounganian et al., 2023). *A. indica* leaf extract exhibited triterpenoid compounds, which shows antifungal properties (Mohanasundaram & Saral, 2023).

2.5.2 In Vivo Models and Toxicological Evaluation

To validate in vitro results, plant extracts are administered in animal to determine pharmacokinetics, toxicity, and therapeutic effect. A recent study with *Curcuma longa* in a murine model of candidiasis reported that curcumin substantially lowered fungal burden in essential organs without hepatotoxicity or nephrotoxicity (Cheraghipour et al., 2021). In another experiment, *Nigella sativa* oil presented rapid elimination of *C. albicans* wild-type (C1 and C2) and URA3 delete (CAF2 and CAI4) strains (Egbe, 2023).

2.5.3 Phytochemical Isolation and Structural Elucidation

Identification and characterization of bioactive compounds are representation of pharmacological validation. For example: Gas chromatography–mass spectrometry (GC-MS), Liquid chromatography–mass spectrometry (LC-MS), and Nuclear magnetic resonance (NMR) spectroscopy have been widely employed. Thymol and carvacrol isolated from *T. vulgaris* were found to inhibit with fungal plasma membranes through ergosterol interaction (Corona-Gómez et al., 2022). Similarly, berberine isolated from *B. aristata* possesses robust antifungal activity through the promotion of apoptosis against *C. albicans* (Huang et al., 2022).

2.5.4 Molecular Docking and Computational Pharmacology

Recent advances in *in silico* tools have enabled researchers to study ligand–target interactions, which can predict pharmacokinetic profiles, and identify drug interaction properties. Molecular docking of allicin from *A. sativum* showed high binding affinity with fungal lanosterol 14α-demethylase, a key enzyme in ergosterol biosynthesis, indicating its potential as a lead antifungal agent (Pereira et al., 2021). Similar computational modeling of eugenol and other demonstrated effective inhibition of fungal virulence factors, providing a predictive framework for drug design (Elsewedy et al., 2024).

2.5.5 Clinical Studies and Human Trials

Despite promising laboratory evidence, relatively few clinical trials have assessed the efficacy of plant-based antifungals. A randomized clinical trial evaluating ajoene cream derived from garlic for treating tinea pedis showed similar efficacy to 1% terbinafine cream with minimal side effects (Feknous et al., 2024). More recently, pomegranate peel extract was tested in oral candidiasis, showing substantial reduction in fungal counts and symptom relief (Gowda et al., 2023).

2.5.6 Standardization, Formulation, and Delivery Systems

One of the major limitations in herbal drug development is variability in the quality of plant extracts. Efforts have been made to standardize formulations using validated analytical methods (Figure 2.2). Novel delivery systems such as nanoemulsions,

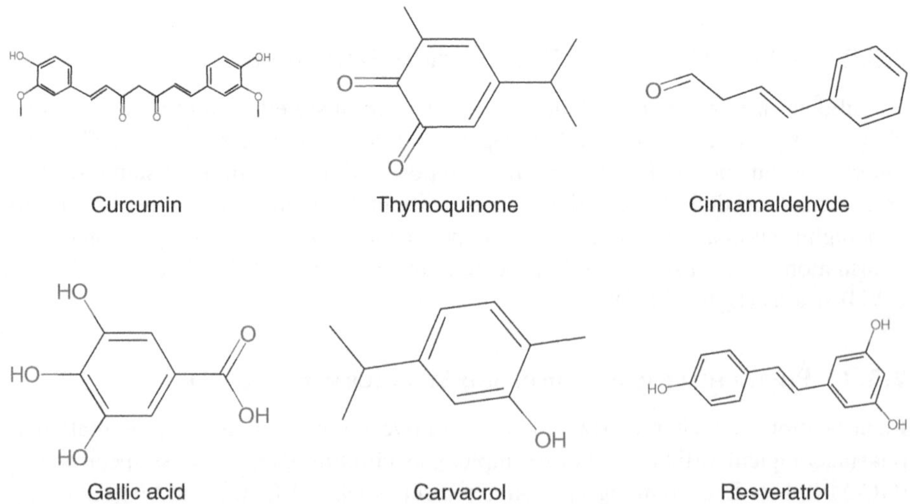

FIGURE 2.2 Structures of recent plant-derived molecules with antifungal/medicinal uses.

liposomes, and hydrogels are also being employed to enhance the solubility, bioavailability, and target specificity of plant-based antifungal agents. For instance, curcumin-loaded liposomes were found to enhance antifungal efficacy and reduce systemic toxicity in an animal model of systemic mycosis (Cheraghipour et al., 2021).

The scientific validation has greatly progressively the acceptance of traditional antifungal remedies. From clinical trials, the journey of medicinal plants into modern pharmacology is supported by evidences. However, large-scale clinical trials, toxicity profiling, regulatory approvals, and standardized formulation strategies are required for more applications.

2.6 CHALLENGES AND FUTURE PROSPECTS

Despite substantial traditional use, the application of medicinal plants as antifungal agents faces several critical challenges. These range from standardization and quality control issues to regulatory, clinical, and environmental concerns, all of which impede their mainstream integration into modern therapeutics.

2.6.1 Standardization and Quality Control

The lack of standardization in extract preparation is one of the major barriers to the global adoption of plant-based antifungal agents, including the amount of active compounds, methods of extraction, and storage stability. Factors such as geographic origin, harvest season, and post-harvest processing significantly influence phytochemical content and bioactivity.

Efforts toward standardization are encouraged worldwide by the World Health Organization (WHO) monographs on selected medicinal plants and regional pharmacopoeias, which provide validated procedures for raw material identification, extract preparation, and quantification of marker compounds. Analytical technologies like HPLC, TLC, and GC-MS are being used more and more to provide batch-to-batch uniformity, reduce adulteration, and check active phytochemical content.

2.6.2 Limited Clinical Evidence

Although many in vitro and in vivo investigations have shown antifungal capability, clinical trials remain rare, sometimes with little sample sizes and without placebo controls (Gowda et al., 2023). Moreover, many studies focus on crude extracts without isolating specific active ingredients, making dose standardization and toxicity assessment difficult.

To speed translation, scientists are using pilot clinical trials and translational animal models with greater similarity to human fungal infections. Such an approach is essential in translating promising in vitro findings into evidence-based antifungal treatments.

2.6.3 Regulatory and Intellectual Property Issues

Medicinal plant products frequently fall into regulatory gray zones, particularly in countries where herbal medicine lacks robust legal frameworks. This leads to inconsistent quality and insufficient clinical validation. In addition, intellectual property rights (IPR) related to traditional knowledge and biodiversity continue to be controversial, raising ethical and legal challenges in the commercialization of ethnomedicine (Timmermans, 2003; Pushpangadan et al., 2016).

2.6.4 Drug Resistance and Biofilm Complexity

Fungal pathogens are evolving resistance even to plant-based compounds. Moreover, fungal biofilms, which are highly resistant to conventional antifungals, also show limited susceptibility to many plant-derived agents, necessitating more advanced formulations such as nanoparticles and synergistic combinations (Ding et al., 2025).

2.6.5 Future Prospects

The following are the areas in which the antifungal role of medicinal plants is promising:

- Nano-formulation and targeted delivery.
- Synergistic therapy (combining plant compounds with existing antifungals).
- Genomics and metabolomics to discover novel compounds and biosynthetic pathways.

Furthermore, collaborative research efforts, global policies, and investment in ethnobotanical databases will quicken the journey from traditional remedy to evidence-based antifungal therapy.

Artificial intelligence platforms like DeepChem and AlphaFold-facilitated docking pipelines have now started screening massive phytochemical libraries for antifungal activity. These platforms enable quick identification of plant-based leads, binding affinity prediction, and compound prioritization for wet-lab confirmation, thereby unifying natural product science with computational pharmacology.

2.7 CONCLUSION

Medicinal plants offer a powerful and large repository of bioactive compounds with potent antifungal activity. As fungal infections increase and become resistant to conventional therapies, there is an urgency for safe and effective antifungal agents. This chapter has traced the ethnobotanical roots, highlighted the pharmacological and biochemical evidence, and analyzed the mechanisms of action of various phytochemicals against fungal pathogens.

Scientific validation through in vitro, in vivo, and clinical studies underscores the relevance of these traditional therapies in modern pharmacology. Bioactive constituents such as alkaloids, flavonoids, terpenoids, and phenolics have emerged as key players in antifungal defense. Nevertheless, a multidisciplinary approach is necessary to overcome challenges related to standardization, safety, regulatory barriers, and large-scale production.

Looking ahead, the integration of advanced technologies such as metabolomics, nanotechnology, and artificial intelligence into phytochemical research holds immense promise. By bridging ancient wisdom with modern science, medicinal plants may serve as the foundation for a new generation of eco-friendly, cost-effective, and highly efficacious antifungal therapies.

REFERENCES

Ahmed, H. F., Seleiman, M. F., Mohamed, I. A., Taha, R. S., Wasonga, D. O., & Battaglia, M. L. (2023). Activity of essential oils and plant extracts as biofungicides for suppression of soil-borne fungi associated with root rot and wilt of marigold (*Calendula officinalis* L.). *Horticulturae*, 9(2), 222. https://doi.org/10.3390/horticulturae9020222

Akwongo, B., Kakudidi, E. K., Nsubuga, A. M., Andama, M., Namaganda, M., Tugume, P., … Katuura, E. (2024). In vitro antifungal activities of medicinal plants used for treatment of candidiasis in Pader district, Northern Uganda. *Tropical Medicine and Health*, 52(1), 84. https://doi.org/10.1186/s41182-024-00628-x

Al Aboody, M. S., & Mickymaray, S. (2020). Anti-fungal efficacy and mechanisms of flavonoids. *Antibiotics*, 9(2), 45. https://doi.org/10.3390/antibiotics9020045

Alajlani, M. (2023). Antifungal property of medicinal plants: A comprehensive review. *International Journal of Herbal Medicine*, 11(4), 51–57. https://www.aspu.edu.sy/SD08/msf/1692526306_.pdf.

Alam, S., Chowdhury, M. N. R., Hossain, M. A., Richi, F. T., Emon, N. U., Mohammad, M., Ahmed, N., & Taher, M.A. (2025). Antifungal potentials of Asian plants: Ethnobotanical insights and phytochemical investigations. *Chemistry & Biodiversity*, 22(5), e202402867. https://doi.org/10.1002/cbdv.202402867

Aliaghazadeh, B., Pascual-Ortiz, M., Martínez, A., Veses, V., & Sheth, C. C. (2025). Unveiling the synergistic effect of icariin and azole drugs on *Candida albicans*. *Frontiers in Oral Health*, 6, 1517210. https://doi.org/10.1016/j.riam.2025.05.001

Barimani, F., Gholami, A. A., & Nabili, M. (2020). In vitro antifungal activity of *Raphanus sativus* L. var. niger (black radish) and Trachyspermum ammi (ajwain) on resistant and susceptible *Aspergillus fumigatus* isolates. *International Journal of Molecular and Clinical Microbiology*, 10(2), 1360–1368. https://www.academia.edu/download/86542408/article_680133_89ca0a33a3f1d6fcb5b7484e9c4b1492.pdf

Cheraghipour, K., Ezatpour, B., Masoori, L., Marzban, A., Sepahvand, A., Rouzbahani, A. K., … Mahmoudvand, H. (2021). Anti-Candida activity of curcumin: A systematic review. *Current Drug Discovery Technologies*, 18(3), 379–390. https://doi.org/10.2174/1570163817666200518074629

Corona-Gómez, L., Hernández-Andrade, L., Mendoza-Elvira, S., Suazo, F. M., Ricardo-González, D. I., & Quintanar-Guerrero, D. (2022). In vitro antimicrobial effect of essential tea tree oil (*Melaleuca alternifolia*), thymol, and carvacrol on microorganisms isolated from cases of bovine clinical mastitis. *International Journal of Veterinary Science and Medicine*, *10*(1), 72–79. https://doi.org/10.1080/23144599.2022.2123082

Dar, U. A., Shahnawaz, M., & Hakeem, K. R. (Eds.). (2024). *Quinone-Based Compounds in Drug Discovery: Trends and Applications*. Elsevier. https://books.google.com/books?hl=en&lr=&id=cIMMEQAAQBAJ&oi=fnd&pg=PP1&dq=Quinone-based+antifungals%2Bmedicinal+plants&ots=nV2H7e-hiK&sig=d47192m9vjDQo5Vz5Dmsgl53PNM

Das, K., Suple, S. D., Hossain, E., Kumar, S., & Kumar, R. (2025). Ethnobotany: A bridge between traditional knowledge and modern pharmacy. *Ethnobotany of India*, *1*, 110–126. https://doi.org/10.5281/zenodo.15868379

Ding, J., Yan, Z., Peng, L., Li, J., Yang, F., & Zheng, D. (2025). Inhibitory effects of berberine on fungal growth, biofilm formation, virulence, and drug resistance as an antifungal drug and adjuvant with prospects for future applications. *World Journal of Microbiology and Biotechnology*, 41(1), 5. https://doi.org/10.3201/eid1001.030119

Dutta, T., Nandy, S., & Dey, A. (2022). Urban ethnobotany of Kolkata, India: A case study of sustainability, conservation and pluricultural use of medicinal plants in traditional herbal shops. *Environment, Development and Sustainability*, 24(1), 1207–1240. https://doi.org/10.1007/s10668-021-01493-y

Egbe, N. (2023). Antifungal effects of *Nigella sativa* L.(Black cumin) seed extracts and seed oil on selected *Candida albicans* strains. *Journal of Current Biomedical Research, 3*(3, May-June), 993–1004. https://doi.org/10.54117/jcbr.v3i3.3

El-Assri, E. M., El Barnossi, A., Chebaibi, M., Hmamou, A., El Asmi, H., Bouia, A., & Eloutassi, N. (2021). Ethnobotanical survey of medicinal and aromatic plants in Taounate, Pre-Rif of Morocco. *Ethnobotany Research and Applications*, 22, 1–23. https://ethnobotanyjournal.org/index.php/era/article/view/3113

Eldesouky, E., Abbas, E., Osmam, A., & Sitohy, M. (2023). Antioxidant and antifungal activity of ginger (*Zingiber officinale*) methanolic extract. *IJCBS*, 24, 482–488. https://www.iscientific.org/wp-content/uploads/2024/05/62-ijcbs-23-24-11-62n.pdf

Elsewedy, H. S., Alshehri, S., Kola-Mustapha, A. T., Genedy, S. M., Siddiq, K. M., Asiri, B. Y., … Ambrose, G. O. (2024). Insights into antibacterial design: Computational modeling of eugenol derivatives targeting DNA gyrase. *Heliyon, 10*(20). https://doi.org/10.1016/j.heliyon.2024.e39394

Feknous, N., Boumendjel, M., & Leblab, F. Z. (2024). Updated insights on the antimicrobial activities of Allium genus (a review). *Russian Journal of Bioorganic Chemistry*, 50(3), 806–823. https://doi.org/10.1134/S1068162024030257

Gowda, G. V., Mamatha, G. P., Usha, V. A., Thimmasetty, J., Rajeshwari, M. R., & Annigeri, R. G. (2023). Efficacy of *Punica granatum* (Pomegranate) in the Management of Oral Candidiasis–A Randomized Controlled Study. *Journal of Indian Academy of Oral Medicine and Radiology*, 35(1), 15–20. https://doi.org/10.4103/jiaomr.jiaomr_174_21

Hassan, M., Fadayomi, V. K., Innocent, I. G., & Suleiman, M. (2022). Antifungal effect of henna (*Lawsonia inermis*) extract on pathogenic fungi. *Microbiology Research Journal International*, 32(3), 15–26. https://doi.org/10.9734/MRJI/2022/v32i330375

Huang, X., Zheng, D., Yong, J., & Li, Y. (2022). Antifungal activity and potential mechanism of berberine hydrochloride against fluconazole-resistant *Candida albicans*. *Journal of Medical Microbiology*, 71(6), 001542. https://doi.org/10.1099/jmm.0.001542

Kedia, A., Prakash, B., Mishra, P. K., & Dubey, N. K. (2014). Antifungal and antiaflatoxigenic properties of *Cuminum cyminum* (L.) seed essential oil and its efficacy as a preservative in stored commodities. *International Journal of Food Microbiology*, 168, 1–7. https://doi.org/10.1016/j.ijfoodmicro.2013.10.008

Khounganian, R. M., Alwakeel, A., Albadah, A., Nakshabandi, A., Alharbi, S., & Almslam, A. S. (2023). The antifungal efficacy of pure garlic, onion, and lemon extracts against *Candida albicans*. *Cureus*, *15*(5). https://www.cureus.com/articles/146915-the-antifungal-efficacy-of-pure-garlic-onion-and-lemon-extracts-against-candida-albicans.pdf

Kiruthik, V., Jeevitha, V., Anu Pragathi, V. S., Janarthanan, P., Hemavathi, M., Ilango, K., … Shanmugapriya, E. (2025). Ethnobotanical survey of anti-mycotic plants indigenous to Tamil Nadu. *Cuestiones de Fisioterapia*, 54(3), 5180–5197. https://cuestionesdefisioterapia.com/index.php/es/article/download/1920/1422

Mahmoudvand, H., Sepahvand, A., Jahanbakhsh, S., Ezatpour, B., & Mousavi, S. A. (2014). Evaluation of antifungal activities of the essential oil and various extracts of *Nigella sativa* and its main component, thymoquinone against pathogenic dermatophyte strains. *Journal de mycologie medicale*, 24(4), e155–e161. https://doi.org/10.1016/j.mycmed.2014.06.048

Memar, M. Y., Raei, P., Alizadeh, N., Aghdam, M. A., & Kafil, H. S. (2017). Carvacrol and thymol: Strong antimicrobial agents against resistant isolates. *Reviews and Research in Medical Microbiology*, 28(2), 63–68. https://doi.org/10.1097/MRM.0000000000000100

Mezzomo, L., Butzge, J., Fiorio, M., Bastiani, C., Mezzari, A., Fuentefria, A., & Apel, M. A. (2025). Origanum essential oil and antifungal activity: A systematic review. *Chemistry & Biodiversity*, e202402296. https://doi.org/10.1002/cbdv.202402296

Mohanasundaram, P., & Saral, A. M. (2023). Phytochemical screening, antibacterial, antifungal, anti-biofilm and antioxidant activity of *Azadiracta indica* A. Juss. Flowers. *Chemistry & Biodiversity*, 20(3), e202201049. https://doi.org/10.1002/cbdv.202201049

do Monte, M. N., de Oliveira, J. R., de Oliveira, L. D., & Jorge, A. O. (2024). *Camellia sinensis* (green tea) promotes antifungal activity against Candida spp. isolated from HIV-positive patients and biocompatibility in murine macrophages. *Research, Society and Development*, 13(2). https://rsdjournal.org/rsd/article/download/45002/35956/470239

Neel, S, Mandal, A., Saha, S., Das, A., Kundu, A., & Singh, A. (2025). *Gymnema sylvestre* saponins for potential antifungal action: *In vitro* and *in silico* perspectives. *Frontiers in Plant Science*, 16, 1508454. https://doi: 10.3389/fpls.2025.1508454

Pereira, R., Mendes, J. D. F. S., dos Santos Fontenelle, R. O., Rodrigues, T. H. S., dos Santos, H. S., Marinho, E. S., … de Morais, S. M. (2021). Antifungal activity, antibiofilm, synergism and molecular docking of *Allium sativum* essential oil against clinical isolates of *C. albicans*. *Research, Society and Development*, 10(12), e313101220457. https://doi.org/10.33448/rsd-v10i12.20457

Pushpangadan, P., George, V., Ijinu, T. P., & Rajasekharan, S. (2016). Ethnopharmacology, traditional knowledge and intellectual property rights. In K.P. Laladhas, Preetha Nilayangode, Oommen V. Oommen (Eds.), *Biodiversity for Sustainable Development* (pp. 97–119). Cham: Springer International Publishing. https://doi.org/10.1007/978-3-319-42162-9_6

Raj, S., Manchanda, R., Bhandari, M., & Alam, M. S. (2022). Review on natural bioactive products as radioprotective therapeutics: Present and past perspective. *Current Topics in Medicinal Chemistry*, 23(14), 1721–1738. https://doi.org/10.2174/1389201023666220110104645

Regar, R. K., Sharma, M., Behera, S., Gupta, P., Lal, R., Prajapati, S., … Kaushik, S. (2025). Exploring the therapeutic potential of *Terminalia chebula* against systemic candidiasis: An in vitro, in vivo, and in silico study. *Fitoterapia*, 184, 106649. https://doi.org/10.1016/j.fitote.2025.106649

Ríos-López, A. L., Dávila-Aviña, J., González, G. M., & Flores-Maldonado, O. (2024). Antifungal and antivirulence activity of vanillin and tannic acid against *Aspergillus fumigatus* and *Fusarium solani*. *Current Microbiology*, 81(6), 156. https://doi.org/10.1007/s00284-024-03678-w

Sarathambal, C., Rajagopal, S., & Viswanathan, R. (2021). Mechanism of antioxidant and antifungal properties of *Pimenta dioica* (L.) leaf essential oil on *Aspergillus flavus*. *Journal of Food Science and Technology*, 58(7), 2497–2506. https://doi.org/10.1007/s13197-020-04756-0

Shah, M., Zia, M., Ahmad, I., Umer Khan, M., Ejaz, H., Alam, M., … Ojha, S. C. (2025). Computational analysis of ayurvedic metabolites for potential treatment of drug-resistant *Candida auris*. *Frontiers in Cellular and Infection Microbiology*, 15, 1537872. https://doi.org/10.3389/fcimb.2025.1537872

Tadesse, D., Lulekal, E., & Masresha, G. (2025). Ethnopharmacological study of traditional medicinal plants used by the people in Metema district, northwestern Ethiopia. *Frontiers in Pharmacology*, 16, 1535822. https://doi.org/10.3389/fphar.2025.1535822

Terras, F. R., Schoofs, H. M., De Bolle, M. F., Van Leuven, F., Rees, S. B., Vanderleyden, J., … Broekaert, W. F. (1992). Analysis of two novel classes of plant antifungal proteins from radish (*Raphanus sativus* L.) seeds. *Journal of Biological Chemistry*, 267(22), 15301–15309. https://doi.org/10.1016/S0021-9258(19)49534-3

Timmermans, K. (2003). Intellectual property rights and traditional medicine: Policy dilemmas at the interface. *Social Science & Medicine*, 57(4), 745–756. https://doi.org/10.1016/S0277-9536(02)00425-2

Tlaamela, D., Mahlo, S. M., & McGaw, L. J. (2024). Ethnobotanical survey and anti-candidal activity of plant species used for oral candidiasis. https://repository.up.ac.za/bitstream/handle/2263/98848/Tlaamela_Ethnobotanical_2024.pdf?sequence=1

Vakili-Ghartavol, M., Arouiee, H., Golmohammadzadeh, S., & Naseri, M. (2024). Improving antifungal effect of peppermint essential oil. *Acta Scientiarum Polonorum. Hortorum Cultus*, 23(2). https://doi.org/10.24326/asphc.2024.5334

Verywell Health. (2024). The surprising medicinal uses of star anise. *Verywell Health*. https://www.verywellhealth.com/star-anise-benefits-uses-side-effects-4588005

Xing, Y., Li, X., Xu, Q., Yun, J., & Lu, Y. (2010). Antifungal activities of cinnamon oil against *Rhizopus nigricans*, *Aspergillus flavus* and *Penicillium expansum* in vitro and in vivo fruit test. *International Journal of Food Science and Technology*, 45(9), 1837–1842. https://doi.org/10.1111/j.1365-2621.2010.02342.x

Yuan, X., Cao, D., Xiang, Y., Jiang, X., Liu, J., Bi, K., … Zhang, Y. (2024). Antifungal activity of essential oils and their potential synergistic effect with amphotericin B. *Scientific Reports*, 14(1), 31125. https://doi.org/10.1038/s41598-024-82380-0

Zhang, C., Zhao, J., Famous, E., Pan, S., Peng, X., & Tian, J. (2021). Antioxidant, hepatoprotective and antifungal activities of black pepper (*Piper nigrum* L.) essential oil. *Food Chemistry*, 346, 128845. https://doi.org/10.1016/j.foodchem.2020.128845

3 Antimicrobial Peptides from Medicinal Plants

New Frontiers in the Management of Multidrug Resistance Fungus

Pesaru Sai Laya, Pisal Ketali Prakash, and Priya Prasad

3.1 INTRODUCTION

3.1.1 An Overview of Fungal Infections

Bacteria, viruses, parasites, fungi, prions, worms and helminths have all been implicated in infectious diseases, with viruses being the most common and bacteria being the most feared until a few decades ago. While methods for controlling bacterial infections have improved, fungi have become the most threatening pathogen (de Pauw, 2011). Yeasts and moulds are currently among the top ten most commonly isolated diseases among patients in Intensive Care Units (Montagna et al., 2013). The prevalence of fungal infections is posing a significant challenge to healthcare professionals. Fungi cause diverse infections in humans, affecting over a billion people worldwide, ranging from minor allergies to potentially threatening invasive fungal infections (Bongomin et al., 2017). Fungal infections, particularly invasive infections, are an increasing public health concern, compounded by the rise of new pathogens, resistant strains and the return of previously rare diseases (Fisher et al., 2022). A recent study estimated that invasive fungal infections affect 6.5 million people and result in 2.5 million deaths, highlighting their significant global concern. However, true prevalence is difficult to measure due to diagnostic problems (Denning, 2024). The growing use of immunosuppressive drugs for organ transplantation and aggressive cancer treatments has improved patient survival but also increased the number of individuals living in immunocompromised states. These individuals are more vulnerable to fungal infections, as fungi are opportunistic pathogens, especially when bacterial competition is reduced by advances in antibiotic therapy (Claeys et al., 2019). Advances in treating non-malignant illnesses such as bronchiectasis and cystic fibrosis have lowered mortality, but persistent risk factors continue to make survivors vulnerable to fungal infections (Sabino & Stevens, 2025). Fungal diseases cause enormous worldwide health costs through

DOI: 10.1201/9781003633952-4

ocular infections that result in blindness, debilitating allergies, hospital-acquired infections, recurring problems such as vulvovaginal candidiasis and other important but non-lethal illnesses. Additionally, mycoses in domesticated animals contribute to this influence. Notably, fungi have been highlighted as possible future bioweapons (Buckley, 2008).

Candidiasis, caused by various *Candida* species, commonly affects the skin, nails and mucous membranes in humans. *Aspergillus* species are responsible for aspergillosis, a fungal infection that primarily targets the respiratory system. Cryptococcosis, caused by *Cryptococcus* species, can lead to serious infections in the lungs and central nervous system. *Histoplasma capsulatum* causes histoplasmosis, a lung infection that may also spread to other organs. Pneumocystis pneumonia, a severe fungal illness, is caused by *Pneumocystis jirovecii* (Babu et al., n.d.). Fungal infections are common in plants and can cause major crop loss. For example, Fusarium Wilt, caused by numerous *Fusarium* species, affects tomatoes, bananas, cotton, etc., causing wilting and death (Rokas, 2022). Fungal diseases are common in animals as well. Ringworm is caused by dermatophyte fungi and affects the skin, hair and nails of many animals, including cats, dogs and livestock (Seyedmousavi et al., 2018). The chytrid fungus *Batrachochytrium dendrobatidis* causes chytridiomycosis, a fungal disease in frogs. It has caused the decline and extinction of many amphibian species around the world (Fisher et al., 2012).

3.1.2 Multidrug Resistance in Fungi

Multidrug resistance (MDR) is defined as the simultaneous acquisition of tolerance to a variety of medications through a restricted or even single genetic alteration (Paul & Moye-Rowley, 2014). MDR in fungi is an emerging global public health threat, leading to treatment failures and elevated mortality rates. Contributing factors include the overuse of antifungal medications, fungi's rapid development of resistance mechanisms and the widespread dissemination of resistant strains in both healthcare settings and the environment. Gaining insight into the molecular mechanisms of resistance and the pathways of fungal transmission is essential for devising effective strategies to manage and combat resistant fungal infections (de Moraes & Ferreira-Pereira, 2024). The antifungal medication includes four classes: azoles, echinocandins, polyenes and pyrimidine analogues. Nonetheless, physicians and mycologists may argue that we are losing the war, with increased drug resistance and tolerance leading to treatment failure (Hui et al., 2024).

Opportunistic fungi, including *Candida albicans*, *Candida auris* and *Aspergillus fumigatus*, have developed advanced resistance mechanisms to many antifungal drugs, limiting treatment options and increasing mortality rates (Perfect, 2017). One of the primary reasons for fungal MDR is overexpression of drug efflux pumps. These include ATP-binding cassette (ABC) transporter family members Cdr1p and Cdr2p, as well as the Mdr1p transporter from the major facilitator superfamily (MFS). These proteins actively remove antifungal drugs from fungal cells, lowering intracellular drug concentrations below therapeutic thresholds. Overexpression of efflux pumps is frequently caused by gain-of-function mutations in transcription factors such as Tac1p and Mrr1p (Prasad Dr & Rawal, 2014).

Alterations of target sites are another significant cause of antifungal resistance. Mutations in the *ERG11* gene, which encodes lanosterol 14α-demethylase, a target of azole medications, can diminish drug binding affinity, resulting in persistent growth despite therapy (Flowers et al., 2015). Resistance to echinocandins is typically caused by mutations in *FKS1* and *FKS2*, which encode subunits of β-1,3-glucan synthase, the drug's molecular target (Zajac et al., 2025).

Biofilm development increases fungus resistance. Biofilms are organized microbial communities embedded in an extracellular matrix that limit drug penetration and promote cell survival during stress. Fungal biofilms are extremely resistant to antifungal medications, requiring greater drug doses or combination treatments (Atriwal et al., 2021). Biofilms also cause persistent infections on medical devices, complicating eradication attempts.

The rise of highly resistant organisms, particularly *C. auris*, has raised global concerns. *C. auris* is commonly resistant to many antifungal classes at the same time, including azoles, echinocandins and polyenes, and has been linked to global healthcare outbreaks (Kim et al., 2024). Its propensity to live on hospital surfaces and colonize humans complicates infection control efforts. Given these challenges, there is an urgent need to create new antifungal medicines that target novel biological pathways. To circumvent current resistance mechanisms, strategies such as combination treatments, efflux pump inhibition and biofilm-specific pathway targeting are being investigated (Perfect, 2017). Furthermore, antifungal stewardship initiatives seek to reduce needless drug exposure, lowering selective pressure for resistance development.

Despite significant advancements, many features of fungal MDR are still poorly understood. Continued research into the regulatory networks that control efflux pump expression, stress responses and biofilm resilience is crucial. Improved diagnostic techniques capable of rapidly identifying resistance at the point of therapy are also critical for better patient outcomes (Almatroudi, 2025). The increasing prevalence of MDR fungi emphasizes the importance of a multidisciplinary strategy that includes molecular research, clinical treatment and public health strategies to manage and prevent these difficult infections successfully.

3.1.3 Need for the Alternative Treatment Strategies

The growing threat of MDR fungal infections has resulted in an urgent need for alternative therapeutic approaches. Traditional antifungal medicines such as azoles, echinocandins and polyenes are becoming increasingly ineffective as different resistance mechanisms emerge. Overexpression of efflux pumps, mutations in drug target genes and biofilm development are all factors that complicate therapy and lead to greater mortality rates (Boakye-Yiadom et al., 2024; Vanden Bossche, 1997). Furthermore, the limited number of existing antifungal medication classes, combined with the slow rate of novel antifungal discovery, exacerbates the situation.

Antimicrobial peptides (AMPs) generated from medicinal plants are one promising development strategy. Unlike conventional antifungals, AMPs frequently target many cellular processes concurrently, such as membrane rupture, metabolic inhibition and modulation of immunological responses, making it more difficult for fungi to acquire resistance (Tam et al., 2015). AMPs, such as defensins, thionins and lipid transfer proteins (LTPs), exhibit broad-spectrum antifungal activity and are effective against resistant strains such as *C. auris* and *Aspergillus* species (De Cesare et al., 2020). The advantages of plant-derived AMPs (pAMPs) go beyond their mode of action. They are typically tiny, heat resistant and can be designed for increased selectivity and lower toxicity to human cells (Yount & Yeaman, 2013). Furthermore, AMPs can work synergistically with existing antifungal medicines, cutting dosages and reducing potential side effects. This combinational approach may also delay the establishment of resistance. Despite their potential, AMPs must overcome several difficulties before they can be used as clinical medicines. Significant challenges remain, including in vivo proteolytic breakdown, possible immunogenicity and cost-effective manufacture. However, breakthroughs in peptide engineering, nanotechnology-based delivery systems and recombinant production methods are quickly overcoming these constraints (Maleki Dizaj et al., 2022; Shriwastav et al., 2025).

Given the limitations of conventional antifungal medications and the increasing spread of MDR fungal diseases, there is an urgent need to investigate and invest in alternative techniques such as pAMPs. Their unique mechanisms, flexibility and synergy with existing treatments make them promising candidates for the next generation of antifungals. The discovery, identification and optimization of therapeutic pAMPs offer the potential to revolutionize the landscape of fungal infection management (Odunitan et al., 2024).

3.2 ANTIMICROBIAL PEPTIDES

Antimicrobial resistance, particularly the emanation of MDR, is one of the major global concerns, creating a public health crisis by threatening the effectiveness of modern medicine. Several bacterial, protozoan and fungal pathogens are developing resistance to the frontline antimicrobial agents. Overuse and misuse of antimicrobials, non-therapeutic use of antimicrobials to prevent disease and promote growth in poultry and livestock and genetic alterations either in the host or pathogens are the key speculations of antimicrobial resistance and MDR (Govindarajan & Kandaswamy, 2023). This is limiting the treatment options and raising mortality rates particularly among immunocompromised individuals. The escalating failure of traditional antibiotics and antifungals has intensified the need for the novel, effective and sustainable treatment strategies. AMPs are one of the potent emerging candidates to combat the disease-causing agents.

AMPs are the naturally occurring, typically cationic, amphipathic molecules, usually 10–50 amino acids long (Govindarajan & Kandaswamy, 2023; Rizzetto et al., 2023). These short stretches of oligopeptides may be small proteins or part of a protein with a broad spectrum of antimicrobial properties. These are a diverse group of molecules present ubiquitously among all organisms, from single-celled bacteria to complex humans. They are an integrative part of the host's first line of defence and contribute to the immune mechanisms. Bacteriocins are the AMPs produced by bacteria as a part of defence to outcompete other microbes (Simons et al., 2020). Plectasin is a fungal defensin produced by *Pseudoplectania nigella* against Gram-positive bacteria (Mygind et al., 2005). Plants possess a wide range of AMPs like defensins, cyclotides and thionins, and these are released in response to infection and environmental stress (Sdvn & Chauhan, 2022). In vertebrates, AMPs are secreted by skin, mucosal surfaces, epithelial cells, immune cells and blood cells. Defensins, cathelicidins, histatins and dermacidins are a few important AMPs in humans (He et al., 2024). Apart from their role in defence, AMPs also has a role in immunomodulation (de la Fuente-Núñez et al., 2017). Although AMPs vary from species to species, they pose a similarity in certain regions of their sequence, structure and charges.

3.2.1 Classification of Antimicrobial Peptides

AMPs are a diverse group of molecules. Due to their vast structural and functional diversity, AMPs can be classified based on various criteria like source, structure, composition of amino acids, mechanism of action, etc.

3.2.1.1 Classification Based on Source

AMPs are all-pervasive and can be found across all kingdoms of life. Categorization of AMPs based on their source encompasses both naturally occurring and synthetic AMPs. Elucidation of AMPs based on their origins contemplates their evolutionary significance and the ecological niches to which they have adapted to.

3.2.1.1.1 Naturally Occurring Antimicrobial Peptides

3.2.1.1.1.1 Animal-Derived Antimicrobial Peptides AMPs of animals are an integrative part of innate immune system and exhibit a potent activity against a broad spectrum of pathogens, including bacteria, viruses, protozoans and fungi (Huan et al., 2020). These are evolutionary conserved sequences, and animal-derived AMPs can be categorized into various classes based on their structural features, functional mechanisms and origin; a few major classes of animal-derived AMPs are:

Defensins: Defensins are the eminent and extensively studied class of AMPs, found in insects, reptiles, birds and mammals. They are characterized by a β-sheet structure stabilized by two or more disulphide linkages. Defensins can be categorized into three groups based on the spatial arrangement of their amino acids, site of secretion and action. They are alpha defensins, beta defensins and theta defensins. Defensins kill microbes by destabilizing the membrane (Lehrer & Ganz, 2002; Reddy et al., 2004)

Cathelicidins: Cathelicidins are the multifunctional peptides exclusively found in vertebrates. These versatile peptides not only serve as antimicrobial agents but also function as immune modulators. Cathelicidins are linear in cytosolic environment and can adopt an α-helical structure while interacting with lipids of membrane. They are present in the inactive precursor form in immune cells and are activated when cleaved upon infection (Kościuczuk et al., 2012; Wang, 2008; Zanetti, 2004)

Cecropins: Cecropins are small peptides typically consisting of 31–39 amino acids and are found in insects. They are first isolated from *Hyalophora cecropia*, the giant silk moth. They are generally linear and adopt α-helical structure in membrane-mimetic environments. They are amphipathic with no disulphide bonds. They exhibit mechanism of action by electrostatically associating with the cell membrane and causing membrane disruption by pore formation. Some cecropins also translocate into the cells and interfere with cellular processes and inhibit cell metabolism (Andreu & Rivas, 1998; Pushpanathan et al., 2013)

Magainins: Magainins are small, cationic, amphipathic peptides, usually 23–27 amino acids long and found in amphibians, extensively in frogs. Like cecropins, magainins are linear in aqueous environments and adopt a stable α-helical conformation on interaction with lipid membranes. Magainins lack cysteine-like residues and disulphide bonds. They target the pathogen cell membrane causing membrane disruption leading to cell death (Pushpanathan et al., 2013; Zasloff, 1987).

Dermicidins: Dermacidins are anionic peptides secreted by human sweat glands, notably eccrine sweat glands. They are initially synthesized as a precursor protein that is 110-amino acids long and 47-amino acids long after proteolytic processing. Dermacidins are exceptionally stable under high-salt conditions and at extreme pH levels. They show extended conformations in aqueous environments and α-helical folded structures at membranes. They kill the cells by pore formation and membrane disruption, but in some cases, they exhibit antibacterial effects by disturbing the membrane potential rather than cell lysis (Paulmann et al., 2012; Schittek et al., 2001)

Histatins: Histatins are histidine-rich, cationic AMPs, present in human saliva and typically 7–38 amino acids long. In aqueous environments, histatins are present as random coils and adopt helical structures on interacting with membranes (Helmerhorst et al., 2001). Histatins display multiple antimicrobial mechanisms like membrane disruption by pore formation, intracellular targeting by reducing ATP production and metal ion (Zn^{+2} and Ca^{+2}) binding to generate reactive oxygen species (Ruissen et al., 2001)

3.2.1.1.1.2 Plant-Derived Antimicrobial Peptides Plants display a wide variety of AMPs. pAMPs are typically small, cationic, amphipathic peptides rich in cysteine motifs stabilized by disulphide linkages and play a vital role in maintaining immune responses in plants. Due to the absence of well-developed immune system like animals, plants secrete a diverse range of AMPs for its defence (Lima et al., 2022). Based on their structure and functions, pAMPs are classified into various families like defensins, thionins, snakins, knotin-like peptides, LTPs, etc. (Li et al., 2021).

3.2.1.1.1.3 Microbial Antimicrobial Peptides Microbial AMPs are the natural defence molecules produced by the microorganisms to enhance their survival in competitive environments. These peptides are highly specialized and play a role in shaping the structure of microbial community by strongly exhibiting activity against closely related microbial species. Based on the source, microbial AMPs are categorized into Bacterial AMPs and Fungal AMPs.

Bacterial AMPs: Bacterial AMPs or bacteriocins are the molecules that are produced under stress or competitive conditions and typically kill or inhibit the growth of closely related bacterial species or other microbial competitors (Alvarez-Sieiro et al., 2016). These are synthesized by ribosomes, may undergo post-translational modifications and are highly specific to target organisms (Yang et al., 2014). Based on their structure and mode of actions, bacteriocins are divided into various classes – Lantibiotics, Non-lantibiotic small peptides and large bacteriocins (Cotter et al., 2005).

Fungal AMPs: Fungal AMPs are the substances produced by fungal species in order to survive microbially competent conditions. They are cysteine-rich peptides with β-sheet structures stabilized by disulphide bonds. They exert potent antifungal and antibacterial properties by disrupting the membrane integrity or by inhibiting the cell wall synthesis compounds (Lacerda et al., 2014; Meyer, 2008).

3.2.1.1.2 Engineered/Synthetic Antimicrobial Peptides

Engineered/Synthetic AMPs are artificially designed AMPs; these are drafted by manipulating the structural and functional properties of naturally occurring AMPs. Even though the naturally occurring AMPs are effective in the defence of the host, their therapeutic potential is held down by their narrow spectrum of activity, cytotoxicity to host cells, proteolytic degradation and difficulty in isolating them in optimal quantities from the host cells. So, by using the techniques in advanced molecular biology, AMPs are tailored to overcome these limitations with increased stability, proper selectivity, reduced toxicity, enhanced efficacy and also combating MDR (Sinha & Shukla, 2019).

3.2.1.1.2.1 Genetically Engineered Antimicrobial Peptides Genetically engineered AMPs are peptides that are designed or intentionally modified from existing or naturally occurring AMPs by using the strategies of molecular biology. These peptides can be synthesized by site-directed mutagenesis, which introduces a specific amino acid alteration to improve their properties; by recombinant DNA technology, which clones and expresses a particular segment of the peptide that is responsible for antimicrobial activity to enhance stable binding and by library screening (Sadeeq et al., 2025; Sinha & Shukla, 2019).

3.2.1.1.2.2 Synthetic Antimicrobial Peptides Synthetic AMPs represent a paradigm shift in the development of AMPs, rather than extracting them from natural sources or engineering them using biotechnology-based approaches. Synthetic AMPs are created either from scratch or by modifying the known AMP sequences and assembling them. Such synthesized peptides can serve as promising agents combating multidrug-resistant organisms. The following are a few strategies used for designing synthetic AMPs (Su et al., 2025):

Solid-phase peptide synthesis: This is a widely used chemical method facilitating efficient, automated and high-throughput synthesis of peptides. In this method, the amino acids are added one after the other in a predetermined order via their C-terminus towards the N-terminus on a solid support, which is usually a resin or polymer bead (Su et al., 2025).

De novo design: This includes the designing of entirely new peptide sequences using computational tools and rational design with well-intended structural and physicochemical properties (Huan et al., 2020).

Template-based design: In this method, the peptide is designed modifying the sequences of known naturally occurring AMPs. The modifications may include amino acid substitutions, cyclization of structure or truncation of a specific segment of the peptide (Huan et al., 2020).

Peptidomimetics: This method includes the designing of small peptide molecules that mimic the structure and function of naturally occurring peptides with altered backbone or functional groups on the side chains (Huan et al., 2020).

3.2.1.2 Classification Based on Charge

Hydrophobicity and charge play an important role in determining how AMPs interact with the negatively charged cell membranes. Most of the AMPs are amphipathic in nature, i.e., having both hydrophilic and hydrophobic regions favouring the interactions and membrane disruption. Thus, AMPs are classified based on their net charge at physiological pH.

3.2.1.2.1 Cationic Antimicrobial Peptides

Most of the AMPs are cationic, carrying a net positive charge ranging from +2 to +9. This positive charge helps to interact strongly with the negatively charged microbial membranes through electrostatic attractions (Hancock & Sahl, 2006). Antibacterial and antifungal AMP Magainin-2 from *Xenopus laevis*, human β-defensin-2 and LL-37, a human cathelicidin, are a few examples for cationic AMPs (Zasloff, 1987).

3.2.1.2.2 Anionic Antimicrobial Peptides

Anionic AMPs are rare, often rich in glutamate and aspartate, carrying a net negative charge ranging from −1 to −7. They challenge microbial survival by interfering with cellular metabolic processes rather than disrupting the cell membrane and also has a role in boosting the host immunity (Hancock & Sahl, 2006). Anionic AMPs generally require cationic cofactors like zinc or calcium to effectively interact with microbial cell membranes (Nguyen et al., 2011). Anionic AMPs are mostly antibacterial with a few with antiviral properties. Dermacidin from humans, Maximin-H5 from *Bombina maxima* and SP-A and SP-D are a few examples of anionic AMPs (Braida et al., 2012; Crouch & Wright, 2001; Schittek et al., 2001).

3.2.1.3 Based on Structure

3.2.1.3.1 α-Helical Peptides

α-Helical AMPs are usually linear without disulphide bonds. They are unstructured in aqueous environment and form helices in hydrophobic environments like membranes. They are characterized by segregated hydrophilic and hydrophobic residues, which favours them in membrane insertion (Shai, 1999; Wimley, 2010). They target the cell via pore formation and carpet-like disruption mechanisms (Brogden, 2005). Examples of α-Helical peptides include magainins of frog, cecropins of insects and melittin of bee venom (Chen et al., 1988; Zasloff, 1987).

3.2.1.3.2 β-Sheet Peptides

β-Sheet AMPs are characterized by two or more β-chains connected by disulphide bonds, stabilizing its structure. In contrast to α-Helical peptides, β-Sheet AMPs maintain a stable structure in both hydrophilic and hydrophobic environments. These peptides target the cell by creating pores and causing leakage of cellular contents. Examples of β-Sheet peptides include protegrins of porcine leukocytes, tachyplesins of horseshoe crab haemocytes and defensins of plants, mammals and insects (Bellm et al., 2000; Nakamura et al., 1988; Selsted & Ouellette, 2005).

3.2.1.3.3 Loop Peptides

A few AMPs attain a loop-like structure due to the formation of a single disulphide bond or a covalent bond between N and C termini of the peptide or via specific cyclization. These structures are highly compact and small with notable stability. Loop-structured AMPs target the cell by disrupting microbial membranes or by interacting with intracellular components (Brogden, 2005; Williams et al., 2020). Examples of loop-structured AMPs are lasioglossin of *Lasioglossum laticeps* and bactenecin from bovine neutrophils (Romeo et al., 1988; Verdon et al., 2008).

3.2.1.3.4 Extended/Linear Peptides

These peptides are rich in specific amino acids like proline, glycine, tryptophan or arginine but lack secondary structures. They do not involve in disrupting membranes but perform antimicrobial functions by interacting with the intracellular components. Upon interacting with membranes (during the process of internalization into cell), these linear peptides fold into specific conformations to attain stability (Brogden, 2005; Nguyen et al., 2011). Examples include histidine-rich histatins, tryptophan-rich Indolicidin and proline- and arginine-rich PR-39 (Holani et al., 2016).

3.2.2 Mechanisms of Action

AMPs, also known as host defence proteins, play a pivotal role in immune regulation, and their mode of action is completely different from antibiotics. Unlike conventional antibiotics, the AMPs target the cellular destruction by primarily disrupting the cell membrane through different mechanisms. This physical mode of action reduces the likelihood of development of resistance, similar to antibiotics. AMPs needle out cells in dual mechanisms, i.e., by direct microbial killing and modulation of host immune responses. This multifaceted activity, broad spectrum efficacy, low resistance potential, and immunomodulatory properties are making AMPs as potential candidates for next-generation antimicrobial therapies to fight against MDR pathogens (Singh et al., 2022).

The mechanisms by which AMPs employ their effect can be broadly classified into two categories: Direct killing mechanism, which involves in targeting the cell by physical or biochemical disruptions, and immunological regulation, which involves in recruitment and activation of immune cells (Figure 3.1).

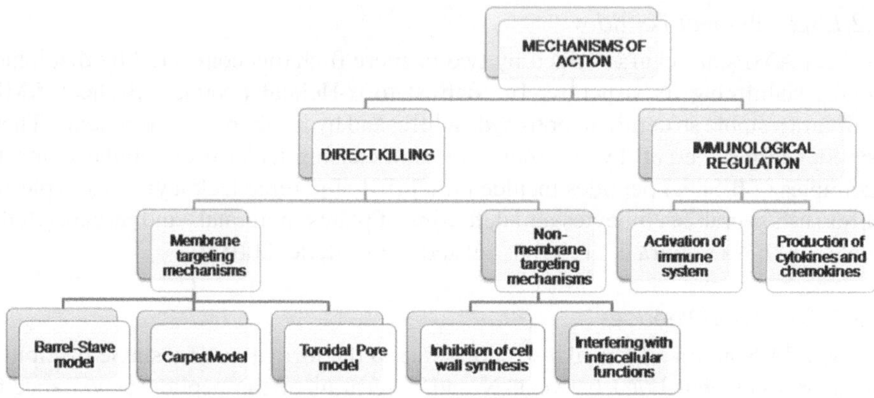

FIGURE 3.1 Classification based on the mechanism of action.

3.2.2.1 Direct Killing Mechanisms

3.2.2.1.1 Membrane Targeting Mechanisms

AMPs often employ their effects by disrupting microbial cellular membranes by binding to them and ultimately causing cell death. The fundamental differences in the membrane composition between host and microbes enable the selective targeting and imposing further reactions. The framework of microbial membranes plays an important role in determining the efficacy of AMPs. The cell membranes of bacteria, fungi and viruses are rich in negatively charged components like phosphatidylglycerol, cardiolipin, lipopolysaccharides and teichoic acids, which strongly attract the cationic AMPs. Through electrostatic interactions, these AMPs target the cell by pore formation or membrane disintegration by binding and integrating into the membrane. The high levels of cholesterol and zwitterionic phospholipids like phosphatidylcholine present in the mammalian cell membranes can reduce the AMP binding, minimizing cytotoxicity. Due to varied membrane interactions and diverse nature of AMPs, various models on the mechanisms of action of AMPs do exist. The membrane targeting mechanisms of AMPs are categorized into receptor-mediated and non-receptor-mediated interactions (Huan et al., 2020; Seyfi et al., 2020; Singh et al., 2022).

3.2.2.1.1.1 Receptor-Mediated Interactions The receptor-mediated mechanism of action is a specialized mode where AMPs bind to specific molecular targets (receptors) present on the surface or within pathogenic cells. Bacteriocins like nicins and mersacidins exert action via specific receptor interactions. Nicin is a lantibiotic AMP produced by *Lactococcus lactis* particularly against Gram-positive bacteria. Nicin binds to the Lipid II molecule of bacterial cell wall, which is involved in peptidoglycan synthesis (an essential component of bacterial cell wall), thereby inhibiting the cell wall biosynthesis and pore formation (Han et al., 2016). Histatin 5 (Hst 5) is an antifungal AMP found in human saliva, and it aims against pathogenic fungi like *C. albicans* (Han et al., 2016). Hst 5 binds to the specific protein Ssa1/2p present on the fungal cell surface, which then internalizes via receptor-mediated endocytosis, thereby disrupting mitochondrial function and accumulating reactive oxygen species (ROS) (Li et al., 2003).

3.2.2.1.1.2 Non-Receptor-Mediated Interactions In non-receptor-mediated mechanism, the AMPs directly interact with microbial membranes without the formation of any AMP–Receptor complexes. This kind of binding is non-specific binding that offers a broad-spectrum activity, enhanced expeditiousness and rapid killing. Different biophysical models (Figure 3.2) have been proposed to explain the membrane disruption by AMPs (Singh et al., 2022).

3.2.2.1.1.2.1 Barrel-Stave Model In Barrel-Stave model, AMPs bind to microbial surfaces via electrostatic interactions. They then undergo conformational changes and enter the membrane, where they aggregate themselves aligning perpendicularly forming a "barrel" (pore)-like structure, and the term Stave indicates peptides that are aligned as vertical slats forming the walls of the pore. The hydrophobic residues face outwards towards the hydrophobic side, and the hydrophilic residues face inwards forming an aqueous transmembrane channel. On triggering, the intracellular contents of the cell are exudated out, thereby leading to the death of cell (De Cesare et al., 2020; Singh et al., 2022). Examples include Alamethicin, an AMP produced by *Trichoderma viride* that forms ion channels by inserting perpendicularly into the membrane, causing membrane depolarization by ion leakage. Pardaxin secreted by *Pardachirus marmoratus* exhibits antibacterial and antitumour activity, targeting the membrane of the pathogen (Nagao et al., 2015).

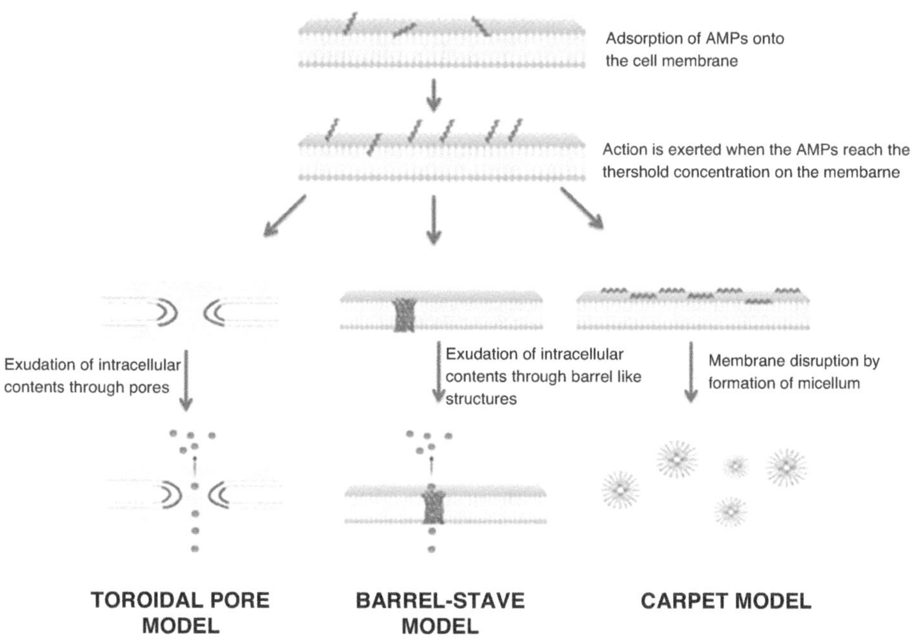

FIGURE 3.2 Different mechanisms of action.

3.2.2.1.1.2.2 Toroidal Pore Model In the toroidal pore model, AMPs align themselves parallel to the membrane of the target cell, and when the threshold is reached, the peptides reorient themselves and get inserted into the membrane causing a curvature and forming lipid–peptide hybrid pores. The lipid head groups and the peptides together form the lining of the pores. Toroidal pores are unstable and short-lived. These pores cause osmotic imbalance, depolarization and cell death (Pouny et al., 1992; Singh et al., 2022). Many AMPs like Magainin2, Melittin, LL-37 and Lacticin Q act via toroidal pore model. Magainin 2 is one of the AMPs present in the skin of the African clawed frog, *X. laevis*. They form toroidal pores, aiding the leakage of intracellular contents (Zasloff, 1987).

3.2.2.1.1.2.3 Carpet Model Carpet model explains membrane disruption without forming well-defined pores. This is a rapid, non-specific, concentration-dependent mechanism that is highly effective against an extensive range of pathogens. According to this model, AMPs align themselves parallel to the membrane (but do not penetrate perpendicularly into the membrane), and when they reach a threshold concentration, i.e., a critical peptide-to-lipid ratio, they start disrupting the membrane by solubilizing lipid bilayers and forming micelle-like aggregates in a detergent-like manner (Zhang et al., 2021). This causes loss of barrier integrity by complete membrane dissolution, thereby resulting in cell death (Nguyen et al., 2011). Cecropin A, Aurein 1,2, Indolicidin and LL-37 are a few examples of AMPs, which exhibit the mode of action through this mechanism (Jenssen et al., 2006).

3.2.2.1.1.2.4 Other Models Apart from the above-discussed ATP-independent cell targeting models, there are other mechanisms like macropinocytosis, which are ATP dependent. In macropinocytosis, AMPs that adhere to the plasma membrane make the membrane to fold inwards along with the peptides to form small vesicle-like structures called macropinosomes. These macropinosomes exudate into the cell and show antimicrobial effect. Sinking raft and electroporation are the other models that explain the mode of action of AMPs. In this method, due to the alteration in charge on both membranous and cytosolic sides, the disposition of AMPs on the membrane aids in the development of unstable holes on the microbial cell membrane leading to the leakage of cellular contents and cell death (Kornspan & Rottem, 2012; Singh et al., 2022).

3.2.2.1.2 Non-Membrane Targeting Mechanisms

While numerous AMPs perform their functions by disrupting microbial membranes, a subset of AMPs operates by targeting the biological functions of the cell instead of directly causing membrane lysis. These AMPs enter the microbial cell without extensively damaging the cell membrane through energy-dependent endocytosis or energy-independent translocation and then target intracellular processes like DNA replication, protein translation, enzyme synthesis, etc., leading to growth inhibition or death (Brogden, 2005; Hale & Hancock, 2007).

3.2.2.1.2.1 Interfering Intracellular Functions Some AMPs can directly bind to nucleic acids inside the pathogen cells and can hamper crucial cellular processes like replication, transcription and translation. Cationic AMPs interact electrostatically with the negatively charged phosphate groups of the nucleic acids, blocking them and effectively halting the microbial proliferation. Cell-Penetrating AMPs (CPAPs) are a group of naturally occurring or synthetic peptide sequences, having the ability to penetrate the cell membranes and accumulate inside the cells. Several CPAPs translocate the plasmalemma by adopting well-defined secondary structures. After entering the cell, they disrupt the cellular homeostasis by increasing ROS, which degrades or blocks the activity of enzymes or proteins taking part in various cellular processes, thereby leading to cell death. Transportan 10 (TP10) is a chimeric hybrid derived from the N-terminus of neuropeptide galanin and the C-terminus of mastoparan, a peptide from wasp venom; it accumulates around the cell surface and internalizes into the cell without actually interacting with the cell membrane, as well as exhibits potential antimicrobial activity against *C. albicans* and *Staphylococcus aureus* (Cruz et al., 2022; Nekhotiaeva et al., 2004).

3.2.2.1.2.2 Inhibiting Cell Wall Biosynthesis Cell wall is crucial for cell survival as it provides structural integrity and aids the organism to tolerate environmental stress and mechanical damage. Some AMPs target the microbial cells by interfering with its cell wall biosynthesis or by remodelling the cell wall. This leads to weakening of structural integrity, osmotic imbalance and, eventually, cell death. Some defensins and AFP-like proteins target biosynthesis of chitin and beta-1,3-glucan synthesis, the key components of fungal cell wall, disrupting the rectitude of cell wall (Aerts et al., 2008; Thevissen et al., 2003).

3.2.2.2 Disruption of Molecular Signalling Pathways

Beyond directly targeting the cells by disrupting the cell membrane and other organelles, AMPs exert profound effects by involving with intracellular signalling pathways. By modulating the key regulatory pathways, AMPs interfere with survival mechanisms by disturbing the organism's metabolism, impacting homeostasis and pushing the cell beyond its capacity to repair.

The Mitogen-Activated Protein Kinase (MAPK) pathway is one of the key pathways targeted by AMPs (mostly in fungal cells), and it plays a vital role in maintaining cell wall integrity, morphogenesis, immunity and stress responses. AMPs, which often interact with the cell wall or cell membrane components, trigger the MAPK pathway, inducing stress responses and leading to cell death. RsAFP2 (Raphanus sativus Antifungal Protein 2), a plant defensin from radish seeds, exhibits antifungal activity by interacting with the glycosylceramides present in the fungal cell membranes. When RsAFP2 binds to glucosylceramides and barges with the integrity of cell membrane, the cell receives the messages of cell wall stress and activates Mkc1-MAPK signalling cascade, which results in the overexpression of cell wall repair genes and generation of excessive ROS, leading to cell death (Aerts et al., 2008).

AMPs, mostly pAMPs, govern the plant defence system by directly targeting the pathogens or by acting as modulators of signalling pathways. Jasmonic acid (JA) pathway, Salicylic acid (SA) pathway and Ethylene (ET) pathway are the important hormone-mediated defence pathways in plants. JA pathway signalling is associated with herbivorous insects like beetles and caterpillars and necrotrophic pathogens like *Alternaria, Botrytis cinerea* and *Sclerotinia sclerotiorum*; and SA pathway signalling is associated with biotrophic pathogens like *Puccinia, Pseudomonas syringae* and *Blumeria graminis* and hemibiotrophic pathogens like *Colletotrichum*. AMPs activate Pattern Recognition Receptors (PRRs) of plant cells and stimulate to trigger various downstream signalling cascades, which lead to the accumulation of defence-related hormones and also the production of other pAMPs. For example, in *Arabidopsis thaliana*, JA and ET response pathways induce the production of PDF1.2 (Plant Defension1.2) upon the infection of *Botrytis* and *Alternaria*. PDF1.2 not only directly involves in the plant defence, but also creates a positive feedback loop by amplifying JA signalling. Similarly, RsAFP2 (a plant defensin from radish) in transgenic tobacco plant targets pathogens directly and also augments the SA pathway. Upon infection with *Phytophthora parasitica*, an oomycete, the triggered SA pathway elevates the expression of pathogenesis-related genes and also favours in systemic acquired resistance (SAR). Plants also exhibit defence responses by the cross-talk between various pathways. Even though the JA and SA pathways are antagonistic to each other, the ET pathway, which is activated parallel to JA pathway, prevents the suppression of SA-mediated pathways favouring robust defence mechanisms (Penninckx et al., 1998; Leon-Reyes et al., 2010).

3.2.2.3 Immunological Regulation of Antimicrobial Peptides

AMPs are not only involved in the conventional microbicidal activity; beyond killing pathogens, they also maintain tissue homeostasis and ensure efficient pathogen clearance. AMPs play a key role as immunomodulators by recruitment and activation of immune cells, modulation of cytokines and chemokines production, neutralization of microbial toxins and involvement in wound healing and tissue regeneration (Hancock & Sahl, 2006; Mookherjee & Hancock, 2007). AMPs also act as alarmins – substances that activate immune system during infection or injury. A human cathelicidin, LL-37, induces the production of IL-8 and orchestrates neutrophils, monocytes and T-cells by acting as "alarmin", and it also facilitates promotion of angiogenesis, wound healing and the elimination of abnormal cells (Mookherjee & Hancock, 2007; Yang et al., 2001)

3.3 PLANT-DERIVED ANTIMICROBIAL PEPTIDES

3.3.1 Plant-Derived Antimicrobial Peptides

Across the centuries, plants have been nature's pharmacy, furnishing with a rich diversity of bioactive compounds with healing properties. The use of medicinal plants for treating various diseases was explained in Atharva Veda, one of the Vedas (a Veda is an Indian scripture of knowledge), and it forms the basis of many traditional and modern therapies (Pattanayak et al., 2010). The plant natural products like alkaloids, flavonoids, terpenes and AMPs show noticeable therapeutic properties against various diseases (Tam et al., 2015). pAMPs are a remarkable group of proteins that act as natural antibiotics and play a vital role in the defence of plants. Being integral to the plant innate immune system, pAMPs provide a rapid and localized response to pathogen invasion. The characteristics like positive charge, high cysteine residues and disulphide bonds endow them to tightly fold and confer to exceptional stability. The biological activities of pAMPs are diverse, and they involve in direct disruption of metabolic membranes as well as interfere with pathogen metabolic processes by altering signalling pathways. In contrast to broad-spectrum chemical defences, pAMPs typically show selective activity targeting a specific group of pathogens and minimizing damage to the host plant's own cells. Most of the pAMPs are pre-synthesized and stored in an inactive form; they become activated when microbial invaders are detected. Due to the escalation in the antimicrobial resistance globally, pAMPs are gaining attention as vital templates for the development of new antibiotics, antifungal and other therapeutic molecules. Their natural origin, structural versatility and reduced potential to trigger resistance offer significant advantages over conventional therapeutic agents (de Oliveira et al., 2024; Li et al., 2021; Tam et al., 2015). The pAMPs are extensively varied and are organized into diverse set of families based on their structure, function and activity.

3.3.2 Classification of Antimicrobial Peptides

A clear understanding of the classification of pAMPs is essential for the rational design of novel therapeutic strategies by correlating structural characteristics with specific antimicrobial activities (Figure 3.3).

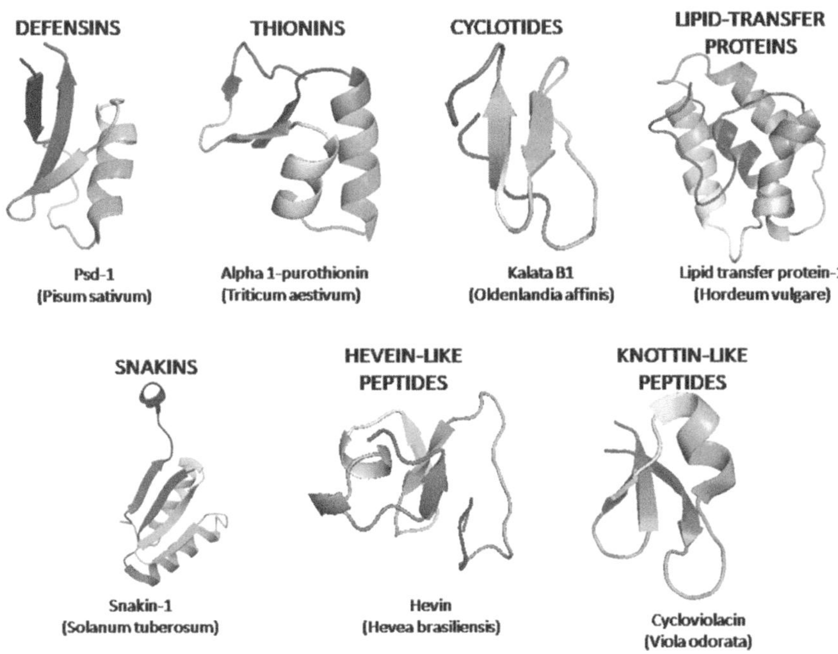

FIGURE 3.3 Three-dimensional structures of antimicrobial peptides from different families. (The structures are retrieved from RCSB-PDB.)

3.3.2.1 Defensins

Defensins are one of the most studied and well characterized families of pAMPs. Plant defensins are small peptides, usually 45–54 amino acids long, mostly cationic, basic, cysteine-rich and amphipathic in nature. They are primarily composed of β-sheets stabilized by 3–4 disulphide linkages, which help them to form a compact, globular structure. The tertiary structure of defensins contains a "Cysteine-stabilized α-helix-β sheet" (CSαβ) motif, which is characterized by one α-helix and three anti-parallel β-sheets typically with eight conserved cysteine residues stabilized with 3–4 disulphide bonds. Defensins can exert antimicrobial activity by membrane permeabilization, binding to cell wall components and inhibiting their synthesis, intracellular targeting by blocking certain processes or molecules that are essential for cell survival, and inhibition of enzyme activity. Binding to the sphingolipids of fungal membrane is a notable characteristic of defensins (Parisi et al., 2019).

3.3.2.2 Thionins

Thionins are basic peptides, typically consisting 45–50 amino acids, carrying a positive charge at physiological pH. Thionins contain two anti-parallel α-helices, a double-stranded β-sheet and a C-terminal coil. They are rich in basic amino acids like arginine and lysine, containing 6–8 cysteine residues favouring the formation of 3–4 disulphide bonds (Florack & Stiekema, 1994; Stec, 2006).

Thionins exert a strong antifungal activity against a wide range of fungal pathogens like *Fusarium, Alternaria and Aspergillus* by disrupting their cell membrane and inhibiting cell wall synthesis and spore formation. Besides being strong antifungals, thionins also perform antibacterial activity against Gram-positive bacteria like *Bacillus* and *Staphylococcus*, as well as Gram-negative bacteria like *Escherichia coli*.

3.3.2.3 Lipid Transfer Proteins

Plant LTPs are small, water soluble peptides, usually 7–10 kDa in size and amphipathic in nature. They are very important pAMPs with a broad range of antibacterial, antiviral and antifungal activities. Plant LTPs are characterized by 4 or 5 α-helices connected by flexible loops and 6–8 conserved cysteine residues conferring to stability facilitating the formation of 4 disulphide bridges. They contain a central hydrophobic pocket that aids in binding and transferring lipids, such as phospholipids, glycerolipids and fatty acids (Edstam et al., 2011).

LTPs perform various physiological functions in plants like lipid transport, pathogen defence, cell membrane biogenesis and stability. LTPs perform antimicrobial activity by binding to lipids (either pathogen membrane lipids or plant defence response lipids like oxylipids) and causing membrane disruption by insertion into microbial membranes either by causing pores or lesions in the membrane, causing leakage of cellular contents or altering membrane lipid composition (Kader, 1996).

3.3.2.4 Cyclotides

Cyclotides are the molecules ranging from 28 to 37 amino acids in length. They are characterized by a cyclic backbone resulted from the formation of peptide bond between N-terminus and C-terminus, which is further stabilized by a knotted arrangement of disulphide bonds at the cysteine motifs. Cyclotides often have a hydrophobic core, which contributes to antimicrobial activity (Strömstedt et al., 2017).

Cyclotides exhibit antimicrobial effects by membrane binding and disruption by pore formation, enzyme inhibition and disruption of microbial replication. They are effective against a wide range of bacterial species like *S. aureus, Bacillus subtilis, E. coli, Pseudomonas aeruginosa* and various *Salmonella* and *Enterococcus* species. They also exert antifungal activity towards various members of *Candida* and *Aspergillus* species.

3.3.2.5 Hevein-Like Peptides

Hevein-like peptides are a distinct class of pAMPs, with a characteristic conserved chitin binding domain. Hevein-like peptides are 30–45 amino acids long, rich in glycine and cysteine residues, forming 3–4 disulphide linkages. These peptides are primarily known to exhibit antifungal activity by disrupting fungal cell wall integrity and inhibiting fungal growth. Hevein-like peptides bind to the chitin domains present in the fungal cell wall, causing structural disruption, osmotic stress and cell lysis (Archer, 1960; Porto et al., 2012).

Hevein-like peptides are effective against various fungal species like *Fusarium, Aspergillus, Botrytis, Candida* and *Trichophyton rubrum*. Besides they also perform antibacterial activity against various Gram-positive and Gram-negative bacteria by causing membrane destabilization (Loo et al., 2021).

3.3.2.6 Snakins

Snakins are a fascinating class of pAMPs with unique structural features and potent antimicrobial properties. These plant defence proteins are structurally similar to the snake venom peptides, thus called snakins. These natural plant defence proteins are 40–60 amino acids long, with high content of basic amino acids, 10–12 cysteine residues forming 6 disulphide bonds and amphipathic in nature. These peptides are characterized by two distinct domains connected by a linker. Snakins are potent antimicrobial, antifungal and antiviral agents. They exhibit antibacterial activity by membrane insertion, pore formation and membrane permeabilization leading to osmotic imbalance and cell lysis, as well as antifungal activity by disrupting cell wall synthesis and inhibiting protein synthesis. In addition to antimicrobial effects, snakins also perform immunomodulatory effects enhancing host's defence mechanism (Berrocal-Lobo et al., 2002; Segura et al., 1999).

3.3.2.7 Knottin-Type Peptides

Knottin-type peptides are characterized by the knot-like structure formed by the disulphide bonds. This kind of architecture gives a compact and resilient shape to the peptides, making them significantly resistant to the proteolytic activity and aiding them to maintain their biological activity even under extremely harsh conditions. Knottin-type peptides are 30–40 amino acids long with 6 cysteine residues and 3 disulphide linkages. They are marked by numerous turns and loops, which offer them a compact structure. They perform antimicrobial activity by interacting with specific cell surface receptors, disrupting membrane and inhibiting enzymes responsible for pathogen survival (Aboye et al., 2015; Hwang et al., 2010).

3.3.3 Impact of Plant-Derived Antimicrobial Peptides on Fungal Metabolism

3.3.3.1 Disruption of Energy Metabolism

The survival and growth of the organism depend on the energy metabolism, i.e., generation of energy currency by glycolysis, Kreb's cycle and oxidative phosphorylation, as well as its storage and utilization by creating macromolecules and maintaining proton gradient across membranes of cell. Disruption in any of these ceases the survival of the organism. A spectrum of pAMPs interfere with energy metabolism by disturbing the membrane potential of plasmalemma, by depolarizing mitochondrial membranes or by directly interfering with energy pathways. ZmD32 is a pAMP that exerts both antimicrobial and antifungal properties. This corn defensin specifically binds to the phosphatidyl phosphates present on the fungal cell membranes and destabilizes the integrity of cell membrane, increases membrane permeability and causes efflux of K^+ ions. This loss in membrane potential impairs the ionic gradient, ATP synthesis and energy metabolism (Kerenga et al., 2019).

3.3.3.2 Effecting Mitochondrial Activity

Mitochondria play a prominent role in energy production, maintaining redox balance, calcium homeostasis and also in regulating cell death by secreting a few important components that mediate cell death. Most of the AMPs directly or indirectly disrupt the functions of mitochondria, leading to the dysregulation of metabolic pathways and causing cell death. A plant defensin, Psd1 (Pisum sativum defensin1), exhibits antifungal activity by targeting the mitochondrial function. Psd1 internalizes the cell by binding to the glucosylceramides on the fungal cell membrane and disrupts the fungal mitochondrial activity by causing hyperpolarization of mitochondrial membrane potential. This interferes with the electron transport chain, thereby affecting ATP synthesis and also the excessive release of ROS, promoting oxidative stress further triggering apoptosis (Lobo et al., 2007; Nicolas, 2009).

3.3.3.3 Disrupting Biosynthetic Pathways

Many pAMPs disturb various biosynthetic pathways by entering into the cell and interfering with cellular processes like DNA replication, protein synthesis, cell wall biosynthesis, lipid biosynthesis and various cell repair mechanisms. They enter into the cell by interacting with the components of cell membrane, by the formation of pores or by disrupting the integrity of the cell membrane and target regulatory proteins or the biosynthetic enzymes by directly interacting with them or inhibiting them. For example, some pAMPs inhibit the synthesis of fungal glucosylceramide, which further affect lipid biosynthesis. Some defensins bind to the intracellular proteins that are involved in cell cycle regulation and cause change in their confirmation, affecting the protein folding, thereby hindering biosynthetic pathways. A spectrum of AMPs target the cell by generating ROS, which damages the enzymes essential for nucleotide, protein and lipid biosynthesis. PnAMP, derived from *Pharbitis* seeds, penetrates the hyphae, disrupts membranes and creates oxidative stress interfering metabolic regulation. However, the exact mechanism by which AMPs interrupt the cellular mechanism is to be explored (Nawrot et al., 2014; van der Weerden et al., 2019).

3.3.3.4 Induction of Autophagy

Autophagy is a cellular mechanism in which the cell breaks down the abnormal cellular components through lysosome-regulated mechanisms. Some pAMPs exert their antifungal activities by inducing excessive autophagy, which results in cell death. HsAFP1, a plant defensin, impairs cell cycle progression and induces cell death in fungal cells. This is evidenced by the upregulation of autophagy-related strains and formation of autophagic bodies in *C. albicans* (Struyfs et al., 2020).

3.3.3.5 Disruption of Vacuolar Function

Vacuoles are the membrane-bound cell organelles that sequester waste products and maintain the osmotic balance of the cell. In addition to other antifungal mechanisms, AMPs also interfere with the functioning of vacuoles. They disrupt the fungal vacuoles by meddling with their function or altering their morphology, thereby further affecting cellular homeostasis, leading to cell death. NbD6 is a plant defensin that interferes with the vacuoles by forming a large vacuole by the fusion of the vacuoles present inside the cells, which is a sign of autophagic response (Struyfs et al., 2021).

3.3.4 Plants-Derived Antimicrobial Peptides as Antifungal Agents: Real-World Applications against Fungal Infections

The escalating challenge of antifungal resistance by fungal pathogens as well as the rising incidence of fungal infections, especially among immunocompromised populations, has created an urgent need for alternative antifungal approaches. pAMPs offer a promising solution for this growing crisis. Their natural origin, unique mechanisms of action, potent antifungal properties and evolutionary refinement as antifungal agents position the pAMPs as valuable candidates in the development of next-generation therapeutics. The following are a few case studies illustrating the role of pAMPs functioning as effective antifungal agents in suppressing or eliminating fungal infections (Table 3.1)

TABLE 3.1
Examples of Some Common Plant-Derived Antimicrobial Peptides, Their Classification, Source and Microbial Activity

pAMP	Class	Source	Microbial Target	References
Fa-AMP1	Plant defensin	*Fagopyrum esculentum*	Antibacterial, antifungal	Fujimura et al. (2003)
Fa-AMP2	Plant defensin	*Fagopyrum esculentum*	Antibacterial, antifungal	Fujimura et al. (2003)
Rs-AFP1	Plant defensin	*Raphanus sativus*	Antifungal	Terras et al. (1995)
Rs-AFP2	Plant defensin	*Raphanus sativus*	Antifungal	Terras et al. (1995)
C.fistula PI	Plant defensin	*Cassia fistula*	Trypsin inhibitor	Wijaya et al. (2000)
Vv-AMP1	Plant defensin	*Vitis vinifera*	Antifungal	De Beer and Vivier (2008)

(Continued)

TABLE 3.1 (*Continued*)
Examples of Some Common Plant-Derived Antimicrobial Peptides, Their Classification, Source and Microbial Activity

pAMP	Class	Source	Microbial Target	References
ZmD32	Plant defensin	*Zea mays*	Antibacterial, antifungal	Kerenga et al. (2019)
NaD1	Plant defensin	*Zea mays*	Antibacterial, antifungal	Kerenga et al. (2019)
Purothionins	Thionin	*Triticum aestivum*	Antibacterial	Fernandez De Caleya et al. (1972)
CaThi	Thionin	*Capsicum annuum*	Antibacterial, antifungal	Taveira et al. (2014)
Cp-thionin II	Thionin	*Vigna unguiculata*	Antifungal	Schmidt et al. (2019)
NTC	Hevein-like peptides	*Hevea brasiliensis*	Antifungal	Parijs et al. (1991)
Pn-AMP1	Hevein-like peptides	*Pharbitis nil*	Antifungal	Choon Koo et al. (1998)
Pn-AMP2	Hevein-like peptides	*Pharbitis nil*	Antifungal	Choon Koo et al. (1998)
pnAMP-h2	Hevein-like peptides	*Pharbitis nil*	Antifungal	Choon Koo et al. (1998)
Ep-AMP1	Knottin-type peptides	*Echinopsis pachanoi*	Antifungal	Aboye et al. (2015)
Psacotheasin	Knottin-type peptides	*Psacothea hilaris*	Antibacterial, antifungal	Hwang et al. (2010)
bevuTI-I	Knottin-type peptides	*Beta vulgaris*	Trypsin inhibitory prolyl oligopeptidase	Retzl et al. (2020)
Snakin-1	Snakins	*Solanum tuberosum*	Antibacterial, antifungal	Segura et al. (1999)
Snakin-2	Snakins	*various flowers*	Antibacterial, antifungal	Tavares et al. (2008)
cyO2	Cyclotides	*Viola odorata*	Antibacterial, anticancer	Pränting et al. (2010)
varv A	Cyclotides	*Viola species*	Anticancer	(Lindholm et al., 2002)
varv F	Cyclotides	*Viola species*	Anticancer	(Lindholm et al., 2002)
kalata B1	Cyclotides	*Oldenlandia affinis*	Insecticidal	Colgrave et al. (2008)
kalata B2	Cyclotides	*Oldenlandia affinis*	Insecticidal	Colgrave et al. (2008)
kalata B6	Cyclotides	*Oldenlandia affinis*	Insecticidal	Colgrave et al. (2008)

(*Continued*)

TABLE 3.1 (*Continued*)
Examples of Some Common Plant-Derived Antimicrobial Peptides, Their Classification, Source and Microbial Activity

pAMP	Class	Source	Microbial Target	References
Poca A	Cyclotides	*Pombalia calceolaria*	Anticancer	Pinto et al. (2018)
Poca B	Cyclotides	*Pombalia calceolaria*	Anticancer	Pinto et al. (2018)
Ps-LTP 1	Lipid transfer protein	*Pisum sativum*	Antibacterial, antifungal	Bogdanov et al. (2016)
McLTP 1	Lipid transfer protein	*Morinda citrifolia*	Antibacterial	Souza et al. (2018)
Lc-LTP1	Lipid transfer protein	*Lens culinaris*	Antibacterial, antifungal	Bogdanov et al. (2015)
Lc-LTP2	Lipid transfer protein	*Lens culinaris*	Antibacterial, antifungal	Bogdanov et al. (2015)
Lc-LTP3	Lipid transfer protein	*Lens culinaris*	Antibacterial, antifungal	Bogdanov et al. (2015)

3.3.4.1 Plant Defensins from *Oryza sativa*

OsAFP1, a plant defensin isolated from *Oryza sativa*, shows a potent antifungal activity inhibiting the growth of *C. albicans*. OsAFP1 is heat resistant, maintains antifungal activity even in harsh conditions and is effective even in very low concentrations. It targets the cell by inducing apoptosis rather than disrupting the fungal cell membrane. The study also reveals that C-terminal ɣ-core motif is critical for its antifungal activity (Ochiai et al., 2018).

3.3.4.2 HsAFP1 from *Heuchera sanguine*

HsAFP1 is a 19-mer plant defensin, shows antifungal activity individually and also when combined with caspofungin. The study revealed that HsAFP1 in combination with caspofungin successfully inhibited the growth of *C. albicans* biofilm formation both in vitro and in vivo in a rat catheter model, by membrane permeabilization and peptide internalization in Candida cells. The study also concluded that this plant defensin also potentiated the activity of caspofungin along with resistant strains (Cools et al., 2017).

3.3.4.3 Nigellothionins from Black Cumin Seeds

NsW1, NsW2 and NsW4 are the Nigellothionins, isolated from black cumin seeds, *Nigella sativa,* which showed a significant inhibition in the growth of *Aspergillus* species compared to the commercial antifungal drug amphotericin B. The study also conveyed that NsW2 was active against azole-resistant clinical strains of *Aspergillus* and *Candida* even in very low concentrations (Barashkova et al., 2021).

3.4 COMPARISON OF PLANT-DERIVED ANTIMICROBIAL PEPTIDES WITH OTHER ANTIFUNGAL TREATMENT STRATEGIES

The growing prevalence of fungal infections is also triggering the rise of antifungal resistance. While the conventional antifungal therapies like the use of azoles, polyenes and echinocandins are still the primary options in treating the fungal infections, the escalated rise in the drug resistance is demanding for the alternative approaches to deal with the MDR pathogens. Even though the immunotherapy approaches are found to be effective, they failed to meet the expectations in immunocompromised individuals. Therefore, pAMPs have gained attention as alternative treatment strategies due to their unique mechanisms of action and potential to combat MDR fungi.

3.4.1 Mechanism of Action

pAMPs perform their antifungal activity by directly disrupting the fungal cell membrane. The cationic and amphipathic nature of these peptides favours them to interact with negatively charged components of the fungal cell membrane, such as phospholipids. After binding, the pAMPs cause cell death by membrane destabilization, pore formation and leakage of cell components. pAMPs can also inhibit the cell growth by interfering with the cell vital processes like DNA replication, transcription, protein translation and enzyme activity (Hancock & Sahl, 2006).

Conventional antifungals like azoles inhibit the fungal growth by targeting an enzyme lanosterol 14α-demethylase, which is responsible for a vital cell membrane component ergosterol. Echinocandins like caspofungin inhibit the synthesis of β-glucan and target the cell wall leading to the cell lysis, and polyenes like amphotericin B and nystatin create pores in the fungal cell membrane altering its dynamics and leading to cell death. But the action of azoles is specific and slower compared to pAMPs; Echinocandins are not effective against moulds, while polyenes are toxic to human cells limiting their use (Baginski & Czub, 2009; Denning, 2003).

3.4.2 Spectrum of Activity

pAMPs exhibit a wide spectrum of activity against various pathogens including yeasts and moulds. Their unique mode of action does not encourage cells to develop resistant mechanisms against them and also confers them to exhibit antibacterial and antiviral properties.

Azoles are mostly potential against *Candida* species; they show limited activity against *Aspergillus* and *Fusarium* and are ineffective on certain drug-resistant strains. Polyenes are effective against a wide range of fungi like *Aspergillus* and *Candida* species, but their prolonged use leads to nephrotoxicity. Echinocandins are at odds with a narrow spectrum of fungal pathogens, and they are ineffective against moulds (Hossain et al., 2022).

3.4.3 Development of Resistance

Lower possibility of resistance development is one of the key advantages of pAMPs. Their nature of interaction, mode of action, non-specificity and broad spectrum of potency contribute to become more difficult to develop resistance. However, due to selective pressure, a few strains may develop resistance towards certain pAMPs, but this might not be as serious as with the conventional drugs. Numerous species of *Candida* and *Aspergillus* are resistant to primary antifungal drugs like azoles, polyenes and echinocandins. The mutations in lanosterol 14α-demethylase and upregulation of efflux pumps in the fungal membranes are the reasons for the emerged azole resistance in *Candida*. The mutations in Fks-1 and change in the membrane dynamics contribute to the echinocandins and polyene resistance (Hossain et al., 2022; Tian et al., 2024).

3.4.4 Toxicity and Side Effects

As pAMPs primarily target the fungal cell membrane, they typically regarded to have minimal toxic effects towards human cells. However, properties like their amino acid sequence, higher hydrophobicity or basic nature of the peptides may affect the mammalian cell membranes leading to cytotoxicity. The long-term use of azoles and echinocandins may exhibit various side effects like liver abnormalities, gastrointestinal disturbances, skin rashes, etc., while polyenes may cause nephrotoxicity (Benitez & Carver, 2019; Marquez & Quave, 2020).

3.5 APPLICATIONS

AMPs generated from medicinal plants are a promising area in the healthcare and agriculture business, providing novel approaches to battle infections, especially those caused by MDR pathogens (Figure 3.4). These tiny, naturally occurring compounds have broad-spectrum antibacterial activity and provide distinct advantages over traditional antibiotics, including quick action, limited resistance development and multifunctional features such as immunomodulation and wound healing (Xuan et al., 2023). The critical need for novel biobased approaches in healthcare is demonstrated by the potential of fungal cell factories and biovalorization processes, underscoring the importance of sustainable biomolecule production for therapeutic purposes (Olaitan Bamidele et al., 2025).

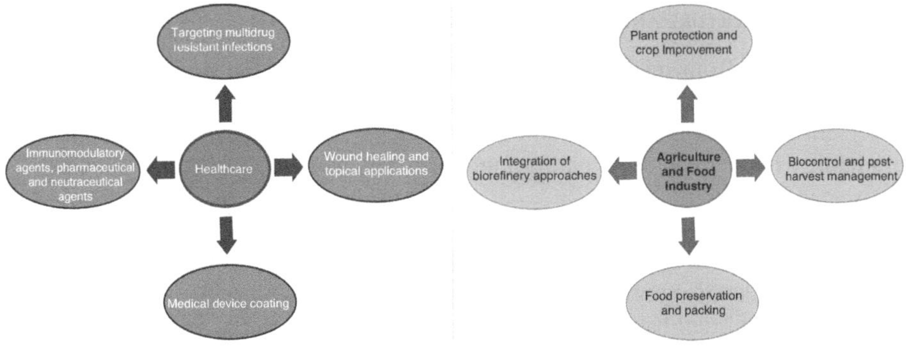

FIGURE 3.4 Applications of antimicrobial peptides in healthcare and agriculture and the food industry.

3.5.1 Healthcare

i. *Treatment of multidrug-resistant infections:* One of the most urgent applications of pAMPs is in the treatment of MDR infections. Traditional antibiotics are increasingly ineffective against resistant bacteria and fungi. AMPs, by targeting microbial membranes and intracellular components simultaneously, reduce the chances of resistance development. Plant defensins and thionins, for instance, have shown potent antifungal and antibacterial activities against *E. coli*, *C. albicans*, *S. aureus* and resistant strains (Tam et al., 2015).
ii. *Wound healing and topical applications:* AMPs found in medicinal plants have an important function in wound healing. Chronic wounds, such as diabetic ulcers and burn injuries, are vulnerable to biofilm-forming, drug-resistant bacteria. Plant-derived peptides not only kill infections but also stimulate tissue regeneration, reduce inflammation and speed up re-epithelialization. Formulations such as AMP-infused hydrogels, lotions and wound dressings are being developed to take advantage of these benefits (Pfalzgraff et al., 2018).
iii. *Medical device coating:* Another application is for coating medical devices, including catheters, implants and prostheses, with AMPs. These surfaces are frequently prone to microbial colonization, resulting in serious illnesses. pAMPs can be fixed on device surfaces, providing ongoing antibacterial protection without systemic toxicity (Costa et al., 2019).
iv. *Immunomodulatory agents, pharmaceutical and nutraceutical products:* Certain AMPs can influence the immune response in the host, making them useful for treating infections and inflammatory illnesses. They can function as immune stimulants or suppressors, balancing immunological responses to prevent chronic inflammation or pathogen evasion (Duarte-Mata & Salinas-Carmona, 2023). pAMPs are excellent candidates for incorporation into pharmaceutical and nutraceutical formulations due to their natural origin and safety features. They can be used as dietary supplements, nasal sprays or topical treatments to improve immunity and prevent infections in susceptible groups (Mahlapuu et al., 2016).

3.5.2 Agriculture and Food Industry

AMPs derived from medicinal plants have shown increasing potential not just in healthcare, but also in agriculture and the food sector (Figure 3.4). Their broad-spectrum antibacterial characteristics, low tendency to produce resistance and biodegradability make them excellent for long-term and environmentally beneficial usage in these industries.

i. *Plant protection and crop improvement:* pAMPs work as a natural defence mechanism in agriculture, and they are increasingly being employed to protect crops from bacterial, fungal and viral infections. These peptides, such as defensins and cyclotides, damage microbial membranes, hinder intracellular activities and cause programmed cell death in pathogens. Unlike synthetic pesticides, AMPs break down spontaneously, which reduces environmental accumulation and toxicity. They also contribute to SAR in crops, improving innate immunity and reducing dependency on chemical treatments (Zhang et al., 2023).

ii. *Biocontrol and post-harvest management:* AMPs can help to significantly reduce post-harvest losses caused by microbial contamination. Edible coatings (EC) and sprays containing AMPs suppress rotting organisms on fruits and vegetables, extending shelf life and maintaining quality. EC are derived from a wide range of agricultural sources such as corn, potatoes, rice, wheat, soybeans and milk. They incorporate various ingredients, including methylcellulose, hydroxypropyl methylcellulose, starches, pectin, chitosan, collagen, gelatin, zein, gluten, casein, soy protein isolate, shellac and beeswax, among others. Coatings incorporating AMPs have been demonstrated to be beneficial at decreasing mould and bacterial contamination while maintaining taste and appearance (Miranda et al., 2024).

iii. *Food preservation and packaging:* In the food business, AMPs are being used in packaging materials to provide active antimicrobial surfaces. These bioactive films limit microbial development on food surfaces, thereby increasing shelf life and maintaining safety. This is especially useful for minimally processed and ready-to-eat foods, when conventional preservatives may be unwanted. Furthermore, the usage of AMPs coincides with consumer demand for natural food preservatives and clean labels (Liu et al., 2021).

Even though AMPs form bioactive films, microplastic contamination in food packaging is a serious threat to public health, food safety and the environment (Chinglenthoiba et al., 2025). Microplastics not only enter food systems, but they also make excellent substrates for biofilm formation due to their hydrophobic and abrasive surface qualities, which promote microbial attachment. These biofilm–microplastic complexes protect embedded bacteria from disinfection efforts and promote the spread of infections and antibiotic resistance genes, posing significant dangers to human health (Rahman et al., 2025). Alternatively, bioplastics are gaining popularity due to their biodegradable and sustainable nature. Global bioplastic production is predicted to increase from 2.4 million tons in 2021 to 7.6 million tons by 2026 (Chandukishore et al., 2025). Many recent studies have discovered that AMPs and antioxidant peptides (AOPs)-functionalized food packaging films/EC have incredible potential in the preservation of various fresh foods. For example, an edible covering containing AOPs derived from fish skin significantly reduces the rotting of fresh fish fillets (Roy et al., 2024).

iv. *Integration of biorefinery approaches:* A forward-thinking method is to include AMPs into biorefinery systems, which convert agricultural waste and hemicellulosic biomass into high-value bioactive chemicals. Several studies have examined how to develop sustainable engineering platforms that use plant and microbial resources. Such platforms allow for the co-production of AMPs with bioethanol, organic acids and other bioactive compounds. Using hemicellulosic waste for AMP synthesis decreases environmental impact and promotes circular bioeconomy models (Sun, 2024; Ulaganathan et al., 2024)

3.6 CHALLENGES AND FUTURE PERSPECTIVES

MDR fungi, such as *C. auris*, *A. fumigatus* and azole-resistant *C. albicans*, have emerged and spread rapidly, posing an urgent worldwide health concern, particularly in immunocompromised patients and hospital settings (Fisher et al., n.d.). pAMPs have sparked interest as potential alternatives to traditional antifungal medications. Despite its potential, various hurdles must be cleared before these peptides may be routinely used in clinical settings.

One of the primary challenges in the application of pAMPs is isolation and purification. AMPs are often expressed in low amounts in plant tissues, making extraction difficult and ineffective (Barashkova & Rogozhin, 2020). Furthermore, AMPs frequently have complicated post-translational modifications, complicating recombinant expression in heterologous systems like *E. coli* or yeast (Tokmakov et al., 2012). This problem not only restricts the scalability of AMP production but it also impacts cost-effectiveness, which is crucial for widespread clinical use. Another major issue is peptide stability and bioavailability. In vivo, AMPs are sensitive to proteolytic breakdown, particularly when supplied systemically. Their amphipathic structure and positive charge may also cause aggregation or non-specific binding to host proteins, lowering their effective concentration at the infection site (Asif et al., 2024). To improve stability, strategies such as nanoparticle encapsulation and chemical changes such as cyclization and PEGylation are being investigated; however, these add complexity and cost to medication development (Suk et al., 2016). Toxicity and selectivity are other causes of concern. While many AMPs primarily target microbial membranes due to their distinct lipid compositions, some pAMPs may be harmful to human cells, especially at higher concentrations (Gagandeep et al., 2024). Furthermore, regulatory and standardization difficulties pose challenges. pAMPs come from a variety of species and might have different sequences and actions based on the plant's environment and developmental stage. The lack of defined methodologies for AMP screening, characterization and quality control makes obtaining regulatory permission for clinical trials and medicinal use challenging (Berida et al., 2024).

Despite these challenges, developments in biotechnology and synthetic biology provide significant opportunities for the synthesis and application of pAMPs. Recombinant DNA technology allows for more efficient AMP generation in microbial or plant expression systems. Transgenic plants or plant cell cultures, for example, can be programmed to produce significant amounts of certain AMPs under regulated conditions (Nazarian-Firouzabadi et al., 2024). Liposomes, dendrimers and polymeric nanoparticles are examples of nanotechnology-based delivery strategies that are proving successful for AMP distribution. These methods can prevent AMP breakdown, improve permeability and targeting and reduce cytotoxicity by regulating the release rate (Fadaka et al., 2021). Combining AMPs with conventional antifungals in such delivery systems may result in synergistic effects, potentially lowering effective doses and minimizing the likelihood of resistance development. The combination of omics technologies, such as genomes, transcriptomics and proteomics, can speed up the discovery of novel pAMPs. High-throughput sequencing and computer modelling enable the screening of hundreds of peptide candidates using structure–activity connections and membrane interaction profiles (Sun, 2024).

Machine learning and artificial intelligence improve predictive models for optimal AMP design. CRISPR-Cas9 gene editing can also be used to boost endogenous AMP production in medicinal plants and even human cells for therapeutic reasons. Furthermore, synthetic biology can be used to create novel AMPs with increased potency, selectivity and stability by integrating patterns from multiple natural peptides. From a clinical standpoint, extensive testing in preclinical models of fungal infection is required to evaluate effectiveness, pharmacokinetics and toxicity. Humanized mouse models, 3D skin infection models and organoids provide more physiologically relevant platforms for evaluating plant-based AMPs against MDR fungi (Zasloff, 2002).

In conclusion, pAMPs represent a compelling frontier in antifungal therapy, particularly in addressing MDR pathogens. Continued interdisciplinary research bridging plant sciences, molecular biology, pharmacology and nanotechnology is crucial for translating these natural molecules from bench to bedside. With sustained investment and innovation, pAMPs could become vital components of next-generation antifungal therapeutics.

REFERENCES

Aboye, T. L., Strömstedt, A. A., Gunasekera, S., Bruhn, J. G., El-Seedi, H., Rosengren, K. J., & Göransson, U. (2015). A cactus-derived toxin-like cystine knot peptide with selective antimicrobial activity. *ChemBioChem*, 16(7), 1068–1077. https://doi.org/10.1002/cbic.201402704

Aerts, A. M., François, I. E. J. A., Cammue, B. P. A., & Thevissen, K. (2008). The mode of antifungal action of plant, insect and human defensins. *Cellular and Molecular Life Sciences*, 65(13), 2069–2079. https://doi.org/10.1007/s00018-008-8035-0

Almatroudi, A. (2025). Biofilm resilience: Molecular mechanisms driving antibiotic resistance in clinical contexts. *Biology*, 14(2). https://doi.org/10.3390/biology14020165

Alvarez-Sieiro, P., Montalbán-López, M., Mu, D., & Kuipers, O. P. (2016). Bacteriocins of lactic acid bacteria: Extending the family. *Applied Microbiology and Biotechnology*, 100(7), 2939–2951. https://doi.org/10.1007/s00253-016-7343-9

Andreu, D., & Rivas, L. (1998). Animal antimicrobial peptides: An overview. *Biopolymers*, 47(6), 415–433. https://doi.org/10.1002/(sici)1097-0282(1998)47:6%3C415::aid-bip2%3E3.0.co;2-d

Archer, B. L. (1960). The proteins of *Hevea brasiliensis* latex. 4. Isolation and characterization of crystalline hevein. *Biochemical Journal*, 75(2), 236–240. https://doi.org/10.1042/bj0750236

Asif, F., Zaman, S. U., Arnab, Md. K. H., Hasan, M., & Islam, Md. M. (2024). Antimicrobial peptides as therapeutics: Confronting delivery challenges to optimize efficacy. *The Microbe*, 2, 100051. https://doi.org/10.1016/j.microb.2024.100051

Atriwal, T., Azeem, K., Husain, F. M., Hussain, A., Khan, M. N., Alajmi, M. F., & Abid, M. (2021). Mechanistic understanding of *Candida albicans* biofilm formation and approaches for its inhibition. *Frontiers in Microbiology*, 12. https://doi.org/10.3389/fmicb.2021.638609

Babu, A., Santosh, R., Muddana, K., Shobha, & Bakki, R. (n.d.). Response to commentary: Fungal infections of oral cavity: Diagnosis, management, and association with COVID-19. SN Comprehensive Clinical Medicine, 3, 2205–2206 https://doi.org/10.1007/s42399-021-01033-9

Baginski, M., & Czub, J. (2009). Amphotericin B and its new derivatives – Mode of action. *Current Drug Metabolism*, 10(5), 459–469. https://doi.org/10.2174/138920009788898019

Barashkova, A. S., & Rogozhin, E. A. (2020). Isolation of antimicrobial peptides from different plant sources: Does a general extraction method exist? *Plant Methods*, 16(1). https://doi.org/10.1186/s13007-020-00687-1

Barashkova, A. S., Sadykova, V. S., Salo, V. A., Zavriev, S. K., & Rogozhin, E. A. (2021). Nigellothionins from black cumin (*Nigella sativa* l.) seeds demonstrate strong antifungal and cytotoxic activity. *Antibiotics*, 10(2), 1–15. https://doi.org/10.3390/antibiotics10020166

Bellm, L., Lehrer, R. I., & Ganz, T. (2000). Protegrins: New antibiotics of mammalian origin. *Expert Opinion on Investigational Drugs*, 9(8), 1731–1742. https://doi.org/10.1517/13543784.9.8.1731

Benitez, L. L., & Carver, P. L. (2019). Adverse effects associated with long-term administration of azole antifungal agents. *Drugs*, 79(8), 833–853. https://doi.org/10.1007/s40265-019-01127-8

Berida, T. I., Adekunle, Y. A., Dada-Adegbola, H., Kdimy, A., Roy, S., & Sarker, S. D. (2024). Plant antibacterials: The challenges and opportunities. *Heliyon*, 10(10). https://doi.org/10.1016/j.heliyon.2024.e31145

Berrocal-Lobo, M., Segura, A., Moreno, M., López, G., García-Olmedo, F., & Molina, A. (2002). Snakin-2, an antimicrobial peptide from potato whose gene is locally induced by wounding and responds to pathogen infection. *Plant Physiology*, 128(3), 951–961. https://doi.org/10.1104/pp.010685

Boakye-Yiadom, E., Odoom, A., Osman, A. H., Ntim, O. K., Kotey, F. C. N., Ocansey, B. K., & Donkor, E. S. (2024). Fungal infections, treatment and antifungal resistance: The Sub-Saharan African context. In *Therapeutic Advances in Infectious Disease* (Vol. 11). SAGE Publications Ltd. https://doi.org/10.1177/20499361241297525

Bogdanov, I. v., Shenkarev, Z. O., Finkina, E. I., Melnikova, D. N., Rumynskiy, E. I., Arseniev, A. S., & Ovchinnikova, T. v. (2016). A novel lipid transfer protein from the pea *Pisum sativum*: isolation, recombinant expression, solution structure, antifungal activity, lipid binding, and allergenic properties. *BMC Plant Biology*, *16*(1), 107. https://doi.org/10.1186/s12870-016-0792-6

Bongomin, F., Gago, S., Oladele, R. O., & Denning, D. W. (2017). Global and multi-national prevalence of fungal diseases—Estimate precision. *Journal of Fungi*, 3(4). https://doi.org/10.3390/jof3040057

Braida, D., Donzelli, A., Martucci, R., Ponzoni, L., Pauletti, A., & Sala, M. (2012). Neurohypophyseal hormones protect against pentylenetetrazole-induced seizures in zebrafish: Role of oxytocin-like and V1a-like receptor. *Peptides*, 37(2), 327–333. https://doi.org/10.1016/j.peptides.2012.07.013

Brogden, K. A. (2005). Antimicrobial peptides: Pore formers or metabolic inhibitors in bacteria? *Nature Reviews Microbiology*, 3(3), 238–250. https://doi.org/10.1038/nrmicro1098

Buckley, M. (2008). *The Fungal Kingdom: Diverse and Essential Roles in Earth's Ecosystem: This Report is Based on a Colloquium, Sponsored by the American Academy of Microbiology, Convened November 2–4, 2007 in Tucson, Arizona*. American Society for Microbiology, Washington, DC. https://europepmc.org/books/NBK559443

Chandukishore, T., Narasimhulu, K., Ledesma-Amaro, R., & Prabhu, A. A. (2025). Metabolic engineering of Yarrowia lipolytica for efficient conversion of valorised food waste into D-lactic acid. *Chemical Engineering Journal*, *519*, 165519. https://doi.org/10.1016/j.cej.2025.165519

Chen, Z., Agerberth, B., Gell, K., Andersson, M., Mutt, V., Östenson, C., Efendić, S., Barros-Söderling, J., Persson, B., & Jörnvall, H. (1988). Isolation and characterization of porcine diazepam-binding inhibitor, a polypeptide not only of cerebral occurrence but also common in intestinal tissues and with effects on regulation of insulin release. *European Journal of Biochemistry*, 174(2), 239–244. https://doi.org/10.1111/j.1432-1033.1988.tb14088.x

Chinglenthoiba, C., Lani, M. N., Anuar, S. T., Amesho, K. T. T., K L, P., & Santos, J. H. (2025). Microplastics in food packaging: Analytical methods, health risks, and sustainable alternatives. *Journal of Hazardous Materials Advances*, 18, 100746. https://doi.org/10.1016/j.hazadv.2025.100746

Claeys, E., Vermeire, K., & Leuven, K. U. (2019). Immunosuppressive drugs in organ transplantation to prevent allograft rejection: Mode of action and side effects. *Journal of Immunological Science*, *3*(4), 14–21.

Colgrave, M. L., Kotze, A. C., Huang, Y.-H., O'Grady, J., Simonsen, S. M., & Craik, D. J. (2008). Cyclotides: natural, circular plant peptides that possess significant activity against gastrointestinal nematode parasites of sheep. *Biochemistry*, *47*(20), 5581–5589. https://doi.org/10.1021/bi800223y

Cools, T. L., Struyfs, C., Drijfhout, J. W., Kucharíková, S., Romero, C. L., Dijck, P. V., Ramada, M. H. S., Bloch, C., Cammue, B. P. A., & Thevissen, K. (2017). A linear 19-mer plant defensin-derived peptide acts synergistically with caspofungin against *Candida albicans* biofilms. *Frontiers in Microbiology*, *8*(OCT). https://doi.org/10.3389/fmicb.2017.02051

Costa, F., Teixeira, C., Gomes, P., & Martins, M. C. L. (2019). Clinical application of AMPs. In K. Matsuzaki (Ed.), *Antimicrobial Peptides: Basics for Clinical Application* (pp. 281–298). Springer Singapore. https://doi.org/10.1007/978-981-13-3588-4_15

Cotter, P. D., Hill, C., & Ross, R. P. (2005). Bacteriocins: Developing innate immunity for food. *Nature Reviews Microbiology*, 3(10), 777–788. https://doi.org/10.1038/nrmicro1273

Crouch, E., & Wright, J. R. (2001). Surfactant proteins A and D and pulmonary host defense. *Annual Review of Physiology*, 63(1), 521–554. https://doi.org/10.1146/annurev.physiol.63.1.521

Cruz, G. S., Santos, A. T. dos, Brito, E. H. S. de, & Rádis-Baptista, G. (2022). Cell-penetrating antimicrobial peptides with anti-infective activity against intracellular pathogens. *Antibiotics*, *11*(12), 1772. https://doi.org/10.3390/antibiotics11121772

de Beer, A., & Vivier, M. A. (2008). Vv-AMP1, a ripening induced peptide from *Vitis vinifera* shows strong antifungal activity. *BMC Plant Biology*, *8*(1), 75. https://doi.org/10.1186/1471-2229-8-75

de Caleya, R. F., Gonzalez-Pascual, B., García-Olmedo, F., & Carbonero, P. (1972). Susceptibility of phytopathogenic bacteria to wheat purothionins in vitro. *Applied Microbiology*, *23*(5), 998–1000. https://doi.org/10.1128/am.23.5.998-1000.1972

De Cesare, G. B., Cristy, S. A., Garsin, D. A., & Lorenz, M. C. (2020). Antimicrobial peptides: A new frontier in antifungal therapy. *mBio*. https://doi.org/10.1128/mbio.02123-20

Denning, D. W. (2003). Echinocandin antifungal drugs. *The Lancet*, 362(9390), 1142–1151. https://doi.org/10.1016/S0140-6736(03)14472-8

Denning, D. W. (2024). Global incidence and mortality of severe fungal disease. *The Lancet Infectious Diseases*, 24(7), e428–e438. https://doi.org/10.1016/S1473-3099(23)00692-8

Duarte-Mata, D. I., & Salinas-Carmona, M. C. (2023). Antimicrobial peptides' immune modulation role in intracellular bacterial infection. *Frontiers in Immunology*, 14. https://doi.org/10.3389/fimmu.2023.1119574

Edstam, M. M., Viitanen, L., Salminen, T. A., & Edqvist, J. (2011). Evolutionary history of the non-specific lipid transfer proteins. *Molecular Plant*, 4(6), 947–964. https://doi.org/10.1093/mp/ssr019

Fadaka, A. O., Sibuyi, N. R. S., Madiehe, A. M., & Meyer, M. (2021). Nanotechnology-based delivery systems for antimicrobial peptides. *Pharmaceutics*, 13(11). https://doi.org/10.3390/pharmaceutics13111795

Fisher, M. C., Alastruey-Izquierdo, A., Berman, J., Bicanic, T., Bignell, E. M., Bowyer, P., Bromley, M., Brüggemann, R., Garber, G., Cornely, O. A., Gurr, S. J., Harrison, T. S., Kuijper, E., Rhodes, J., Sheppard, D. C., Warris, A., White, P. L., Xu, J., Zwaan, B., & Verweij, P. E. (2022). Tackling the emerging threat of antifungal resistance to human health. *Nature Reviews Microbiology*, 20(9), 557–571. https://doi.org/10.1038/s41579-022-00720-1

Fisher, M. C., Hawkins, N. J., Sanglard, D., & Gurr, S. J. (n.d.). *Worldwide emergence of resistance to antifungal drugs challenges human health and food security*. https://www.science.org

Fisher, M. C., Henk, D. A., Briggs, C. J., Brownstein, J. S., Madoff, L. C., McCraw, S. L., & Gurr, S. J. (2012). Emerging fungal threats to animal, plant and ecosystem health. *Nature*, 484(7393), 186–194. https://doi.org/10.1038/nature10947

Florack, D. E. A., & Stiekema, W. J. (1994). Thionins: Properties, possible biological roles and mechanisms of action. *Plant Molecular Biology*, 26(1), 25–37. https://doi.org/10.1007/BF00039517

Flowers, S. A., Colón, B., Whaley, S. G., Schuler, M. A., & Rogers, P. D. (2015). Contribution of clinically derived mutations in *ERG11* to azole resistance in *Candida albicans*. *Antimicrobial Agents and Chemotherapy*, 59(1), 450–460. https://doi.org/10.1128/AAC.03470-14

de la Fuente-Núñez, C., Silva, O. N., Lu, T. K., & Franco, O. L. (2017). Antimicrobial peptides: Role in human disease and potential as immunotherapies. *Pharmacology & Therapeutics*, 178, 132–140. https://doi.org/10.1016/j.pharmthera.2017.04.002

Fujimura, M., Minami, Y., Watanabe, K., & Tadera, K. (2003). Purification, characterization, and sequencing of a novel type of antimicrobial peptides, *Fa*-AMP1 and *Fa*-AMP2, from Seeds of Buckwheat (*Fagopyrum esculentum* Moench.). *Bioscience, Biotechnology, and Biochemistry*, *67*(8), 1636–1642. https://doi.org/10.1271/bbb.67.1636

Gagandeep, K. R., Balenahalli Narasingappa, R., & Vishnu Vyas, G. (2024). Unveiling mechanisms of antimicrobial peptide: Actions beyond the membranes disruption. *Heliyon*, 10(19). https://doi.org/10.1016/j.heliyon.2024.e38079

Govindarajan, D. K., & Kandaswamy, K. (2023). Antimicrobial peptides: A small molecule for sustainable healthcare applications. *Medicine in Microecology*, 18. https://doi.org/10.1016/j.medmic.2023.100090

Grover, V., Jain, A., Bhardwaj, A., Mehta, M., Arora, S., Algarni, Y. A., Bavabeedu, S. S., Das, G., Ali, A. B. M., Ahmed, N., & Heboyan, A. (2025). Development of synthetic antimicrobial peptides based on genomic analysis of *Streptococcus salivarius*. *Journal of Clinical Laboratory Analysis*, *39*(4). https://doi.org/10.1002/jcla.25156

Hale, J. D., & Hancock, R. E. (2007). Alternative mechanisms of action of cationic antimicrobial peptides on bacteria. *Expert Review of Anti-Infective Therapy*, 5(6), 951–959. https://doi.org/10.1586/14787210.5.6.951

Han, J., Jyoti, Md. A., Song, H.-Y., & Jang, W. S. (2016). Antifungal activity and action mechanism of histatin 5-halocidin hybrid peptides against Candida ssp. *PLoS ONE*, *11*(2), e0150196. https://doi.org/10.1371/journal.pone.0150196

Hancock, R. E. W., & Sahl, H.-G. (2006). Antimicrobial and host-defense peptides as new anti-infective therapeutic strategies. *Nature Biotechnology*, 24(12), 1551–1557. https://doi.org/10.1038/nbt1267

He, X., Zhang, Y., Yuan, S., & Mao, M. (2024). A comprehensive review of antimicrobial peptides: Mechanisms, classifications and designations. *Theoretical and Natural Science*, 46(1), 131–141. https://doi.org/10.54254/2753-8818/46/20240614

Helmerhorst, E. J., Troxler, R. F., & Oppenheim, F. G. (2001). The human salivary peptide histatin 5 exerts its antifungal activity through the formation of reactive oxygen species. *Proceedings of the National Academy of Sciences*, 98(25), 14637–14642. https://doi.org/10.1073/pnas.141366998

Holani, R., Shah, C., Haji, Q., Inglis, G. D., Uwiera, R. R. E., & Cobo, E. R. (2016). Proline-arginine rich (PR-39) cathelicidin: Structure, expression and functional implication in intestinal health. *Comparative Immunology, Microbiology and Infectious Diseases*, 49, 95–101. https://doi.org/10.1016/j.cimid.2016.10.004

Hossain, C. M., Ryan, L. K., Gera, M., Choudhuri, S., Lyle, N., Ali, K. A., & Diamond, G. (2022). Antifungals and drug resistance. *Encyclopedia*, 2(4), 1722–1737. https://doi.org/10.3390/encyclopedia2040118

Huan, Y., Kong, Q., Mou, H., & Yi, H. (2020). Antimicrobial Peptides: Classification, Design, Application and Research Progress in Multiple Fields. *Frontiers in Microbiology, 11*. https://doi.org/10.3389/fmicb.2020.582779

Hui, S. T., Gifford, H., & Rhodes, J. (2024). Emerging antifungal resistance in fungal pathogens. *Current Clinical Microbiology Reports*, 11(2), 43–50. https://doi.org/10.1007/s40588-024-00219-8

Hwang, B., Hwang, J. S., Lee, J., & Lee, D. G. (2010). Antifungal properties and mode of action of psacotheasin, a novel knottin-type peptide derived from *Psacothea hilaris*. *Biochemical and Biophysical Research Communications*, 400(3), 352–357. https://doi.org/10.1016/j.bbrc.2010.08.063

Jenssen, H., Hamill, P., & Hancock, R. E. W. (2006). Peptide antimicrobial agents. *Clinical Microbiology Reviews*, 19(3), 491–511. https://doi.org/10.1128/CMR.00056-05

Kader, J.-C. (1996). Lipid-transfer proteins in plants. *Annual Review of Plant Physiology and Plant Molecular Biology*, 47(1), 627–654. https://doi.org/10.1146/annurev.arplant.47.1.627

Kerenga, B. K., McKenna, J. A., Harvey, P. J., Quimbar, P., Garcia-Ceron, D., Lay, F. T., Phan, T. K., Veneer, P. K., Vasa, S., Parisi, K., Shafee, T. M. A., van der Weerden, N. L., Hulett, M. D., Craik, D. J., Anderson, M. A., & Bleackley, M. R. (2019). Salt-tolerant antifungal and antibacterial activities of the corn defensin ZmD32. *Frontiers in Microbiology, 10*. https://doi.org/10.3389/fmicb.2019.00795

Kim, J. S., Cha, H., & Bahn, Y. S. (2024). Comprehensive overview of *Candida auris*: An emerging multidrug-resistant fungal pathogen. *Journal of Microbiology and Biotechnology*, 34(7), pp. 1365–1375. https://doi.org/10.4014/jmb.2404.04040

Koo, J. C., Lee, S. Y., Chun, H. J., Cheong, Y. H., Choi, J. S., Kawabata, S., Miyagi, M., Tsunasawa, S., Ha, K. S., Bae, D. W., Han, C., Lee, B. L., & Cho, M. J. (1998). Two hevein homologs isolated from the seed of *Pharbitis nil L.* exhibit potent antifungal activity. *Biochimica et Biophysica Acta (BBA) – Protein Structure and Molecular Enzymology, 1382*(1), 80–90. https://doi.org/10.1016/S0167-4838(97)00148-9

Kornspan, J. D., & Rottem, S. (2012). The phospholipid profile of mycoplasmas. *Journal of Lipids, 2012*, 1–8. https://doi.org/10.1155/2012/640762

Kościuczuk, E. M., Lisowski, P., Jarczak, J., Strzałkowska, N., Jóźwik, A., Horbańczuk, J., Krzyżewski, J., Zwierzchowski, L., & Bagnicka, E. (2012). Cathelicidins: Family of antimicrobial peptides. A review. *Molecular Biology Reports*, 39(12), 10957–10970. https://doi.org/10.1007/s11033-012-1997-x

Lacerda, A. F., Vasconcelos, Ã. A. R., Pelegrini, P. B., & Grossi de Sa, M. F. (2014). Antifungal defensins and their role in plant defense. *Frontiers in Microbiology, 5*. https://doi.org/10.3389/fmicb.2014.00116

Lehrer, R. I., & Ganz, T. (2002). Defensins of vertebrate animals. *Current Opinion in Immunology*, 14(1), 96–102. https://doi.org/10.1016/S0952-7915(01)00303-X

Leon-Reyes, A., Du, Y., Koornneef, A., Proietti, S., Körbes, A. P., Memelink, J., Pieterse, C. M. J., & Ritsema, T. (2010). Ethylene signaling renders the jasmonate response of *Arabidopsis* insensitive to future suppression by salicylic acid. *Molecular Plant-Microbe Interactions®, 23*(2), 187–197. https://doi.org/10.1094/MPMI-23-2-0187

Li, J., Hu, S., Jian, W., Xie, C., & Yang, X. (2021). Plant antimicrobial peptides: Structures, functions, and applications. *Botanical Studies*, 62(1), 5. https://doi.org/10.1186/s40529-021-00312-x

Li, X. S., Reddy, M. S., Baev, D., & Edgerton, M. (2003). *Candida albicans* Ssa1/2p is the cell envelope binding protein for human salivary histatin 5. *Journal of Biological Chemistry*, 278(31), 28553–28561. https://doi.org/10.1074/jbc.M300680200

Lima, A. M., Azevedo, M. I. G., Sousa, L. M., Oliveira, N. S., Andrade, C. R., Freitas, C. D. T., & Souza, P. F. N. (2022). Plant antimicrobial peptides: An overview about classification, toxicity and clinical applications. *International Journal of Biological Macromolecules*, 214, 10–21. https://doi.org/10.1016/j.ijbiomac.2022.06.043

Lindholm, P., Göransson, U., Johansson, S., Claeson, P., Gullbo, J., Larsson, R., Bohlin, L., & Backlund, A. (2002). Cyclotides: a novel type of cytotoxic agents. *Molecular Cancer Therapeutics*, *1*(6), 365–369.

Liu, Y., Sameen, D. E., Ahmed, S., Dai, J., & Qin, W. (2021). Antimicrobial peptides and their application in food packaging. *Trends in Food Science and Technology*, 112, 471–483. https://doi.org/10.1016/j.tifs.2021.04.019

Lobo, D. S., Pereira, I. B., Fragel-Madeira, L., Medeiros, L. N., Cabral, L. M., Faria, J., Bellio, M., Campos, R. C., Linden, R., & Kurtenbach, E. (2007). Antifungal *Pisum sativum* defensin 1 interacts with *Neurospora crassa* cyclin F related to the cell cycle. *Biochemistry*, 46(4), 987–996. https://doi.org/10.1021/bi061441j

Loo, S., Tay, S. V., Kam, A., Tang, F., Fan, J.-S., Yang, D., & Tam, J. P. (2021). Anti-fungal hevein-like peptides biosynthesized from quinoa cleavable hololectins. *Molecules*, 26(19), 5909. https://doi.org/10.3390/molecules26195909

Mahlapuu, M., Håkansson, J., Ringstad, L., & Björn, C. (2016). Antimicrobial peptides: An emerging category of therapeutic agents. *Frontiers in Cellular and Infection Microbiology*, 6(DEC). https://doi.org/10.3389/fcimb.2016.00194

Maleki Dizaj, S., Salatin, S., Khezri, K., Lee, J. Y., & Lotfipour, F. (2022). Targeting multidrug resistance with antimicrobial peptide-decorated nanoparticles and polymers. *Frontiers in Microbiology*, 13. https://doi.org/10.3389/fmicb.2022.831655

Marquez, L., & Quave, C. L. (2020). Prevalence and therapeutic challenges of fungal drug resistance: Role for plants in drug discovery. *Antibiotics*, 9(4). https://doi.org/10.3390/antibiotics9040150

Meyer, V. (2008). Genetic engineering of filamentous fungi—Progress, obstacles and future trends. *Biotechnology Advances*, 26(2), 177–185. https://doi.org/10.1016/j.biotechadv.2007.12.001

Miranda, M., Bai, J., Pilon, L., Torres, R., Casals, C., Solsona, C., & Teixidó, N. (2024). Fundamentals of edible coatings and combination with biocontrol agents: A strategy to improve postharvest fruit preservation. *Foods*, 13(18). https://doi.org/10.3390/foods13182980

Montagna, M. T., Caggiano, G., Lovero, G., De Giglio, O., Coretti, C., Cuna, T., Iatta, R., Giglio, M., Dalfino, L., Bruno, F., & Puntillo, F. (2013). Epidemiology of invasive fungal infections in the intensive care unit: Results of a multicenter Italian survey (AURORA Project). *Infection*, 41(3), 645–653. https://doi.org/10.1007/s15010-013-0432-0

Mookherjee, N., & Hancock, R. E. W. (2007). Cationic host defence peptides: Innate immune regulatory peptides as a novel approach for treating infections. *Cellular and Molecular Life Sciences*, 64(7–8), 922–933. https://doi.org/10.1007/s00018-007-6475-6

de Moraes, D. C., & Ferreira-Pereira, A. (2024). Multidrug-resistant fungi. *Journal of Fungi*, 10(10). https://doi.org/10.3390/jof10100686

Mygind, P. H., Fischer, R. L., Schnorr, K. M., Hansen, M. T., Sönksen, C. P., Ludvigsen, S., Raventós, D., Buskov, S., Christensen, B., De Maria, L., Taboureau, O., Yaver, D., Elvig-Jørgensen, S. G., Sørensen, M. V., Christensen, B. E., Kjærulff, S., Frimodt-Moller, N., Lehrer, R. I., Zasloff, M., & Kristensen, H.-H. (2005). Plectasin is a peptide antibiotic with therapeutic potential from a saprophytic fungus. *Nature*, 437(7061), 975–980. https://doi.org/10.1038/nature04051

Nagao, T., Mishima, D., Javkhlantugs, N., Wang, J., Ishioka, D., Yokota, K., Norisada, K., Kawamura, I., Ueda, K., & Naito, A. (2015). Structure and orientation of antibiotic peptide alamethicin in phospholipid bilayers as revealed by chemical shift oscillation analysis of solid state nuclear magnetic resonance and molecular dynamics simulation. *Biochimica et Biophysica Acta (BBA) - Biomembranes*, 1848(11), 2789–2798. https://doi.org/10.1016/j.bbamem.2015.07.019

Nakamura, T., Furunaka, H., Miyata, T., Tokunaga, F., Muta, T., Iwanaga, S., Niwa, M., Takao, T., & Shimonishi, Y. (1988). Tachyplesin, a class of antimicrobial peptide from the hemocytes of the horseshoe crab (*Tachypleus tridentatus*). Isolation and chemical structure. *Journal of Biological Chemistry*, 263(32), 16709–16713. https://doi.org/10.1016/S0021-9258(18)37448-9

Nawrot, R., Barylski, J., Nowicki, G., Broniarczyk, J., Buchwald, W., & Goździcka-Józefiak, A. (2014). Plant antimicrobial peptides. *Folia Microbiologica*, *59*(3), 181–196. https://doi.org/10.1007/s12223-013-0280-4

Nazarian-Firouzabadi, F., Torres, M. D. T., & de la Fuente-Nunez, C. (2024). Recombinant production of antimicrobial peptides in plants. *Biotechnology Advances*, 71. https://doi.org/10.1016/j.biotechadv.2023.108296

Nekhotiaeva, N., Elmquist, A., Rajarao, G. K., Hällbrink, M., Langel, U., & Good, L. (2004). Cell entry and antimicrobial properties of eukaryotic cell-penetrating peptides. *FASEB Journal : Official Publication of the Federation of American Societies for Experimental Biology*, *18*(2), 394–396. https://doi.org/10.1096/fj.03-0449fje

Nguyen, L. T., Haney, E. F., & Vogel, H. J. (2011). The expanding scope of antimicrobial peptide structures and their modes of action. *Trends in Biotechnology*, 29(9), 464–472. https://doi.org/10.1016/j.tibtech.2011.05.001

Nicolas, P. (2009). Multifunctional host defense peptides: intracellular-targeting antimicrobial peptides. *The FEBS Journal*, 276(22), 6483–6496. https://doi.org/10.1111/j.1742-4658.2009.07359.x

Ochiai, A., Ogawa, K., Fukuda, M., Ohori, M., Kanaoka, T., Tanaka, T., Taniguchi, M., & Sagehashi, Y. (2018). Rice defensin OsAFP1 is a new drug candidate against human pathogenic fungi. *Scientific Reports*, *8*(1). https://doi.org/10.1038/s41598-018-29715-w

Odunitan, T. T., Oyaronbi, A. O., Adebayo, F. A., Adekoyeni, P. A., Apanisile, B. T., Oladunni, T. D., & Saibu, O. A. (2024). Antimicrobial peptides: A novel and promising arsenal against methicillin-resistant *Staphylococcus aureus* (MRSA) infections. *Pharmaceutical Science Advances*, 2, 100034. https://doi.org/10.1016/j.pscia.2023.100034

Olaitan Bamidele, M., Bola Bamikale, M., Adepeju Bamidele, M., Cardenas Hernandez, E., Castillo Olvera, G., Sandoval Cortes, J., & Noe Aguilar, C. (2025). Microbiological technology to extract bioactive molecules from agricultural and industrial waste. In K. Sharma (Ed.), *Bio-prospecting of Novel Microbial Bioactive Compounds for Sustainable Development: Management of Natural Resources through Microbial Conversion into Valuable Products* (pp. 121–138). Springer Nature Switzerland. https://doi.org/10.1007/978-3-031-82178-3_8

de Oliveira, S. S. S., Cherene, M. B., Taveira, G. B., de Oliveira Mello, É., de Oliveira Carvalho, A., & Gomes, V. M. (2024). Plant antimicrobial peptides and their main families and roles: A review of the literature. *Current Issues in Molecular Biology*, *47*(1), 1. https://doi.org/10.3390/cimb47010001

Parisi, K., Shafee, T. M. A., Quimbar, P., van der Weerden, N. L., Bleackley, M. R., & Anderson, M. A. (2019). The evolution, function and mechanisms of action for plant defensins. *Seminars in Cell & Developmental Biology*, 88, 107–118. https://doi.org/10.1016/j.semcdb.2018.02.004

Pattanayak, P., Behera, P., Das, D., & Panda, S. (2010). *Ocimum sanctum* Linn. A reservoir plant for therapeutic applications: An overview. *Pharmacognosy Reviews*, 4(7), 95. https://doi.org/10.4103/0973-7847.65323

Paul, S., & Moye-Rowley, W. S. (2014). Multidrug resistance in fungi: Regulation of transporter-encoding gene expression. *Frontiers in Physiology*, 5. https://doi.org/10.3389/fphys.2014.00143

Paulmann, M., Arnold, T., Linke, D., Özdirekcan, S., Kopp, A., Gutsmann, T., Kalbacher, H., Wanke, I., Schuenemann, V. J., Habeck, M., Bürck, J., Ulrich, A. S., & Schittek, B. (2012). Structure-activity analysis of the dermcidin-derived peptide DCD-1L, an anionic antimicrobial peptide present in human sweat. *Journal of Biological Chemistry*, 287(11), 8434–8443. https://doi.org/10.1074/jbc.M111.332270

de Pauw, B. E. (2011). What are fungal infections? *Mediterranean Journal of Hematology and Infectious Diseases*, 3(1). https://doi.org/10.4084/MJHID.2011.001

Penninckx, I. A. M. A., Thomma, B. P. H. J., Buchala, A., Métraux, J.-P., & Broekaert, W. F. (1998). concomitant activation of jasmonate and ethylene response pathways is required for induction of a plant defensin gene in arabidopsis. *The Plant Cell*, *10*(12), 2103–2113. https://doi.org/10.1105/tpc.10.12.2103

Perfect, J. R. (2017). The antifungal pipeline: A reality check. *Nature Reviews Drug Discovery*, 16(9), 603–616. https://doi.org/10.1038/nrd.2017.46

Pfalzgraff, A., Brandenburg, K., & Weindl, G. (2018). Antimicrobial peptides and their therapeutic potential for bacterial skin infections and wounds. *Frontiers in Pharmacology*, 9(MAR). https://doi.org/10.3389/fphar.2018.00281

Pinto, M. E. F., Najas, J. Z. G., Magalhães, L. G., Bobey, A. F., Mendonça, J. N., Lopes, N. P., Leme, F. M., Teixeira, S. P., Trovó, M., Andricopulo, A. D., Koehbach, J., Gruber, C. W., Cilli, E. M., & Bolzani, V. S. (2018). Inhibition of breast cancer cell migration by cyclotides isolated from *Pombalia calceolaria*. *Journal of Natural Products*, *81*(5), 1203–1208. https://doi.org/10.1021/acs.jnatprod.7b00969

Porto, W. F., Souza, V. A., Nolasco, D. O., & Franco, O. L. (2012). In silico identification of novel hevein-like peptide precursors. *Peptides*, 38(1), 127–136. https://doi.org/10.1016/j.peptides.2012.07.025

Pouny, Y., Rapaport, D., Mor, A., Nicolas, P., & Shai, Y. (1992). Interaction of antimicrobial dermaseptin and its fluorescently labeled analogs with phospholipid membranes. *Biochemistry*, 31(49), 12416–12423. https://doi.org/10.1021/bi00164a017

Pranting, M., Loov, C., Burman, R., Goransson, U., & Andersson, D. I. (2010). The cyclotide cycloviolacin O_2 from Viola odorata has potent bactericidal activity against Gram-negative bacteria. *Journal of Antimicrobial Chemotherapy*, *65*(9), 1964–1971. https://doi.org/10.1093/jac/dkq220

Prasad Dr, R., & Rawal, M. K. (2014). Efflux pump proteins in antifungal resistance. *Frontiers in Pharmacology*, 5. https://doi.org/10.3389/fphar.2014.00202

Pushpanathan, M., Gunasekaran, P., & Rajendhran, J. (2013). Antimicrobial peptides: Versatile biological properties. *International Journal of Peptides*, *2013*, 1–15. https://doi.org/10.1155/2013/675391

Rahman, Md. A., Akter, S., Md. Ashrafudoulla, Chowdhury, M. A. H., Yoon, H. J., & Ha, S.-D. (2025). Biofilm–microplastic interactions in food safety: mechanisms, risks, and control strategies. *Critical Reviews in Food Science and Nutrition*, 1–19. https://doi.org/10.1080/10408398.2025.2517825

Reddy, K. V. R., Yedery, R. D., & Aranha, C. (2004). Antimicrobial peptides: Premises and promises. *International Journal of Antimicrobial Agents*, 24(6), 536–547. https://doi.org/10.1016/j.ijantimicag.2004.09.005

Retzl, B., Hellinger, R., Muratspahić, E., Pinto, M. E. F., Bolzani, V. S., & Gruber, C. W. (2020). Discovery of a beetroot protease inhibitor to identify and classify plant-derived cystine knot peptides. *Journal of Natural Products*, *83*(11), 3305–3314. https://doi.org/10.1021/acs.jnatprod.0c00648

Rizzetto, G., Gambini, D., Maurizi, A., Candelora, M., Molinelli, E., Cirioni, O., Brescini, L., Giacometti, A., Offidani, A., & Simonetti, O. (2023). Our experience over 20 years: antimicrobial peptides against gram positives, gram negatives, and fungi. *Pharmaceutics*, 15(1). https://doi.org/10.3390/pharmaceutics15010040

Rokas, A. (2022). Evolution of the human pathogenic lifestyle in fungi. *Nature Microbiology*, 7(5), 607–619. https://doi.org/10.1038/s41564-022-01112-0

Romeo, D., Skerlavaj, B., Bolognesi, M., & Gennaro, R. (1988). Structure and bactericidal activity of an antibiotic dodecapeptide purified from bovine neutrophils. *Journal of Biological Chemistry*, 263(20), 9573–9575. https://doi.org/10.1016/S0021-9258(19)81553-3

Roy, S., Ramakrishnan, R., Afzia, N., Ghosh, T., & Zhang, W. (2024). Recent progress in the antimicrobial and antioxidant peptide activated film/coatings for food packaging applications: a Review. *Food Bioscience*, 62, 105288. https://doi.org/10.1016/j.fbio.2024.105288

Ruissen, A. L. A., Groenink, J., Helmerhorst, E. J., Walgreen-Weterings, E., van't Hof, W., Veerman, E. C. I., & Nieuw Amerongen, A. V. (2001). Effects of histatin 5 and derived peptides on *Candida albicans*. *Biochemical Journal*, 356(2), 361–368. https://doi.org/10.1042/bj3560361

Sabino, R., & Stevens, D. A. (2025). Biology, immunology, epidemiology, and therapy of fungal infections: A themed issue dedicated to Professor David A. Stevens: Editorial. *Journal of Fungi*, 11(3). https://doi.org/10.3390/jof11030179

Sadeeq, M., Li, Y., Wang, C., Hou, F., Zuo, J., & Xiong, P. (2025). Unlocking the power of antimicrobial peptides: advances in production, optimization, and therapeutics. *Frontiers in Cellular and Infection Microbiology*, *15*, 1528583. https://doi.org/10.3389/fcimb.2025.1528583

Schittek, B., Hipfel, R., Sauer, B., Bauer, J., Kalbacher, H., Stevanovic, S., Schirle, M., Schroeder, K., Blin, N., Meier, F., Rassner, G., & Garbe, C. (2001). Dermcidin: a novel human antibiotic peptide secreted by sweat glands. *Nature Immunology*, 2(12), 1133–1137. https://doi.org/10.1038/ni732

Schmidt, M., Arendt, E. K., & Thery, T. L. C. (2019). Isolation and characterisation of the antifungal activity of the cowpea defensin Cp-thionin II. *Food Microbiology*, *82*, 504–514. https://doi.org/10.1016/j.fm.2019.03.021

Sdvn, P., & Chauhan, S. (2022). Alternative against multi drug resistance. *Journal of Bacteriology & Mycology*, 9(2). www.austinpublishinggroup.com

Segura, A., Moreno, M., Madueño, F., Molina, A., & García-Olmedo, F. (1999). Snakin-1, a peptide from potato that is active against plant pathogens. *Molecular Plant-Microbe Interactions®*, 12(1), 16–23. https://doi.org/10.1094/MPMI.1999.12.1.16

Selsted, M. E., & Ouellette, A. J. (2005). Mammalian defensins in the antimicrobial immune response. *Nature Immunology*, 6(6), 551–557. https://doi.org/10.1038/ni1206

Seyedmousavi, S., Bosco, S., De Hoog, S., Ebel, F., Elad, D., Gomes, R. R., Jacobsen, I. D., Martel, A., Mignon, B., Pasmans, F., Piecková, E., Rodrigues, A. M., Singh, K., Vicente, V. A., Wibbelt, G., Wiederhold, N. P., & Guillot, J. (2018). Fungal infections in animals: A patchwork of different situations. *Medical Mycology*, 56, S165–S187. https://doi.org/10.1093/mmy/myx104

Seyfi, R., Kahaki, F. A., Ebrahimi, T., Montazersaheb, S., Eyvazi, S., Babaeipour, V., & Tarhriz, V. (2020). Antimicrobial peptides (AMPs): Roles, functions and mechanism of action. *International Journal of Peptide Research and Therapeutics*, 26(3), 1451–1463. https://doi.org/10.1007/s10989-019-09946-9

Shai, Y. (1999). Mechanism of the binding, insertion and destabilization of phospholipid bilayer membranes by α-helical antimicrobial and cell non-selective membrane-lytic peptides. *Biochimica et Biophysica Acta (BBA) - Biomembranes*, *1462*(1–2), 55–70. https://doi.org/10.1016/S0005-2736(99)00200-X

Shriwastav, S., Kaur, N., Hassan, M., Ahmed Mohammed, S., Chauhan, S., Mittal, D., Aman, S., & Bibi, A. (2025). Antimicrobial peptides: A promising frontier to combat antibiotic resistant pathogens. *Annals of Medicine & Surgery*, 87(4), 2118–2132. https://doi.org/10.1097/MS9.0000000000003106

Simons, A., Alhanout, K., & Duval, R. E. (2020). Bacteriocins, antimicrobial peptides from bacterial origin: Overview of their biology and their impact against multidrug-resistant bacteria. *Microorganisms*, 8(5), 639. https://doi.org/10.3390/microorganisms8050639

Singh, T., Choudhary, P., & Singh, S. (2022). Antimicrobial peptides: Mechanism of action. In Enany, S., Masso-Silva, J. & Savitskaya, A. (Eds.), *Insights on Antimicrobial Peptides*. IntechOpen. https://doi.org/10.5772/intechopen.99190

Sinha, R., & Shukla, P. (2019). Antimicrobial peptides: Recent insights on biotechnological interventions and future perspectives. *Protein & Peptide Letters*, *26*(2), 79–87. https://doi.org/10.2174/0929866525666181026160852

Souza, A. A., Costa, A. S., Campos, D. C. O., Batista, A. H. M., Sales, G. W. P., Nogueira, N. A. P., Alves, K. M. M., Coelho-de-Souza, A. N., & Oliveira, H. D. (2018). Lipid transfer protein isolated from noni seeds displays antibacterial activity in vitro and improves survival in lethal sepsis induced by CLP in mice. *Biochimie*, *149*, 9–17. https://doi.org/10.1016/j.biochi.2018.03.011

Stec, B. (2006). Plant thionins – the structural perspective. *Cellular and Molecular Life Sciences*, 63(12), 1370–1385. https://doi.org/10.1007/s00018-005-5574-5

Strömstedt, A. A., Park, S., Burman, R., & Göransson, U. (2017). Bactericidal activity of cyclotides where phosphatidylethanolamine-lipid selectivity determines antimicrobial spectra. *Biochimica et Biophysica Acta (BBA) – Biomembranes*, *1859*(10), 1986–2000. https://doi.org/10.1016/j.bbamem.2017.06.018

Struyfs, C., Cools, T. L., de Cremer, K., Sampaio-Marques, B., Ludovico, P., Wasko, B. M., Kaeberlein, M., Cammue, B. P. A., & Thevissen, K. (2020). The antifungal plant defensin HsAFP1 induces autophagy, vacuolar dysfunction and cell cycle impairment in yeast. *Biochimica et Biophysica Acta (BBA) - Biomembranes*, *1862*(8), 183255. https://doi.org/10.1016/j.bbamem.2020.183255

Struyfs, C., Cammue, B. P. A., & Thevissen, K. (2021). Membrane-interacting antifungal peptides. *Frontiers in Cell and Developmental Biology*, *9*. https://doi.org/10.3389/fcell.2021.649875

Su, Z., Yu, H., Lv, T., Chen, Q., Luo, H., & Zhang, H. (2025). Progress in the classification, optimization, activity, and application of antimicrobial peptides. *Frontiers in Microbiology*, *16*, 1582863. https://doi.org/10.3389/fmicb.2025.1582863

Suk, J. S., Xu, Q., Kim, N., Hanes, J., & Ensign, L. M. (2016). PEGylation as a strategy for improving nanoparticle-based drug and gene delivery. *Advanced Drug Delivery Reviews*, 99, 28–51. https://doi.org/10.1016/j.addr.2015.09.012

Sun, S. (2024). Progress in the identification and design of novel antimicrobial peptides against pathogenic microorganisms. In Liu, S. (Ed.), *Probiotics and Antimicrobial Proteins* (pp. 918–936). Springer. https://doi.org/10.1007/s12602-024-10402-4

Tam, J. P., Wang, S., Wong, K. H., & Tan, W. L. (2015). Antimicrobial peptides from plants. *Pharmaceuticals*, 8(4), 711–757. https://doi.org/10.3390/ph8040711

Tavares, L. S., Santos, M. de O., Viccini, L. F., Moreira, J. S., Miller, R. N. G., & Franco, O. L. (2008). Biotechnological potential of antimicrobial peptides from flowers. *Peptides*, *29*(10), 1842–1851. https://doi.org/10.1016/j.peptides.2008.06.003

Taveira, G. B., Mathias, L. S., da Motta, O. v., Machado, O. L. T., Rodrigues, R., Carvalho, A. O., Teixeira-Ferreira, A., Perales, J., Vasconcelos, I. M., & Gomes, V. M. (2014). Thionin-like peptides from *Capsicum annuum* fruits with high activity against human pathogenic bacteria and yeasts. *Peptide Science*, *102*(1), 30–39. https://doi.org/10.1002/bip.22351

Terras, F. R., Eggermont, K., Kovaleva, V., Raikhel, N. v, Osborn, R. W., Kester, A., Rees, S. B., Torrekens, S., van Leuven, F., & Vanderleyden, J. (1995). Small cysteine-rich antifungal proteins from radish: their role in host defense. *The Plant Cell*, *7*(5), 573–588. https://doi.org/10.1105/tpc.7.5.573

Thevissen, K., Ferket, K. K. A., François, I. E. J. A., & Cammue, B. P. A. (2003). Interactions of antifungal plant defensins with fungal membrane components. *Peptides*, 24(11), 1705–1712. https://doi.org/10.1016/j.peptides.2003.09.014

Tian, S., Wu, Y., Li, H., Rong, C., Wu, N., Chu, Y., Jiang, N., Zhang, J., & Shang, H. (2024). Evolutionary accumulation of *FKS* 1 mutations from clinical echinocandin-resistant *Candida auris. Emerging Microbes & Infections*, *13*(1). https://doi.org/10.1080/22221751.2024.2377584

Tokmakov, A. A., Kurotani, A., Takagi, T., Toyama, M., Shirouzu, M., Fukami, Y., & Yokoyama, S. (2012). Multiple post-translational modifications affect heterologous protein synthesis. *Journal of Biological Chemistry*, 287(32), 27106–27116. https://doi.org/10.1074/jbc.M112.366351

Ulaganathan, N., Dhodduraj, K., Chandukishore, T., & Prabhu, A. A. (2024). Production of thermostable recombinant cutinase using mixed food waste: A sustainable approach toward environmental remediation. *Chemical Engineering Journal*, 496. https://doi.org/10.1016/j.cej.2024.153892

Vanden Bossche, H. (1997). Mechanisms of antifungal resistance. *Revista Iberoamericana de Micologia*, 14(2), 44–49.

van Parijs, J., Broekaert, W. F., Goldstein, I. J., & Peumans, W. J. (1991). Hevein: an antifungal protein from rubber-tree (*Hevea brasiliensis*) latex. *Planta*, *183*(2), 258–264. https://doi.org/10.1007/BF00197797

Verdon, J., Berjeaud, J.-M., Lacombe, C., & Héchard, Y. (2008). Characterization of anti-Legionella activity of warnericin RK and delta-lysin I from *Staphylococcus warneri. Peptides*, 29(6), 978–984. https://doi.org/10.1016/j.peptides.2008.01.017

Wang, G. (2008). Structures of human host defense cathelicidin LL-37 and its smallest antimicrobial peptide KR-12 in lipid micelles. *Journal of Biological Chemistry*, 283(47), 32637–32643. https://doi.org/10.1074/jbc.M805533200

Wijaya, R., Neumann, G. M., Condron, R., Hughes, A. B., & Polya, G. M. (2000). Defense proteins from seed of *Cassia fistula* include a lipid transfer protein homologue and a protease inhibitory plant defensin. *Plant Science*, *159*(2), 243–255. https://doi.org/10.1016/S0168-9452(00)00348-4

Williams, R. M., Yates, L. A., & Zhang, X. (2020). Structures and regulations of ATM and ATR, master kinases in genome integrity. *Current Opinion in Structural Biology*, 61, 98–105. https://doi.org/10.1016/j.sbi.2019.12.010

Wimley, W. C. (2010). Describing the mechanism of antimicrobial peptide action with the interfacial activity model. *ACS Chemical Biology*, 5(10), 905–917. https://doi.org/10.1021/cb1001558

Xuan, J., Feng, W., Wang, J., Wang, R., Zhang, B., Bo, L., Chen, Z. S., Yang, H., & Sun, L. (2023). Antimicrobial peptides for combating drug-resistant bacterial infections. In Zhe-Sheng (Jason) C. (Ed.), *Drug Resistance Updates* (Vol. 68). Churchill Livingstone. https://doi.org/10.1016/j.drup.2023.100954

Yang, D., Chertov, O., & Oppenheim, J. J. (2001). Participation of mammalian defensins and cathelicidins in anti-microbial immunity: receptors and activities of human defensins and cathelicidin (LL-37). *Journal of Leukocyte Biology*, 69(5), 691–697. https://doi.org/10.1189/jlb.69.5.691

Yang, S.-C., Lin, C.-H., Sung, C. T., & Fang, J.-Y. (2014). Antibacterial activities of bacteriocins: Application in foods and pharmaceuticals. *Frontiers in Microbiology*, *5*. https://doi.org/10.3389/fmicb.2014.00241

Yount, N. Y., & Yeaman, M. R. (2013). Peptide antimicrobials: Cell wall as a bacterial target. *Annals of the New York Academy of Sciences*, 1277(1), 127–138. https://doi.org/10.1111/nyas.12005

Zajac, C., Scott, N. E., Kline, S., Erayil, S. E., & Selmecki, A. (2025). Hotspot gene conversion between FKS1 and FKS2 in echinocandin resistant *Candida glabrata* serial isolates. *Npj Antimicrobials and Resistance*, 3(1), 31. https://doi.org/10.1038/s44259-025-00102-6

Zanetti, M. (2004). Cathelicidins, multifunctional peptides of the innate immunity. *Journal of Leukocyte Biology*, 75(1), 39–48. https://doi.org/10.1189/jlb.0403147

Zasloff, M. (1987). Magainins, a class of antimicrobial peptides from Xenopus skin: isolation, characterization of two active forms, and partial cDNA sequence of a precursor. *Proceedings of the National Academy of Sciences*, 84(15), 5449–5453. https://doi.org/10.1073/pnas.84.15.5449

Zasloff, M. (2002). Antimicrobial peptides of multicellular organisms. *Nature*, 415. www.nature.com

Zhang, Q.-Y., Yan, Z.-B., Meng, Y.-M., Hong, X.-Y., Shao, G., Ma, J.-J., Cheng, X.-R., Liu, J., Kang, J., & Fu, C.-Y. (2021). Antimicrobial peptides: Mechanism of action, activity and clinical potential. *Military Medical Research*, 8(1), 48. https://doi.org/10.1186/s40779-021-00343-2

Zhang, Y. M., Ye, D. X., Liu, Y., Zhang, X. Y., Zhou, Y. L., Zhang, L., & Yang, X. L. (2023). Peptides, new tools for plant protection in eco-agriculture. *Advanced Agrochem*, 2(1), 58–78. https://doi.org/10.1016/j.aac.2023.01.003

4 Microbial Allies
The Role of Endophytes in the Production of Antifungal Compounds

R. Nischitha

4.1 INTRODUCTION

Endophytic fungi are microorganisms that live in the internal tissues of apparently healthy plants without causing any visible harm (Petrini, 1991). In recent decades, endophytic fungi have gained scientific attention due to their unique association with host plant and their potential to produce diverse bioactive metabolites (Decunta et al., 2021). As studies reported, these compounds have broader ecological functions and significant applications in antimicrobial, antioxidant, and stress migration functions (Rigobelo, 2025). One of the modern medical challenges is the spread of antimicrobial resistance (AMR) over several effective standard treatments, and global estimates indicate that AMR is responsible for the death of over 1.2 million people per year. Moreover, this high-impact burden is estimated to increase further in the coming decades. Therefore, it is imperative to search for new antimicrobial compounds from natural sources. Endophytic fungi from unexplored host plants are considered to be the sources of bioactive compounds (Rajakaruna et al., 2024). In addition, their vast production of diverse bioactive functional compounds makes them alternative sources for developing therapies (Muhammad et al., 2024; Vishwakarma et al., 2024).

A range of fungal endophytes are found in medicinal plants and tree species, but those associated with perennial grass species represent specific attention. Grasses are diverse and ecologically important; their beneficial fungal communities are often adapted to challenging stress conditions. For several decades, grass endophytic fungal symbiosis has been well known for its biotic and abiotic stress mitigation by producing diverse secondary metabolites. Whereas in the recent study, grass-endophytic fungal metabolites were also reported in the pharmaceutical field (Nischitha, 2024). Such symbiotic fungal endophytes may also influence crop improvement and soil health approaches in the conservation of endangered medicinal plant species (Mishra et al., 2024).

This chapter highlights the diverse endophytic fungal bioactive metabolites and their antimicrobial potential isolated from the grass species. It explains endophytic fungal metabolite profiling with types of bioactive functional groups and their mechanisms of antimicrobial actions, as well as their applications in the agricultural and

DOI: 10.1201/9781003633952-5

pharmaceutical fields. By linking the endophytic fungal ecological functions and biological values, this chapter aims to emphasize the role of grass-endophytic fungi in both environmental sustainability and public health innovation.

4.1.1 Habitat-Specific and Host Specificity of Endophytic Fungi

The assemblages of endophytic fungi vary according to host and habitat. The specificity of unique fungal diversity is influenced by the plant genotype, geography, and environmental conditions. For example, assemblages of endophytic fungi with host plants in drought or nutrient-poor soil frequently produce a potent metabolite involved in drought tolerance; meanwhile, during this biotic and abiotic stress, endophytes may tend to produce specific metabolites, which may also reflect high potency in other biological activities. The specificity of the host plants majorly influenced the production of diverse secondary metabolites in the endophytic fungi. Meanwhile, the endophytic fungi associated with ethnobotanically significant medicinal plant or tree species harbor structurally complex metabolites and mimic those of the host plant. In the case of *Taxus brevifolia, Azadirachta indica,* or *Catharanthus roseus,* endophytic fungi isolated from these plants showed plant-like biosynthesis of metabolites (camptothecin and taxol) with anticancer and antimicrobial potency by highlighting the co-evolutionary potential of the symbionts (Kharwar et al., 2011; Nischitha et al., 2025).

The endophytic fungal isolation and identification follows a multistep protocol, where the apparently healthy plant tissues were subjected to surface sterilization using ethanol, hydrogen peroxide, and sodium hypochlorite, followed by a sterilized water wash to eradicate epiphytic contamination (Schulz et al., 1993). Segmented plant tissues are aseptically inoculated on nutrient and non-nutrient media like potato dextrose agar (PDA), malt extract agar (MEA), Sabouraud dextrose agar (SDA), water agar (WA), or standard blotter method and incubated under 12/12h light/nUV light (350–400 nm) cycle at 25 ± 2°C for 5–7 days (Nischitha et al., 2025). Molecular analysis of these isolates is most commonly performed by amplification and sequencing initially by rDNA gene region internal transcribed spacer (ITS) and multigene analysis like β-tubulin, small subunit (SSU), large subunit (LSU), and translation elongation factor 1-α (TEF1-α), followed by comparison of sequenced data with the reference database (e.g., NCBI GenBank or UNITE) for taxonomical identity (White et al., 1990; Zhu et al., 2024). The endophyte diversity at the ecosystem level is evaluated using phylogenetic analyses and community outlining based on either metagenomics or high-throughput sequencing. Such strategies facilitate the discovery of cryptic diversity and increase the information on habitat and host interaction, which are paramount in guided bioprospection of new bioactive metabolites.

4.1.2 Biosynthetic Potential of Endophytic Fungi

Endophytic fungi have been reported to produce a broad set of structurally diverse secondary metabolites (polyketides, terpenoids, alkaloids, and non-ribosomal peptides). Such bio-metabolites tend to involve in the disruption of cell membranes, the inhibition of crucial enzymes, and the creation of oxidative stress by production of reactive oxygen species (ROS) (Wadhwa et al., 2024). Host–fungus interactions, environmental factors, and the genetic composition of the organism play a vital role in the regulation of metabolite biosynthesis (Balkrishna et al., 2022). New discoveries of genome mining and synthetic biology have occurred, resulting in a more detailed understanding of underlying biosynthetic pathways, which provides a potential way to target antimicrobial agent discovery and optimization in endophytic sources.

4.1.3 Characterization of Antimicrobial Agents

Antimicrobial compounds produced by endophytic fungi have been revealed to be structurally and functionally diverse; however, their bioactive properties can be distributed into major, representative chemical groups. These are polyketides, terpenoids, alkaloids, and non-ribosomal peptides. Members of the groups are considered by different modes of action against bacterial and fungal pathogens, which also complements their potential as new agents.

4.1.3.1 Hydroxylated Compounds

Organic molecules with one or more hydroxyl (–OH) groups are known as hydroxylated compounds. The hydroxyl groups impart excessive chemical reactivity and biological activity. The traces of these compounds are maximally identified in the natural and those based on endophytic fungi, and these groups take part in various physiological, pharmacological, and ecological processes.

The enzymatic addition of hydroxyl groups through hydroxylation increases the level of solubility and chemical reactivity and increases the functional activity (Caruso et al., 2022). In fungi, the hydroxylation reaction is often facilitated by hydroxylation enzymes (e.g., cytochrome P450 monooxygenases), and this process helps to diversify fungal metabolites and has an excessive effect on their pharmacological value (Toghueo, 2020). Hydroxylated flavonoids, alkaloids, terpenoids, and polyketides produced by endophytic fungi associated with medicinal plants have been reported, and these are connected with intense indicators of antimicrobial activity through interruption of vital microbial functions (Figure 4.1; Table 4.1).

The endophytic fungi could create hydroxylated compounds that have an antimicrobial action as a result of the multi-targeted activity, damaging the membrane of a microbe, inhibiting the action of enzymes, causing oxidative stress, or damaging either the DNA or the RNA.

Their hydroxyl groups enhance hydrogen bonding with biological targets, increase water solubility, and improve membrane permeability, making them highly effective agents for antimicrobial intervention. Several endophytic fungi have been reported to produce structurally diverse hydroxylated metabolites with notable antimicrobial activities. For example, *Phomopsis* sp. isolated from *Terminalia chebula* produces phomalactone and hydroxylated phenylpropanoids, which significantly inhibit the growth of *Bacillus subtilis* and *Candida albicans* (Kumari and Kumar, 2025). *Diaporthe* sp. isolated from *Morinda officinalis* has been shown to produce hydroxylated polyketides such as diaporthin and orthosporin, which are effective against both Gram-positive and Gram-negative bacteria (Wei et al., 2023).

Hydroxylated tetrahydrofuran derivatives from *Diaporthe* spp. isolated from *C. roseus* demonstrated strong antifungal activity, particularly against *Alternaria*

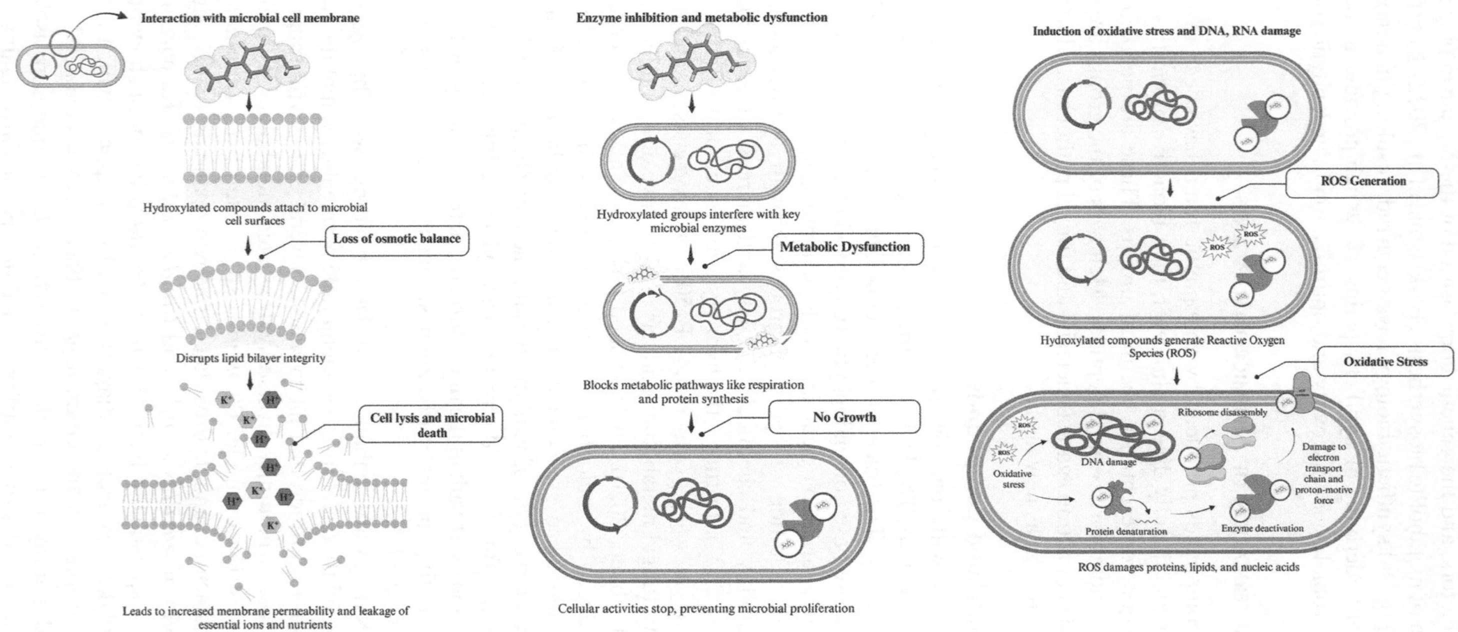

FIGURE 4.1 Antimicrobial mechanism of endophytic fungal compounds on microbial growth inhibition.

TABLE 4.1
Biological Significance of Hydroxylated Compounds of Grass Endophytic Fungi

Endophytic Fungi	Host	Compound	Biological Significance	References
Aspergillus chevalieri	*Cynodon dactylon*	5-Hydroxymethylfurfural	Antifungal	Kamunkar and Nischitha (2025)
Aspergillus fumigatus	*Cynodon dactylon*	5α,8α-epidioxy-ergosta-6,22-diene-3β-ol	Antifungal	Zhang et al. (2016)
		Ergosta-14,22-diene-3β-ol	Antifungal	
Aspergillus sp.	*Cynodon dactylon*	3β-hydroxy-5α,8α-epidioxy-ergosta-6,22-diene	Antibacterial	Li et al. (2005)
Aspergillus stellatus	*Cynodon dactylon*	Isovaline, 3-hydroxyisovaline	Antifungal	Kamunkar and Nischitha (2025)
Chaetomium subaffine	*Heteropogon contortus*	Hydroxypyridone	Antifungal	Nischitha and Shivanna (2022a)
Curvularia protuberata	*Paspalidium flavidum*	4-Hydroxycoumarin	Antimicrobial	Nischitha and Shivanna (2021)
		7-Hydroxycoumarin	Antimicrobial	
		4-Methoxycinnamaldehyde	Nematicidal	
Curvularia tsudae	*Cynodon dactylon*	Umbelliferone	Antimicrobial	Nischitha et al. (2020)
		4-Hydroxycoumarin	Antimicrobial	
Hypoxylon sp.	*Cynodon dactylon*	Propanoic acid, 3-hydroxy-, hydrazide	Antifungal	Kamunkar and Nischitha (2025)
Penicillium pinophilum	*Alloteropsis cimicina*	4-Hydroxycoumarin Umbelliferone 4-Methoxycinnamic acid	Antimicrobial	Nischitha and Shivanna (2021)
Pestalotiopsis guepinii	*Ischaemum ciliare*	4-Hydroxycoumarin 4-Methoxycinnamic acid	Antimicrobial	Shivanna and Nischitha (2024)

alternata, Botrytis cinerea, Colletotrichum gloeosporioides, Fusarium graminearum, and *Phytophthora cinnamomi* (Yan et al., 2018). Additionally, hydroxylated xanthones and lactones produced by *Aspergillus nidulans* and *Xylaria feejeensis* exhibited activity against multidrug-resistant bacteria, functioning via inhibition of DNA gyrase and interference with bacterial replication (Elmaidomy et al., 2022). Advances in metabolomic screening and biosynthetic pathway elucidation have helped to identify key fungal hydroxylases, particularly P450 monooxygenases, that are involved in tailoring steps of hydroxylated metabolite production. The ability to bioengineer these enzymatic steps offers opportunities to enhance the production of hydroxylated antimicrobial agents for potential application in medical, agricultural, and industrial settings.

4.1.3.2 Carboxylic Acids and Derivatives

Endophytic fungi produce a significant group of bioactive secondary metabolites, carboxylic acids, and derivatives. These compounds are structurally diverse and are well-known biologically active when defined by the presence of one or more carboxyl (–COOH) groups. They play a vital role in the fitness of host plants in plant-fungal symbiosis; in addition, they could also have useful potential in the pharmaceutical, agricultural, and industrial markets. One of the most outstanding functions of this group is antimicrobial activity. The action of these metabolites is to lesion the cells of microbes, disrupt the constraints of the microbial cell membranes, and balance the interior cell pH, further inducing the permeability of the cell membranes, resulting in leakages of ions and microbial cell disruption. In addition, they disrupt central metabolic pathways containing glycolysis and the tricarboxylic acid (TCA) cycle, blocking the synthesis of ATP and eventually leading to depletion of energy. Another group of derivatives causes oxidative stress through ROS, producing cytotoxicity that damages proteins, lipids, and nucleic acids. Some aromatic acids also reduce the virulence of microbes, interfering with quorum sensing and the pathogens' biofilm formation process as a means of colonizing and survival. Indicatively, when under acidic conditions, their effectiveness tends to be more effective, which enhances both enzyme inhibition and membrane destabilization.

Endophytic fungi synthesize their aliphatic and aromatic fatty acids due to a variety of biosynthetic pathways, such as polyketide and shikimate pathways, and intermediates of the TCA cycle (Kharwar et al., 2011). Examples include benzoic acid, cinnamic acid, salicylic acid and its methyl ester (methyl salicylate), dicarboxylic acids, citric acid, succinic acid, and fumaric acid (Table 4.2). *C. gloeosporioides* endophytic in *Artemisia annua* is reported to produce benzoic acid, which has broad-spectrum antifungal and antibacterial activity (Zhang et al., 2006). Several endophytic fungi associated with medicinal plants have been reported to produce carboxylic acid derivatives with therapeutic potential. One study reported about a *Phomopsis* species of *A. indica* that synthesizes salicylic acid and methyl salicylate products of metabolism that serve as plant defense and wound concentrates and have anti-inflammatory and antimicrobial effects (Kharwar et al., 2011). Similarly, fungal endophytes like *Aspergillus niger* produce citric acid with its chelation and acidifying properties, which is utilized in both the pharmaceutical and food sectors. Succinic acid and fumaric acid, produced by the genera *Fusarium* and *Penicillium*, have found industrial use (Zhang et al., 2006).

These carboxyl group compounds have ecological and physiologic effects on plant development in addition to their antimicrobial potential. In one instance, *Nigrospora oryzae* obtained in *C. roseus* was found to produce phenylacetic acid, which has an auxin-like activity, and it further affects the growth of plants (Verma et al., 2009). Moreover, endophytic aromatic carboxylic acids have been demonstrated to suppress quorum sensing in plant and human pathogens, thereby increasing their prospects in biological control measures. Biosynthetically, aromatic acids including benzoic acid and cinnamic acid would generally originate as products of the shikimate pathway, with citric acid and succinic acid as TCA cycle intermediates, providing intermediates to this pathway; recent breakthroughs in genome mining, metabolic engineering,

TABLE 4.2
Biological Significance of Carboxylic Acids, Flavonoids, and Phenolics of Grass Endophytic Fungi

Endophytic Fungi	Host	Compound	Biological Significance	References
		Carboxylic Acids		
Aspergillus chevalieri	*Cynodon dactylon*	Phthalic acid, butyl undecyl ester Pentadecanoic acid Octadecanoic acid 22-Tricosenoic acid	Antifungal	Kamunkar and Nischitha (2025)
Aspergillus fumigatus	*Cynodon dactylon*	9-Octadecenoic acid, (E)- 9-Octadecenoic acid, methyl ester 9,12-Octadecadienoic acid, ethyl ester 9,12,15-Octadecatrienoic acid 9,12,15-Octadecatrienoic acid, ethyl ester 14-Pentadecenoic acid n-Hexadecanoic acid Hexadecanoic acid, methyl ester Hexadecanoic acid, ethyl ester Hexadecenoic acid, Z-11- Heptadecanoic acid Linoleic acid Linoleic acid ethyl ester Octadecanoic acid Tetradecanoic acid, ethyl ester	Antibacterial and antifungal	Zhang et al. (2016)
Aspergillus sp.	*Cynodon dactylon*	Helvolic acid	Antifungal	Li et al. (2005)
Aspergillus stellatus	*Cynodon dactylon*	Tetracosanoic acid	Antifungal	Kamunkar and Nischitha (2025)
		Pentadecanoic acid Cis-10-Heptadecenoic acid	Antifungal	Kamunkar and Nischitha (2025)
Chaetomium subaffine	*Heteropogon contortus*	Sorbic acid	Antimicrobial	Nischitha and Shivanna (2022a)
Curvularia protuberata	*Paspalidium flavidum*	Azelaic Acid Perfluorobutanoic acid	Antibacterial Antimicrobial	Nischitha and Shivanna (2021)
Digitaria bicornis	*Penicillium citrinum*	Mycophenolic acid	Antimicrobial	Nischitha and Shivanna (2022b)
Hypoxylon sp.	*Cynodon dactylon*	Propanoic acid, 3-hydroxy-, hydrazide	Antifungal	Kamunkar and Nischitha (2025)
Penicillium pinophilum	*Alloteropsis cimicina*	Phthalic acid Cinnamic acid	Antimicrobial Antimicrobial	Nischitha and Shivanna (2021)

(Continued)

TABLE 4.2 (*Continued*)
Biological Significance of Carboxylic Acids, Flavonoids, and Phenolics of Grass Endophytic Fungi

Endophytic Fungi	Host	Compound	Biological Significance	References
Pestalotiopsis guepinii	*Ischaemum ciliare*	Caffeic acid	Antimicrobial	Nischitha and Shivanna (2024)
		Flavonoids and Phenolic		
Aspergillus chevalieri	*Cynodon dactylon*	5-Hydroxymethylfurfural	Antifungal	Kamunkar and Nischitha (2025)
		Phthalic acid, butyl undecyl ester		
Curvularia protuberata	*Paspalidium flavidum*	4-Hydroxycoumarin	Antimicrobial	Nischitha and Shivanna (2021)
		7-Hydroxycoumarin	Antimicrobial	
		Vanillyl alcohol	Antimicrobial	
Curvularia tsudae	*Cynodon dactylon*	Umbelliferone	Antimicrobial	Nischitha et al. (2020)
		Berberine	Antimicrobial	
		Esculetin	Antibacterial	
		4-Hydroxycoumarin	Antimicrobial	
Penicillium pinophilum	*Alloteropsis cimicina*	Scopoletin	Antimicrobial	Nischitha and Shivanna (2021)
		4-Methoxycinnamic acid	Antimicrobial	
		4-Hydroxycoumarin	Antimicrobial	
		Umbelliferone	Antimicrobial	
		Eugenol	Antimicrobial	
Pestalotiopsis guepinii	*Ischaemum ciliare*	Resveratrol	Antioxidant	Nischitha and Shivanna (2024)
		4-Hydroxycoumarin	Antimicrobial	
		4-Methoxycinnamic acid	Antimicrobial	

and precursor feeding approaches have increased knowledge of these pathways and provide opportunities to intensify the quantity and variety of carboxylic acid derivatives produced by fungal endophytes (Strobel, 2003; Kusari and Spiteller, 2014).

4.1.3.3 Flavonoids and Phenolics

Other secondary metabolic products of endophytic fungi are flavonoids and phenolic compounds, which are known to be an essential source of biological activities. Flavonoids are polyphenols possessing a C_6-C_3-C_6 carbon skeleton structure, whereas phenolic substances are compounds with one or more hydroxyl groups on each aromatic ring. These metabolites are mostly biosynthesized in the polyketide and shikimate pathways, which the fungal endophytes and most of their host plants share (Hashem et al., 2023). Recent studies have shown the occurrence of numerous flavonoids and phenolic acids produced by the endophytic fungi. Quercetin, kaempferol, catechin, gallic acid, caffeic acid, ferulic acid, and chlorogenic acid are various metabolites reported from various endophytic fungal strains (Tan et al., 2018). For example, *Penicillium chrysogenum* growing in *T. brevifolia* was identified to synthesize quercetin and kaempferol components that are proven antioxidants, anti-inflammatory, and antimicrobial (Yadav and Meena, 2021). Similarly, *F. oxysporum* of *C. roseus* produced caffeic and chlorogenic acids, which are powerful free radical scavengers and antimicrobial agents (Tu et al., 2021).

The antimicrobial properties of fungal-derived phenolic acids like ferulic acid and gallic acid result in cross-breaking of the microbial cell wall as well as annihilation of the microbial biofilm. Besides acting as antimicrobial agents, these metabolites have been examined for their anticancer activity as well (Singh et al., 2021). Quercetin, rutin, and endophytic fungus-derived related flavonoids can interrupt signaling pathways like NF-κB and MAPK resulting in apoptosis and prevention of cancer cell growth (Rai et al., 2022). On the other hand, the study also reported that rutin and gallic acid produced by an *Aspergillus* species endophytic in *Ginkgo bilba* showed great antioxidant and antimicrobial activity (Wu et al., 2022). Besides the proposed antimicrobial and anticancer activity, the endophytic flavonoids and phenolics are found to be effective against inflammation and cardioprotection. These compounds inhibit the activity of pro-inflammatory mediators and vascular tissue vulnerability to oxidative harm, which makes them suitable for use in chronic inflammatory and cardiovascular disease (Gouda et al., 2016). Moreover, derivatives including caffeic and chlorogenic acids also show a quorum-sensing inhibitory effect that decreases the pathogenic bacterial virulence.

The reason behind these multiple and varied bioactivities of these metabolites lies in the processes that involve radical scavenging, inhibitory effect on microbial enzymes, modulation of intercellular signalling, intercalation of DNA, and inhibitory effect on microbial communication systems. These are complex activities that render flavonoids and functional food uses. Table 4.2 presents the bioactive flavonoid and phenolic compounds selected with respect to grass endophytes. New metabolic engineering and synthetic biology tools are opening up possibilities for improved synthesis of these metabolites. Co-cultivation and techniques that involve epigenetic modulation and activation of silent biosynthetic gene clusters (BGCs) are investigated to discover new flavonoid and phenolic compounds in endophytic fungi (Kusari and Spiteller, 2014).

4.1.3.4 Terpenoids and Steroids

A well-recognized type of secondary metabolite produced by endophytic fungi is terpenoids and steroids, whose extensive biological activities are well established. Isoprenoids, or terpenoids, are built up using isoprene (C_5H_8) units and have a great structural variety, ranging from simple monoterpenes (C_{10}) to complex triterpenes (C_{30}) and others. Triterpenoid-based steroids play the most significant role in modulating cellular signalling, cell membrane functions, and immune-regulating agents, and they usually have a tetracyclic structure that serves as the backbone of the steroid. Endophytic fungi use the mevalonate (MVA) pathway to biosynthesize all terpenoids and steroidal molecules (Zhao et al., 2020; Kusari et al., 2014). On the one hand, certain species use the methylerythritol phosphate (MEP) pathway (Zhao et al., 2020; Kusari et al., 2014). These metabolites take part in the defense of the host and give it selective benefits within the ecological niches. Certain endophytic fungal terpenes with potent biological activities are represented in Table 4.3.

As an example, *Annulohypoxylon* sp. isolated from *Taxus wallichiana* produces paclitaxel (taxol), a di-terpene with anticancer and antifungal effects (Gauchan et al.,

2021). Similarly, *Pestalotiopsis microspora*, endophytic in *T. brevifolia*, has been known to produce 7-epi-10-deacetyltaxol (EDT), thus there is a possibility that endophytes could reproduce or supplement plant metabolism of the host plant (Subban et al., 2017). Likewise, *Curvularia pallescens*, isolated from *Humboldtia brunonis*, produces strong-acting antifungal curvularin (Sheik and Chandrashekar, 2018), and *Trichoderma* spp. are recognized to produce antimicrobial and plant growth-promoting harziane diterpenoids (Li et al., 2020).

Steroidal metabolites extracted from the endophytic fungi are ergosterol derivatives, beta-sitosterol, and stigmasterol. They are linked with a range of pharmacological effects, such as antimicrobial, anti-inflammatory, anticancer, and immunostimulatory properties (Hashem et al., 2023). As an evidence, ergosterol peroxide extracted by *P. chrysogenum* showed considerable antibacterial and toxicity activity (Shaaban et al., 2023); on the other hand, *C. gloeosporioides* isolated from *Hevea brasiliensis* is reported to produce β-sitosterol, which has antioxidant, antimicrobial, and cholesterolemic properties (Praptiwi et al., 2020).

Fungal terpenoids commonly confront antimicrobial mechanisms, including interruption of microbial cell membrane integrity, cell leakage of contents, failure to produce ergosterol in fungal pathogens, and application of key bacterial enzymes. Steroid-type compounds, especially the ergosterol derivatives, alter the stability of membranes and can also alter the virulence and survival signaling pathways of the microbes. Furthermore, a significant number of fungal steroids would have therapeutic value beyond anti-inflammatory therapy, which would allow them to be used as anti-inflammatory and immunomodulators (Gouda et al., 2016).

The increasing pharmaceutical potential of fungal terpenoids and steroids has led to more investigation on how to improve their production. Biotechnological approaches, including precursor supplying, fermentation, and engineering genetic approaches, are being used to raise metabolite production (Hashem et al., 2023). Also, there is increasing access to novel biosynthetic gene clusters, which will expand new initiatives in self-sustaining production of high-value terpenoid and steroid natural products using endophytic fungi through the development of genome mining and metabolomics.

4.1.3.5 Alkaloids and Amides

Endophytic fungi produced certain alkaloids and amides that have become of great interest due to their strong antimicrobial properties. All these are nitrogen-based secondary metabolites, which can inhibit a wide range of bacterial and fungal pathogens by various mechanisms. These are the formation of cell membranes against them, the disruption of nucleic acid synthesis, and the inhibition of quorum-sensing systems of microbes (Hashem et al., 2023; Gouda et al., 2016).

Some of the new antimicrobial alkaloids identified in recent years are from endophytes. Chaetoglobosins are cytochalasin alkaloids generated by *Chaetomium globosum* derived from *G. biloba*, which have a high level of antifungal properties against plant and human infections (Li et al., 2014). On the same note, terretonin and other indole alkaloids with a large spectrum of antibacterial actions against *Staphylococcus aureus* and *Escherichia coli* were reported to be synthesized by a species of *Aspergillus* isolated by a marine sponge (El-Demerdash et al., 2020). *Xylaria* sp. isolated from *Camptotheca acuminata* yields 3-nitropropionic acid

TABLE 4.3
Biological Significance of Terpenoids, Steroids, Alkaloids, Amides, Sulfonamides, and Organosulfur Compounds of Grass Endophytic Fungi

Endophytic Fungi	Host	Compound	Biological Significance	References
		Terpenoids and Steroids		
Aspergillus chevalieri	*Cynodon dactylon*	7,9-Di-tert-butyl-1-oxaspiro(4,5) deca-6,9-diene	Antifungal	Kamunkar and Nischitha (2025)
		22-Tricosenoic acid		
Aspergillus flavus	*Cynodon dactylon*	Fumigaclavine A	Antibacterial	Kalimuthu et al. (2022)
		Fumigaclavine E		
		γ-Tocopherol	Antifungal	Zhang et al. (2016)
		Fumigaclavine C		
		Fumigaclavine D		
		α-Tocopherol		
Aspergillus fumigatus	*Cynodon dactylon*	Fumigaclavine F		
		Fumigaclavine G		
		Fumigaclavine H		
		Festuclavine		
		Ergosta-14,22-diene-3β-ol		
		5α,8α-Epidioxy-ergosta-6,22- diene-3β-ol		
Aspergillus sp.	*Cynodon dactylon*	Helvolic acid	Antifungal	Li et al. (2005)
		Ergosterol		
		Ergosterol	Antibacterial	Ma et al. (2004)
Festuca rubra	*Epichloë festucae*	Cyclonerodiol	Antifungal	Yue et al. (2000)
Penicillium citrinum	*Digitaria bicornis*	Ergosterol peroxide	Antiviral	Nischitha and Shivanna (2022b)

(Continued)

TABLE 4.3 (*Continued*)
Biological Significance of Terpenoids, Steroids, Alkaloids, Amides, Sulfonamides, and Organosulfur Compounds of Grass Endophytic Fungi

Endophytic Fungi	Host	Compound	Biological Significance	References
Penicillium pinophilum	*Alloteropsis cimicina*	(22E)-Stigmasta-5,22-dien-3-ol	Antimicrobial	Nischitha and Shivanna (2021)
		Alkaloids and Amides		
Aspergillus chevalieri	*Cynodon dactylon*	Pyrrolo[1,2-a]pyrazine-1,4-dione,hexahydro-3-	Antifungal	Kamunkar and Nischitha (2025)
Aspergillus fumigatus	*Cynodon dactylon*	Fumigaclavine A	Antibacterial	Zhang et al. (2016)
		Fumigaclavine E		
		Fumigaclavine C	Antifungal	
		Fumigaclavine D		
		Fumigaclavine F		
		Fumigaclavine G		
		Fumigaclavine H		
		Festuclavine		
Aspergillus stellatus		2-Methylamino-N-phenyl-acetamide	Antifungal	Kamunkar and Nischitha (2025)
Chaetomium subaffine	*Heteropogon contortus*	N-isobutyl-decanamide	Antifungal	Nischitha and Shivanna (2022a)
Curvularia tsudae	*Cynodon dactylon*	Berberine	Antimicrobial	Nischitha et al. (2020)
Digitaria bicornis	*Penicillium citrinum*	Stearamide	Antimicrobial	Nischitha and Shivanna (2022b)
		Oleamide	Antibacterial	
Epichloë coenophiala	*Lolium arundinaceum*	Lolines	Anti-herbivory	Realini et al. (2024)
Epichloë occultans	*Lolium multiflorum/Lolium rigidum*	Lolines	Anti-herbivory	Ueno et al. (2020)
Epichloë uncinata	*Festuca pratensis*	Lolines	Anti-herbivory	Vikuk et al. (2019)

(Continued)

TABLE 4.3 (*Continued*)
Biological Significance of Terpenoids, Steroids, Alkaloids, Amides, Sulfonamides, and Organosulfur Compounds of Grass Endophytic Fungi

Endophytic Fungi	Host	Compound	Biological Significance	References
Festuca rubra	*Epichloë festucae*	*N,N*-diacetamide	Antifungal	Yue et al. (2000)
Hypoxylon sp.	*Cynodon dactylon*	Propanoic acid, 3-hydroxy-, hydrazide	Antifungal	Kamunkar and Nischitha (2025)
		Sulfonamides and Organosulfur		
Curvularia protuberata	*Paspalidium flavidum*	4-Methoxycinnamaldehyde	Nematicidal	Nischitha and Shivanna (2021)
	N-Butylbenzenesulfonamide			
		Antimicrobial		
Curvularia tsudae	*Cynodon dactylon*	Sulfadimidine	Antifungal	Nischitha et al. (2020)
Epichloë festucae	*Festuca rubra*	Cyclonerodiol	Antifungal	Yue et al. (2000)
Epichloë coenophiala	*Lolium arundinaceum*	Ergovaline	Anti-herbivory	Cibils-Stewart et al. (2022)
	Peramine			
Epichloë festucae	*Festuca rubra*	Ergovaline	Anti-herbivory	Vassiliadis et al. (2023)
	Peramine	antimammalian and insecticidal alkaloids	Fernando et al. (2021)	
	Lolitrem B			
Epichloë festucae var. *lolii*	*Lolium perenne*	Ergovaline	Anti-herbivory	Vassiliadis et al. (2023)
	Peramine			
	Lolitrem B			
Penicillium pinophilum	*Alloteropsis cimicina*	Sulfamethazine	Antimicrobial	Nischitha and Shivanna (2021)
	Eugenol	Antimicrobial		
	Dextramycin	Antibacterial		

compounds, which can inhibit fungal pathogens due to their cellular respiration pathway disruption ability (Nouh et al., 2021). *Talaromyces purpureogenus* produces the antibacterial agent purpurealidin E, an alkaloid that has been found to demonstrate highly antibacterial effects against various clinical isolates (Hasanien et al., 2022). Amide compounds have proved to be some of the endophytic fungi metabolites with important antimicrobial activity. As an example, *C. gloeosporioides* can be isolated from *Cinnamomum malabatrum* and used to generate amide derivatives of clerodin; these substances displayed good antibacterial action against the Gram-positive and Gram-negative bacteria (Packiaraj et al., 2016).

These discoveries show the biochemical diversity of endophytic fungi and the potential that could not yet be exhausted in the production of new antimicrobial agents on the basis of nitrogen-derived compounds. The ongoing initiative in metabolomics, fermentation optimization, and genome mining can also increase the discovery of such valuable metabolites of interest in pharmaceutical and agricultural science. Likewise, *Alternaria alternata*, endophytic in *A. indica*, produced piperamide-like compound that revealed a high antibacterial effect especially against multi-bacterial resistant strains (Chatterjee et al., 2019). Certain antimicrobial functions of alkaloids produced by fungi are the inhibition of the topoisomerase enzyme, disturbance of bacterial DNA replication, and destabilization of the membrane. Amides usually disrupt the biofilm biosynthesis of microbial cell walls, hindrance of quorum-sensing mechanisms, which play an essential role in the expression of virulence factors, and, in addition, efflux pump inhibitors increase the susceptibility of antibiotics. Recent advances in genome mining and metabolic profiling are facilitating the discovery of new antimicrobial alkaloidal and amide metabolites from endophytic fungi (Table 4.3). Strategies such as co-cultivation with competing microorganisms and epigenetic modulation are being employed to activate cryptic biosynthetic gene clusters, leading to the enhanced production of novel antimicrobial compounds (Zhu et al., 2023).

4.1.3.6 Sulfonamides and Organosulfur Compounds

Sulfonamides and organosulfur compounds produced by endophytic fungi have demonstrated notable antimicrobial properties against a broad range of pathogens. These natural products exert their antimicrobial effects primarily by interfering with vital microbial processes such as cell wall synthesis, membrane integrity, protein biosynthesis, and oxidative stress regulation. Natural sulfonamide-like compounds isolated from *A. alternata* associated with *A. indica* exhibited potent antibacterial activity by inhibiting folate biosynthesis, a mechanism similar to synthetic sulfonamide drugs but with enhanced efficacy against resistant strains (Chatterjee et al., 2019). Similarly, *Lasiodiplodia pseudotheobromae* obtained from *Illigera rhodantha* produced sulfur-rich secondary metabolites that displayed significant inhibition against Gram-positive and Gram-negative bacteria, including *S. aureus* and *E. coli* (Wei et al., 2014).

In addition, endophytic fungi such as *Bionectria* sp. isolated from *Huperzia serrata* have been reported to synthesize sulfur-containing peptides with strong antimicrobial effects against both human and plant pathogenic fungi (Yang et al., 2019). Organosulfur compounds from *Xylaria* sp., including sulfur-containing diketopiperazines, have shown pronounced antifungal activity, particularly against *C. albicans* and *A. niger* (Fan et al., 2022). Mechanistically, these compounds disrupt bacterial and fungal cell membranes, inhibit key metabolic enzymes involved in folate or amino acid synthesis, and induce oxidative stress by altering cellular redox balance. Sulfur atoms in these molecules enhance their reactivity with biological nucleophiles, contributing to membrane destabilization and enzyme inactivation. The antimicrobial potency of sulfonamides and organosulfur compounds derived from endophytes highlights their potential as promising candidates for the development of novel anti-infective agents (Table 3). Their structural diversity and broad-spectrum activity provide an important alternative to conventional antibiotics, particularly in the face of rising AMR.

4.2 BIOSYNTHETIC PATHWAYS OF ANTIMICROBIAL COMPOUNDS

Antimicrobial compounds are essential in combating microbial infections and are widely used across the medical, agricultural, and food industries. These bioactive molecules originate from a range of natural and synthetic sources, including plants and fungi. A deeper understanding of their biosynthetic pathways is critical not only for improving production efficiency but also for enhancing their efficacy and guiding the discovery of novel therapeutic agents.

Among microbial sources, endophytic fungi have gained recognition as prolific producers of structurally diverse antimicrobial metabolites. Such organisms have well-developed BGCs, which allow them to produce secondary metabolites based on particular pathways. Of the above, the most prominent are the polyketide synthase (PKS) and the non-ribosomal peptide synthetase (NRPS) pathways, which were used to make most of the known antibiotics and antifungal compounds. Moreover, endophytic fungi utilize the biosynthetic pathway of ribosomally synthesized and post-translationally modified peptides (RiPPs), terpenoid pathway, and alkaloid biosynthesis. Such pathway plasticity and modularity are connected with diverse profiles of antimicrobial compounds in endophytic fungi, thus being potential targets of bioprospecting and biotechnological exploitation due to high variability.

4.2.1 POLYKETIDE SYNTHASE (PKS) PATHWAY

Polyketides, a large chemical family, are natural products whose biosynthesis is by the iterative process of acyl-CoA condensation catalyzed by PKS. PKS pathways play a significant role in synthesizing bioactive compounds in endophytic fungi that have antifungal and antibacterial efficacy. Important ones may be lovastatin, with main interest based on cholesterol-lowering properties; lovastatin also has antifungal properties and griseofulvin, which disrupts fungal mitosis by blocking microtubule polymerization. PKS enzymes are widely divided into type I (modular; e.g., macrolides), type II (iterative; e.g., anthracyclines), and type III (non-modular; e.g., chalcones). *Aspergillus* spp. demonstrated the ability to biosynthesize polyketide derivatives with marked activity against *S. aureus*, marking the promise of PKS gene clusters as an antimicrobial discovery platform (Bai et al., 2023).

4.2.2 Non-Ribosomal Peptide Synthetase (NRPS) Pathway

NRPS systems perform the synthesis of metabolites based on peptides utilizing standard and non-proteinogenic amino acids. These few modular enzymes have ATP-dependent mechanisms and involve the production of well-known antimicrobials (echinocandins (β-glucan synthesis) and cyclosporin (an antifungal immunosuppressant)). Endophytic fungi like *Penicillium* spp. have the NRPS gene cluster that encodes the cyclic peptides that have robust antifungal actions against plant pathogens (Niu et al., 2020). The systems have a flourishing platform where new classes of peptide-based antibiotics with extraordinary bioactivities could be discovered.

4.2.3 Ribosomally Synthesized and Post-translationally Modified Peptides (RiPPs)

RiPPs include bacteriocins, lantibiotics, and others. Alterations such as cyclization, glycosylation, methylation, and oxidation take place to enhance the stability, structural enrichment, and biological activity of peptides. Although the study of RiPP biosynthesis has been largely reduced to bacterial systems, recent genomic efforts have the potential to reveal that RiPP-type biosynthetic pathways also occur in endophytic fungi. Despite the lack of direct isolation of RiPPs produced by endophytic fungi, there are some well-documented examples of RiPPs among Basidiomycota and plant pathogenic fungi. Genome mining studies are discovering a potentially expansive field of fungal secondary metabolic systems in fungal genome studies, the RiPP biosynthetic gene clusters. Fungal origin RiPPs demonstrate that fungi of basidiomycota synthesize amatoxins, which are highly toxic cyclic peptides that go through very extensive post-translational modifications (PTMs) including oxidation and cyclization. Omphalotins, belonging to basidiomycota too, undergo methylation and oxidation, which vitally changes their bioactivity and stabilizes them as chemicals (Ozaki et al., 2023).

The overall process used to produce RiPPs in fungal systems is a conserved pathway in which the synthesis of RiPP precursor proteins occurs on the ribosome and precursor peptides are successively modified by enzymes. PTMs to proteins occur via enzymes like methyltransferases, cyclases, and oxidases, which convert linear precursors into other complex, bioactive forms. This modified variability enables RiPPs to target particular biological receptors, such as microbial membrane enzymes and signalling. The advent of improved technology in genome sequencing and mining has recently led to the identification of many uncharacterized RiPP clusters of genes in fungal endophytes. The typification of those clusters may find earlier unknown RiPPs of distinctive structure and potentially valuable in pharmaceuticals. The potential of fungal RiPPs as a future research and drug development source in natural products and the field of antimicrobial agents is shown by their modular biosynthesis structure and chemical variability.

4.2.4 Terpenoid Biosynthesis

Terpenoid biosynthesis in endophytic fungi follows conserved pathways in other organisms, primarily the MVA and MEP pathways. These routes produce isoprene units, which act as building blocks in the generation of an enormous range of terpenoid structures. Classification established some of the producers of terpenoids, and endophytic fungi have been known to produce the classes of terpenoids: monoterpenes, sesquiterpenes, diterpenes, and triterpenes with specific ecological and biological functions of antimicrobial, anticancer, and anti-inflammatory. For example, *Taxomyces andreanae* synthesizes taxol, a well-characterized diterpene with strong anticancer and antifungal activity. Similarly, *P. microspora* produces antimicrobial sesquiterpenes that have exhibited to be active against various pathogens (Pandy et al., 2023). The formation of these terpenoids is usually a product of enzyme-catalyzed adjustments that include cyclases, oxidation, and methylation processes, which help to add structural variance and biological activity of the final metabolites.

Other examples are fusicoccins and diterpenoid compounds, which affect plant pathogen interaction by influencing stomatal behavior and making plants resistant to the disease (Heiser, 2004). Antimicrobial potential against *Trichoderma* species was also shown to hold potent potential due to sesquiterpenoids, once again contributing to the importance of endophytic fungi in protecting plants and antagonizing microorganisms (Leylaie and Zafari, 2018). Terpenoid profiling is a unique feature of these fungi, especially isolated ones within plant hosts, which orchestrates its potential use in both the domains of medicine and agriculture, especially the development of agents with antimicrobial and anticancer effects (Luo et al., 2024).

4.2.5 Alkaloid Biosynthesis

In endophytic fungi, biosynthesis of alkaloids occurs in the production of the nitrogen-bearing secondary metabolite. In most cases, these alkaloids are formed on the basis of an amino acid such as tryptophan, tyrosine, or ornithine. The metabolites play a key role in fungal defense and symbiosis as well as in the resistance of host plants to pests and pathogens. Fungal alkaloids have quite a broad range of activity, which includes antimicrobial, anticancer, insecticidal, and neuroactive activities (Linh et al., 2021). Some structural classes of alkaloid have been identified that endophytic fungi are known to produce indole, isoquinoline, and pyrroloquinoline alkaloids. To illustrate the matter, species of *Penicillium* and *Aspergillus* can produce ergotamine and fumonisin, which exhibit biological properties as well as engage in controlling the growth and stress responses in plants (Moussa et al., 2025). The biosynthetic response regularly entails cyclization, methylation, and oxidation reactions mediated by enzymes that are coded by distinct biosynthetic gene clusters.

Recently, it has been reported that new alkaloid biosynthesis pathways previously not characterized exist among endophytic fungi following the development of genomic sequencing, metabolic profiling, and bioinformatics. Such findings present new forms of biotechnological applications, especially in the area of drug discovery and the development of neutral products (Meghana et al., 2025). The chemical diversity and often unique structural features of fungal alkaloids make endophytic fungi an important and sustainable resource for generating novel therapeutic agents.

4.3 REGULATION OF BIOSYNTHETIC PATHWAYS

Endophytic fungi that produce antimicrobial secondary metabolites have a rigorously regulated production, which is determined by genetic, epigenetic, and environmental controls. This regulation combination makes the production of secondary metabolites closely in line with ecological needs, host interactions, and external cues. Control of BGCs at the genetic level is centered on global transcriptional regulators such as LaeA (Latent secondary metabolites activator A), VeA (Velvet A), and the broad velvet complex. The master regulator of fungal secondary metabolism, a methyltransferase-like protein LaeA, restructures the chromatin structure, and this correlates with the ability to produce an entire cluster of secondary metabolites at once because one protein can affect the expression of various BGCs (Wang et al., 2025). LaeA overexpression, or a phenotype leading to LaeA dysfunction, has been related to useful increases in the production of antimicrobial polyketides and non-ribosomal peptides in endophytic species of the genera *Aspergillus* and *Penicillium* (Hou et al., 2024). A significant part of the velvet complex is VeA, which coordinates signals that provide fungal growth and secondary metabolite production. Deletion of VeA homologs in *Fusarium* endophytes resulted in the downregulation of antimicrobial pathways, whereas in the case of the overexpression of the same route, cryptic pathways became accessible (Cai et al., 2022).

4.4 ENGINEERING BIOSYNTHETIC PATHWAYS FOR ENHANCED PRODUCTION

Epigenetically, chromatin remodeling plays a vital role in the expression of silent gene clusters or low-expression gene clusters that are mostly not expressed in normal laboratory conditions. Novel antimicrobial activity has been induced in endophytic fungi by using the histone deacetylase inhibitors (HDACi) including suberoylanilide hydroxamic acid (SAHA) and sodium butyrate (Xue et al., 2023). As an illustration, it was found that HDACi treatment of *Colletotrichum* species induces previously silent gene clusters, thus allowing the discovery of new antibacterial polyketide derivatives (Sharma et al., 2017). Equally, secondary metabolism is regulated by DNA methylation status. The removal of demethylation in the *Diaporthe* endophytes triggered new pathways that led to the production of synthesized unique antimicrobial compounds (Alanzi et al., 2023). Therefore, biosynthesis of antimicrobial compounds in endophytic fungi is under the multifactorial control of transcriptional regulators and their interactions with enzymes, alterations of chromatin, and environmental signals. To produce opportunities with the silent biosynthetic pathways and improve the production of the new antimicrobial agents available for pharmaceutical and agricultural purposes, it is encouraging to proceed with studying this network of regulation.

The field of bioengineering and the production of valuable antimicrobial compounds have become potential solutions for endophytic fungi. This is because conventional fermentation procedures usually give low metabolite titers as a result of the existence of silent or weakly expressed BGCs. In order to address these challenges, recent breakthroughs in the fields of synthetic biology, genetic engineering, and omics-driven optimization have enabled marked increases in the availability and scalability of these bioactive secondary metabolites. CRISPR-Cas9 (Clustered Regularly Interspaced Short Palindromic Repeats-related protein 9) genome editing is one of the excellent tools in this industry. The technology allows the specific gene of BGCs to be activated, deleted, or altered specifically. The recent use of endophytic fungi to trigger a silent PKS and NRPS hybrid cluster and to allow large-scale yields of such novel polyketides in *A. nidulans* has been demonstrated through the use of CRISPR-Cas9 (Alanzi et al., 2023). Omics technologies have also developed, which has given a higher pace to pathway engineering.

- Transcriptomics would offer expression profiles of genes under varying growth changes to assist in identifying important BGCs that can be upregulated in a selective manner.
- Proteomics can be used to locate rate-limiting enzymes in a biosynthetic pathway and influence enzyme engineering strategies.
- Metabolomics can provide a detailed overview of metabolic activity and the secondary metabolite production, enabling optimization of media and the feeding of precursors (Chowdhary et al., 2022).

Epigenetic reprogramming is another important field. SAHA, which is an inhibitor of histone deacetylase (HDACi), has been shown to activate cryptic BGCs of endophytic fungi. As an example, exposure of *Xylaria* isolates to HDACi not only leads to the increased concentrations of the already discovered antibacterial compounds (Verma et al., 2023; Harshitha et al., 2023). Co-cultivation techniques where endophytic fungi are grown with bacterial competitors have proved helpful as a strategy to create analogs of natural stress situations, resulting in secondary metabolism. Similarly, co-cultivation of *Colletotrichum* with *B. subtilis* resulted in the high-level production of new hydroxylated terpenoids with potent antimicrobial activity (Alanzi et al., 2023). In addition, other strategies have been developed, which include metabolic pathway refactoring, modular biosynthetic assembly, and computational pathway modeling, which further refine control over the production of complex natural products. The strategies hold an optimistic approach toward future production and optimization of industrial-scale production of antimicrobial products of endophytic fungi.

4.5 COMPARATIVE ANALYSIS OF BIOACTIVITY

A comparative examination of antimicrobial compounds of the endophytic fungus shows great differences in terms of activity and efficacy, range of activity, and action. The chief distinction has been said to be among their chemical structure, the set of functional groups (e.g., hydroxyl, carboxyl, and amine functional groups), molecular size, and their biosynthetic origin like polyketides, non-ribosomal peptides, terpenoids, alkaloids, and hydroxylated derivatives.

4.5.1 Potency and Spectrum of Activity

Endophytic fungi produce polyketides and non-ribosomal peptides, usually with broad-spectrum antimicrobial activity and bactericidal against both Gram-negative and Gram-positive bacteria and against pathogenic fungi. An example of a mega-event polyketide is griseofulvin, which exists as a polyketide belonging to species of endophytic *Pencillium* and exhibits viable antifungal actions, especially against dermatophytes (Enyi et al., 2025). In addition, the synthesis of cycle peptides by *P. chrysogenum* has demonstrated significant bactericidal action on multi-drug-resistant *S. aureus* (Zhang et al., 2019). Terpenoids and alkaloids are, however, more likely to have narrower spectrums of action and may be selective against certain groups of microbes. To illustrate, artemisinin analogs of *C. pallescens* have considerable antifungal and antimalarial activity with moderate effect against Gram-negative bacteria (Hussain et al., 2017). Similarly, other derivatives of the alkaloid, such as those produced in swainsonine and purpurealidin, are also bacteria-positive (Hasanien et al., 2022).

Antimicrobial effects associated with hydroxylated molecules are well known: hydroxylated polyketides and flavonoids often have antimicrobial activities based on membrane disruption and oxidative stress. They tend to augment membrane permeability and restrain the growth of bacteria by the mediation of ROS. Whereas, hydroxylated derivatives obtained in *Phomopsis* or *Diaporthe* species demonstrated antimicrobial activity against both *E. coli* and *S. aureus* (Zhou and Hou, 2019), which indicates the presence of a dual-activity against Gram-positive and Gram-negative bacteria.

4.5.2 Mechanism of Action

Antimicrobial mechanisms of endophytic fungal metabolites are considerably different among the various chemical groups, as these compounds comprise different structures and ecological roles (Eshboev et al., 2024):

- Polyketides usually act either to inhibit microtubule assembly (examples include griseofulvin) or to interfere with nucleic acid synthesis.
- The typical mechanism of action of non-ribosomal peptides (NRPS-derived) consists of interfering with the cell membrane integrity of the fungi or interrupting the activity of enzymes that are vital in the synthesis of fungal cells, such as the β-glucan synthase.
- The action of terpenoids is based upon the interruption of microbial membrane structure and biosynthesis of ergosterol, which weakens and ruins the membrane structure and functioning.
- Alkaloids block the enzyme topoisomerase, which impedes DNA replication.
- Hydroxylated compounds cause oxidative stress, enzyme inactivation of microbial respiration, and membrane disruption.

This diversity of endophytic fungi mechanisms helps them to take many ecological niches and compete with other forms of microorganisms, leading to increased survival and defenses of the host plants.

4.5.3 Structure–Activity Relationship (SAR)

To gain insight into the role of particular structural characteristics of bioactive compounds on their antimicrobial efficacy and selectivity, SAR analysis is vital. Hydroxyl moieties, hydroxylated phenolic functional groups, are also functional groups that augment aqueous solubility, along with facilitating greater biological target engagement via hydrogen bond. In NRPS peptides, large macrocyclic elements led to insertion and disruption of membranes to help in broad-spectrum antimicrobial activity. This lipophilic character of terpenoids allows insertions into the membranes of microorganisms and changes the fluidity and permeability of microbial membranes, causing cell lysis. Meanwhile, the aromatic systems, which are planar, are present in polyketides and alkaloids and induce intercalation with the DNA and enzyme inhibition, interrupting critical cell processes. Notably, SAR studies have demonstrated that small chemical changes that include methylation, hydroxyl, or glycosylation can significantly change the potency, selectivity and action spectrum of these antimicrobial agents (Niu et al., 2020).

4.5.4 Comparative Performance against Resistant Strains

It has been found that endophytic fungal metabolites exhibit significant results in combating antibiotic-resistant pathogens in situations where they usually perform better compared to traditional antibiotics. A number of endophytic fungal-derived polyketides and cyclic peptides have also proved to be active against multidrug-resistant strains of microorganisms such as methicillin-resistant *Staphylococcus aureus* (MRSA) and vancomycin-resistant Enterococcus (VRE). These results emphasize the possibility of endophyte-derived natural products as a new scaffold to overcome the global pressure of AMR and align with their plausibility in pursuing next-generation therapeutic agents.

4.5.5 Potential for Drug Development

Despite being a small fraction of the total chemical classes of fungal secondary metabolites, polyketides and non-ribosomal peptides represent by far the most promising development in new broad-spectrum antibiotics because of their outstanding structural variety and highly effective bioactivity. Terpenoids, which are lipophilic with membrane-destroying properties, have high potential as antifungals. In the meantime, alkaloids and their hydroxylated web products are understandable scaffolds for designing functioning antibacterial agents specifically targeting antibiotic-resistant pathogens.

Recent improvements in metabolic engineering and synthetic biology have also supported the range of potential applications in clinical practice, since it is now possible to rationally optimize the biosynthetic pathway and improve the yield of a metabolite, as well as achieve a better pharmacological profile of this compound. Such technological advances are leading the way toward the translational processing of products based on endophyte-derived natural products as effective therapeutic agents.

4.5.6 Mechanisms of Antimicrobial Action

The antimicrobial action of secondary metabolites by endophytic fungi is diverse in that it interferes with crucial functions in the pathogenic microbe. These compounds contain a wide variety of structural elements (polyketides, non-ribosomal peptides, terpenoids, alkaloids, and hydroxylated derivatives), which are the sources of a wide range of biological activity, providing valuable targets in combating drug-resistant organisms.

One of the mechanisms of action is cell membrane integrity permeation. Terpenoids, cyclic peptides, and some polyketides such as camptothecin incorporate into microbial membranes, disrupting structure and causing the extracellular loss of ions and metabolites. For example, harziane diterpenoids of *Trichoderma* species lead to cytoplasmic leakage via disruption of the membrane (Li et al., 2020). *Penicillium* species produce cyclic lipopeptides, which disrupt the formation of mitotic spindles, thus interfering with DNA replication.

An additional antimicrobial approach is to block protein synthesis. The *Phomopsis* species hydroxylated xanthones disrupt the assembly of 70s ribosome, effectively inhibiting the translation in bacteria (Wu et al., 2024). Other metabolites also produce oxidative stress, producing ROS, which induce lipid peroxidation, DNA strand fragmentation, and inactivation of enzymes. Case in point: hydroxylated polyketides of *Diaporthe* endophytes also show *S. aureus* and *Pseudomonas aeruginosa* killing properties with the help of ROS (Saxena et al., 2024). Also, antivirulence and quorum-sensing inhibitory activities are reported. Some of the hydroxylated flavonoids in endophytic fungi inhibit the quorum-sensing pathways in *P. aeruginosa*, spreading biofilm and virulence factors production (Bajpai et al., 2020). Lastly, efflux pump inhibitors are a new form of antimicrobials. Endophytic fungi-derived specific alkaloids have been demonstrated to inhibit multidrug efflux pumps in Gram-negative bacteria, which improves the activity of traditional antibiotics like tetracycline and ciprofloxacin (Saxena et al., 2024). Taken together, these various and complementary mechanisms support the fact that endophytic fungal metabolites have therapeutic potential as a reservoir of novel antimicrobials in an era of increasing AMR.

4.6 LIMITATIONS IN THE UTILIZATION OF ENDOPHYTIC FUNGI

Although endophytic fungi have unlimited potential, their commercial and biotechnological usage is subject to a number of challenges. The main burden is the ease of growing and sustaining pure fungal isolates. Most species of endophytic fungi are slow growing, have specialized nutritional requirements, or require growth factors of the host in order to grow, which are thus recalcitrant to culturing in a typical laboratory setting. Also, some strains are non-viable or lose their bioactivity when subjected to repeated sub-culturing, an issue that interferes with the uniform production of metabolites. The second highest challenge is the degree of variability and instability in the expression of secondary metabolites. Production of antimicrobial compounds is very susceptible to environmental conditions, cues received by the host, and epigenetic regulation. As a result, compounds observed in plants may not be reliably reproduced in vitro. Fluctuations in metabolite yield and composition between batches complicate downstream processes such as extraction, purification, standardization, and quality control.

Moreover, there is a lack of comprehensive genomic and biosynthetic data for many endophytic fungi, limiting the capacity to predict or manipulate their metabolic pathways. A significant proportion of BGCs remain silent under laboratory conditions, requiring advanced techniques such as co-cultivation, epigenetic modulation, or genome editing to activate them, methods that are often labor-intensive and not always successful. These limitations underscore the need for integrated approaches involving synthetic, systems biology, and metabolomics to unlock the full potential of endophytic fungi for sustainable applications in pharmaceutical and agriculture.

4.7 APPLICATIONS AND FUTURE PERSPECTIVES

This is because endophytic fungi (the fungi that live asymptomatically in the tissues of plants) have attracted the attention of many scientists owing to their ability to yield a diverse secondary metabolite repertoire, especially antimicrobial compounds. These fungi enter into mutualistic relationships with their host plants, which in many cases lead to the production of one-of-a-kind secondary metabolites not synthesized by free-living relatives. Endophytic fungi represent a new and environmentally friendly source of novel antibiotics, antifungals, and antivirals as the world faces an ever-growing threat of AMR.

Endophyte-derived antimicrobial compounds have a wide range of activity against pathogens, including drug-resistant bacteria and fungi, and thus represent interesting prospects as new antimicrobials applicable in medicine and agriculture. These fungi have the potential to become an environmentally friendly alternative to the use of synthetic pesticides and antibiotics as natural biocontrol agents in agricultural systems. Moreover, they can be used in the promotion of sustainable agriculture since they have applications in the growth promotion of plants, stress tolerance, and biofertilization. It is likely that, as genomics, metabolomics, and synthetic biology move rapidly forward, new biosynthetic pathways and regulatory mechanisms will be opened up that can be used to provide increased flux to key features of the production process. Artificial intelligence (AI) and machine learning (ML) are, in this regard, emerging to become extremely useful in the speedier process of discovery. These technologies are applicable in predicting biosynthetic gene clusters and modeling of the metabolic pathways, which helps to prioritize chemical compounds that possess a lot of antimicrobial potential. AI-based exploratory analysis of large metabolomic datasets also allows greater efficiency, e.g., by facilitating dereplication, identifying novel bioactive metabolites, and predicting SAR, which can also be applied to drug discovery pipelines based on endophytes.

Ecological and environmental research will also be significant in determining the biosafety and ecological effects of procedures for implementing endophyte technology. Further bioprospecting of less-discovered or endemic plant species has the potential of discovering new fungal strains that can be used in the production of certain compounds that have never been discovered but can be used therapeutically. Long-term endophytic fungi hold great promise in how next-generation antimicrobial treatments are developed, and sustainable agricultural technologies can be achieved to meet and fill previous gaps in the management of infectious diseases and environmental sustainability, which are two of the most problematic issues of our time.

4.8 CONCLUSION

Endophytes are a profitable source of antimicrobial compounds allied with a cross-utilization of them in areas of medicine and agriculture. Their ability to make multiple bioactive metabolites holds potential answers to the growing menace of AMR or to curb reliance on using chemical pesticides. As genomic, metabolomic, and synthetic biological techniques continue to improve, there is a potential to effectively harness all capabilities of these symbiotic microorganisms to optimize their production and discovery of metabolites. In the future, the prospects of endophytic fungi in biotechnological innovation are very bright, thanks to new strain discoveries, increasing knowledge on plant-fungal symbiosis, and adherence to sustainability and environmentally healthy uses in pharmaceutical product innovation and ecosystem agricultural management.

REFERENCES

Alanzi, A., Elhawary, E. A., Ashour, M. L., & Moussa, A. Y. (2023). *Aspergillus* co-cultures: A recent insight into their secondary metabolites and microbial interactions. *Archives of Pharmacal Research*, 46(4), 273–298.

Bai, X., Sheng, Y., Tang, Z., Pan, J., Wang, S., Tang, B., & Zhang, H. (2023). Polyketides as secondary metabolites from the genus *Aspergillus*. *Journal of Fungi*, 9(2), 261.

Bajpai, R., Yusuf, M. A., Upreti, D. K., Gupta, V. K., & Singh, B. N. (2020). Endolichenic fungus, *Aspergillus quandricinctus* of *Usnea longissima* inhibits quorum sensing and biofilm formation of *Pseudomonas aeruginosa* PAO1. *Microbial Pathogenesis*, 140, 103933.

Balkrishna, A., Joshi, R., & Arya, A. D. V. (2022). Endophytic fungi: A comprehensive review on their secondary metabolites, pharmacological interventions and host plant interactions. Wu, Q.-S., Zou, Y.-N., Xu, Y.-J., Eds, *Endophytic Fungi: Biodiversity, Antimicrobial Activity and Ecological Implications*; pp. 1–62. Nova Science Publisher, New York.

Cai, L., Wang, J., Li, Y., Qin, M., Yin, X., He, Z., & Kang, J. (2022). Global regulatory factor VeA upregulates the production of antitumor substances in endophytic *Fusarium solani*. *Antonie van Leeuwenhoek*, 115(8), 1085–1100.

Caruso, D. J., Palombo, E. A., Moulton, S. E., & Zaferanloo, B. (2022). Exploring the promise of endophytic fungi: A review of novel antimicrobial compounds. *Microorganisms*, 10(10), 1990.

Chatterjee, S., Ghosh, R., & Mandal, N. C. (2019). Production of bioactive compounds with bactericidal and antioxidant potential by endophytic fungus *Alternaria alternata* AE1 isolated from *Azadirachta indica* A. Juss. *Plos One*, 14(4), e0214744.

Chowdhary, K., Arora, H., & Sharma, S. (2022). CRISPR/Cas9-based genome editing as a way ahead for inducing production of bioactive metabolites in endophytes. *National Academy Science Letters*, 45(3), 275–280.

Decunta, F. A., Pérez, L. I., Malinowski, D. P., Molina-Montenegro, M. A., & Gundel, P. E. (2021). A systematic review on the effects of *Epichloë* fungal endophytes on drought tolerance in cool-season grasses. *Frontiers in Plant Science*, 12, 644731.

Elmaidomy, A. H., Shady, N. H., Abdeljawad, K. M., Elzamkan, M. B., Helmy, H. H., Tarshan, E. A., & Abdelmohsen, U. R. (2022). Antimicrobial potentials of natural products against multidrug resistance pathogens: A comprehensive review. *RSC Advances*, 12(45), 29078–29102.

El-Demerdash, A., Kumla, D., & Kijjoa, A. (2020). Chemical diversity and biological activities of meroterpenoids from marine derived-fungi: comprehensive update. *Marine Drugs*, *18*(6), 317.

Enyi, E. O., Chigozie, V. U., Okezie, U. M., Udeagbala, N. T., & Oko, A. O. (2025). A review of the pharmaceutical applications of endophytic fungal secondary metabolites. *Natural Product Research*, *39*(11), 3295–3311.

Eshboev, F., Mamadalieva, N., Nazarov, P. A., Hussain, H., Katanaev, V., Egamberdieva, D., & Azimova, S. (2024). Antimicrobial action mechanisms of natural compounds isolated from endophytic microorganisms. *Antibiotics*, 13(3), 271.

Fan, Y., Ma, Z., Zhang, Y., Wang, Y., Ding, Y., Wang, C., & Cao, S. (2022). Sulfur-containing compounds from endophytic fungi: sources, structures and bioactivities. *Journal of Fungi*, 8(6), 628.

Fernando, K., Reddy, P., Spangenberg, G. C., Rochfort, S. J., & Guthridge, K. M. (2021). Metabolic potential of Epichloë endophytes for host grass fungal disease resistance. *Microorganisms*, *10*(1), 64.

Gauchan, D. P., Vélëz, H., Acharya, A., Östman, J. R., Lundén, K., Elfstrand, M., & Garcia-Gil, M. R. (2021). *Annulohypoxylon* sp. strain MUS1, an endophytic fungus isolated from *Taxus wallichiana* Zucc., produces taxol and other bioactive metabolites. *3 Biotech*, 11, 1–16.

Gouda, S., Das, G., Sen, S. K., Shin, H. S., & Patra, J. K. (2016). Endophytes: A treasure house of bioactive compounds of medicinal importance. *Frontiers in Microbiology*, 7, 1538.

Harshitha, K., Nair, A. R., & Pillai, P. P. (2023). Overview of bioactive metabolite (s) produced by endophytes and future perspectives on epigenetic modification/regulation of cryptic biosynthetic pathways. *Phytochemistry Letters*, 53, 116–131.

Hasanien, Y. A., Nassrallah, A. A., Zaki, A. G., & Abdelaziz, G. (2022). Optimization, purification, and structure elucidation of anthraquinone pigment derivative from *Talaromyces purpureogenus* as a novel promising antioxidant, anticancer, and kidney radio-imaging agent. *Journal of Biotechnology*, 356, 30–41.

Hashem, A. H., Attia, M. S., Kandil, E. K., Fawzi, M. M., Abdelrahman, A. S., Khader, M. S., Khodaira, M.A., Emam, A.E., Goma, M.A., & Abdelaziz, A. M. (2023). Bioactive compounds and biomedical applications of endophytic fungi: A recent review. *Microbial Cell Factories*, 22(1), 107.

Heiser, I. (2004). Interactions between host plants and fungal and bacterial pathogens. In Bertold H., & Erich F. Elstner (Eds.), *Plant Toxicology* (pp. 569–610). CRC Press.

Hou, X., Liu, L., Xu, D., Lai, D., & Zhou, L. (2024). Involvement of LaeA and velvet proteins in regulating the production of mycotoxins and other fungal secondary metabolites. *Journal of Fungi*, 10(8), 561.

Hussain, M. A., Mahajan, V., Rather, I. A., Awasthi, P., Chouhan, R., Dutt, P., Sharma, Y. P., Bedi, Y. S., & Gandhi, S. G. (2017). Isolation and identification of growth promoting endophytes from *Artemisia annua L.* and its effects on artemisinin content. *Trends in Phytochemical Research*, 1(4), 207–214.

Kalimuthu, A. K., Pavadai, P., Panneerselvam, T., Babkiewicz, E., Pijanowska, J., Mrówka, P., Rajagopal, G., Deepak, V., Sundar, K., Maszczyk, P., & Kunjiappan, S. (2022). Cytotoxic potential of bioactive compounds from *Aspergillus flavus*, an endophytic fungus isolated from *Cynodon dactylon*, against breast cancer: Experimental and computational approach. *Molecules*, 27(24), 8814.

Kamunkar, A. K., & Nischitha, R. (2025). Exploring endophytic fungi from *Cynodon dactylon*: GC–MS profiling and biological activity. *Fungal Genetics and Biology*, *176*, 103959.

Kharwar, R. N., Mishra, A., Gond, S. K., Stierle, A., & Stierle, D. (2011). Anticancer compounds derived from fungal endophytes: Their importance and future challenges. *Natural Product Reports,* 28(7), 1208–1228.

Kumari, R., & Kumar, S. (2025). Endophytic fungi from indian medicinal plants: A goldmine of novel biomolecules with potential biological applications. In Vineet M., Kamlesh K. S., Mahiti G., Nadeem A. (Eds.), *Natural Bioactives from the Endophytes of Medicinal Plants* (pp. 86–99). CRC Press.

Kusari, S., & Spiteller, M. (2012). Metabolomics of endophytic fungi producing associated plant secondary metabolites: Progress, challenges and opportunities. In Ute, R. (Ed.), *Metabolomics* (pp. 241–266). In-Tech.

Leylaie, S., & Zafari, D. (2018). Antiproliferative and antimicrobial activities of secondary metabolites and phylogenetic study of endophytic *Trichoderma* species from *Vinca* plants. *Frontiers in Microbiology*, 9, 1484.

Li, W. Y., Liu, Y., Lin, Y. T., Liu, Y. C., Guo, K., Li, X. N., & Li, S. H. (2020). Antibacterial harziane diterpenoids from a fungal symbiont *Trichoderma atroviride* isolated from *Colquhounia coccinea* var. mollis. *Phytochemistry*, 170, 112198.

Li, Y., Song, Y. C., Liu, J. Y., Ma, Y. M., & Tan, R. X. (2005). Anti-Helicobacter pylori substances from fungal cultures. *World Journal of Microbiology and Biotechnology*, 21, 553–558.

Linh, T. M., Mai, N. C., Hoe, P. T., Ngoc, N. T., Thao, P. T. H., Ban, N. K., & Van, N. T. (2021). Development of a cell suspension culture system for promoting alkaloid and *vinca* alkaloid biosynthesis using endophytic fungi isolated from local *Catharanthus roseus*. *Plants*, 10(4), 672.

Luo, P., Huang, J. H., Lv, J. M., Wang, G. Q., Hu, D., & Gao, H. (2024). Biosynthesis of fungal terpenoids. *Natural Product Reports*, 41(5), 748–783.

Li, H., Xiao, J., Gao, Y. Q., Tang, J. J., Zhang, A. L., & Gao, J. M. (2014). Chaetoglobosins from Chaetomium globosum, an endophytic fungus in Ginkgo biloba, and their phytotoxic and cytotoxic activities. *Journal of Agricultural and Food Chemistry*, 62(17), 3734–3741.

Ma, Y. M., Li, Y., Liu, J. Y., Song, Y. C., & Tan, R. X. (2004). Anti-Helicobacter pylori metabolites from Rhizoctonia sp. Cy064, an endophytic fungus in Cynodon dactylon. *Fitoterapia*, 75(5), 451–456.

Meghana, S. B., Sukrutha, S. K., Swamy, M. K., Purushotham, B., & Akhtar, M. S. (2025). Endophytes-mediated secondary metabolites synthesis in plants: An omic approach. In Akhtar, S. & Hakeem, K. R. (Eds.), *Agricultural Biotechnology* (pp. 217–230). Apple Academic Press.

Mishra, S., Srivastava, A., Singh, A., Pandey, G. C., & Srivastava, G. (2024). An overview of symbiotic and pathogenic interactions at the fungi-plant interface under environmental constraints. *Frontiers in Fungal Biology*, 5, 1363460.

Moussa, A. Y., Albelbisy, M. A., & Singab, A. N. B. (2025). The underrepresented quinolinone alkaloids in genera *Penicillium* and *Aspergillus*: Structure, biology, and biosynthetic machinery. *Chemistry & Biodiversity*, 22(3), e202402218.

Muhammad, M., Basit, A., Ali, K., Ahmad, H., Li, W. J., Khan, A., & Mohamed, H. I. (2024). A review on endophytic fungi: A potent reservoir of bioactive metabolites with special emphasis on blight disease management. *Archives of Microbiology*, 206(3), 129.

Nischitha, R. (2024). Role of grass endophytic fungi as a natural resource of bioactive metabolites. *Archives of Microbiology*, 206(10), 418.

Nischitha, R., & Shivanna, M. B. (2021). Antimicrobial activity and metabolite profiling of endophytic fungi in Digitaria bicornis (Lam) Roem. and Schult. and Paspalidium flavidum (Retz.) A. Camus. *3 Biotech*, 11(2), 53.

Nischitha, R., & Shivanna, M. B. (2022a). Diversity and in silico docking of antibacterial potent compounds in endophytic fungus Chaetomium subaffine Sergeeva and host Heteropogon contortus (L.) P. Beauv. *Process Biochemistry*, 112, 124–138.

Nischitha, R., & Shivanna, M. B. (2022b). Screening of secondary metabolites and antioxidant potential of endophytic fungus Penicillium citrinum and host Digitaria bicornis by spectrophotometric and electrochemical methods. *Archives of Microbiology*, 204(4), 206.

Nischitha, R., Vasanthkumari, M. M., Kumaraswamy, B. E., & Shivanna, M. B. (2020). Antimicrobial and antioxidant activities and chemical profiling of Curvularia tsudae endophytic in Cynodon dactylon (L.) Pers. *3 Biotech*, 10(7), 300.

Nischitha, R., Suryavamshi, L. G., Ashoka, G. B., & Shivanna, M. B. (2025). Spotlight on the hidden treasure of endophytic fungal bioactive metabolites. In Manoharachary, C., Singh, H. B., Singh, S. K., & Sharma, Y. P. (Eds.), *Biodiversity, Bioengineering, and Biotechnology of Fungi* (pp. 129–150). Academic Press.

Niu, X., Thaochan, N., & Hu, Q. (2020). Diversity of linear non-ribosomal peptide in biocontrol fungi. *Journal of Fungi*, 6(2), 61.

Nouh, F. A. A., Gezaf, S. A., & Abdel-Azeem, A. M. (2021). Recent advances in fungal antimicrobial. *Industrially Important Fungi for Sustainable Development, Volume 2: Bioprospecting for Biomolecules*, 177.

Ozaki, T., Minami, A., & Oikawa, H. (2023). Recent advances in the biosynthesis of ribosomally synthesized and posttranslationally modified peptides of fungal origin. *The Journal of Antibiotics*, 76(1), 3–13.

Packiaraj, R., Jeyakumar, S., Ayyappan, N., Adhirajan, N., Premkumar, G., Rajarathinam, K., & Muthuramkumar, S. (2016). Antimicrobial and cytotoxic activities of endophytic fungus *Colletotrichum gloeosporioides* isolated from endemic tree *Cinnamomum malabatrum*. *Studies in Fungi*, 1(1), 104–113.

Pandy, R., Kumar, S. S., Suresh, P., Annaraj, J., Pandi, M., Vellasamy, S., & Sagadevan, S. (2023). Screening and characterization of fungal Taxol-producing endophytic fungi for evaluation of antimicrobial and anticancer activities. *Open Chemistry*, 21(1), 20220344.

Petrini, O. (1991). Fungal endophytes of tree leaves. In Andrews, J. H. & Hirano, S. S. (Eds.), *Microbial Ecology of Leaves* (pp. 179–197). Springer New York.

Praptiwi, P., Fathoni, A., & Ilyas, M. (2020). Diversity of endophytic fungi from *Vernonia amygdalina*, their phenolic and flavonoid contents and bioactivities. *Biodiversitas Journal of Biological Diversity*, 21(2), 436–441.

Rai, N., Gupta, P., Keshri, P. K., Verma, A., Mishra, P., Kumar, D., Kumar, A., Singh, S.K., & Gautam, V. (2022). Fungal endophytes: An accessible source of bioactive compounds with potential anticancer activity. *Applied Biochemistry and Biotechnology*, 194(7), 3296–3319.

Rajakaruna, O., Wijayawardene, N. N., Udagedara, S., Jayasinghe, P. K., Gunasekara, S. S., Boonyuen, N., & Ariyawansa, K. G. (2024). Exploring fungal diversity in seagrass ecosystems for pharmaceutical and ecological insights. *Journal of Fungi*, 10(9), 627.

Realini, F. M., Escobedo, V. M., Ueno, A. C., Bastías, D. A., Schardl, C. L., Biganzoli, F., & Gundel, P. E. (2024). Anti-herbivory defences delivered by Epichloë fungal endophytes: a quantitative review of alkaloid concentration variation among hosts and plant parts. *Annals of Botany*, 133(4), 509–520.

Rigobelo, E. C. (2025). Endophytic filamentous fungi as sources of metabolites for agricultural applications. In *Fungal Metabolites for Agricultural Applications: Biostimulation and Crop Protection by Fungal Biotechnology* (pp. 1–20). Springer Nature Switzerland.

Saxena, S., Dufossé, L., Deshmukh, S. K., Chhipa, H., & Gupta, M. K. (2024). Endophytic fungi: A treasure trove of antifungal metabolites. *Microorganisms*, 12(9), 1903.

Schulz, B., Wanke, U., Draeger, S., & Aust, H. J. (1993). Endophytes from herbaceous plants and shrubs: Effectiveness of surface sterilization methods. *Mycological Research*, 97(12), 1447–1450.

Shaaban, R., Elnaggar, M. S., Khalil, N., & Singab, A. N. B. (2023). A comprehensive review on the medicinally valuable endosymbiotic fungi *Penicillium chrysogenum*. *Archives of Microbiology*, 205(6), 240.

Sharma, V. K., Kumar, J., Singh, D. K., Mishra, A., Gond, S. K., Verma, S. K., & Kharwar, R. N. (2017). Induction of cryptic metabolite production through epigenetic tailoring in *Colletotrichum gloeosporioides* isolated from *Syzygium cumini*. In de Azevedo, J. L., Quecine & M. C. (Eds.), *Diversity and Benefits of Microorganisms from the Tropics* (pp. 169–184). Springer International Publishing.

Shivanna, M. B., & Nischitha, R. (2024). Diversity of endophytic fungi of *Ischaemum ciliare* Retz. as a potential source of antimicrobial and antioxidant agents. In A. M. Singh, & A. K. Roy (Eds.), *Biodiversity conservation, present scenario and future prospect*. 1st ed. (pp. 156–170). Walnut Publication, India. (ISBN: 978-93-5911-652-5/eBook ISBN: 978-93-5911-577-1).

Singh, A., Singh, D. K., Kharwar, R. N., White, J. F., & Gond, S. K. (2021). Fungal endophytes as efficient sources of plant-derived bioactive compounds and their prospective applications in natural product drug discovery: Insights, avenues, and challenges. *Microorganisms*, 9(1), 197.

Strobel, G. A. (2003). Endophytes as sources of bioactive products. *Microbes and Infection*, 5(6), 535–544.

Subban, K., Singh, S., Subramani, R., Johnpaul, M., & Chelliah, J. (2017). Fungal 7-epi-10-deacetyltaxol produced by an endophytic *Pestalotiopsis microspora* induces apoptosis in human hepatocellular carcinoma cell line (HepG2). *BMC Complementary and Alternative Medicine*, 17, 1–16.

Tan, X. M., Zhou, Y. Q., Zhou, X. L., Xia, X. H., Wei, Y., He, L. L., ang, H.Z. & Yu, L. Y. (2018). Diversity and bioactive potential of culturable fungal endophytes of Dysosma versipellis; a rare medicinal plant endemic to China. *Scientific Reports*, 8(1), 5929.

Toghueo, R. M. K. (2020). Bioprospecting endophytic fungi from *Fusarium* genus as sources of bioactive metabolites. *Mycology*, 11(1), 1–21.

Tu, N. H. K., Thuy, N. T. T., Vinh, D. T. T., Anh, N. P. Q., & Ly, H. D. (2021). Detection of antimicrobial, antioxidant and cytotoxicity activities of Fusarium oxysporum F01 isolated from *Catharanthus roseus* collected in Vietnam. *Journal of Pure Applied Microbiology*, 15(3), 1643–1654.

Ueno, A. C., Gundel, P. E., Seal, C. E., Ghersa, C. M., & Martínez-Ghersa, M. A. (2020). The negative effect of a vertically-transmitted fungal endophyte on seed longevity is stronger than that of ozone transgenerational effect. *Environmental and Experimental Botany*, 175, 104037.

Vassiliadis, S., Reddy, P., Hemsworth, J., Spangenberg, G. C., Guthridge, K. M., & Rochfort, S. J. (2023). Quantitation and distribution of Epichloë-derived alkaloids in perennial ryegrass tissues. *Metabolites*, 13(2), 205.

Verma, A., Tiwari, H., Singh, S., Gupta, P., Rai, N., Kumar Singh, S., Singh, B.P., Rao, S., & Gautam, V. (2023). Epigenetic manipulation for secondary metabolite activation in endophytic fungi: Current progress and future directions. *Mycology*, 14(4), 275–291.

Verma, V. C., Gond, S. K., Kumar, A., Mishra, A., & Kharwar, R. N. (2009). Endophytic actinomycetes from *Azadirachta indica* A. Juss.: Isolation, diversity, and anti-microbial activity. *Microbial Ecology*, 57(4), 749–756.

Vishwakarma, S., Chaudhry, V., Chand, S., Sagar, K., Gupta, K. K., Bhardwaj, N., Prasad, R., Kumar, P., & Chandra, H. (2024). The potential of fungal endophytes in plants: Sources of bioactive compounds. *Indian Journal of Microbiology*, 65, 1–15.

Wadhwa, K., Kapoor, N., Kaur, H., Abu-Seer, E.A., Tariq, M., Siddiqui, S., Yadav, V.K., Niazi, P., Kumar, P., & Alghamdi, S. (2024). A comprehensive review of the diversity of fungal secondary metabolites and their emerging applications in healthcare and environment. *Mycobiology*, 52(60), 335–387

Wang, M., Liang, X., Ma, R., Lin, W., Fang, Z., Zhang, L., Chen, G., He, J., Bai, E., Pi, E., & Shen, C. (2025). Correlation analysis of secondary metabolism and endophytic fungal assembles provide insights into screening efficient taxol-related fungal elicitors. *Plant, Cell & Environment*, 48(6), 4196–4211.

Wei, W., Jiang, N., Mei, Y. N., Chu, Y. L., Ge, H. M., Song, Y. C., Ng, S.W., & Tan, R. X. (2014). An antibacterial metabolite from *Lasiodiplodia pseudotheobromae* F2. *Phytochemistry*, 100, 103–109.

Wei, W., Khan, B., Dai, Q., Lin, J., Kang, L., Rajput, N. A., & Liu, G. (2023). Potential of secondary metabolites of *Diaporthe* species associated with terrestrial and marine origins. *Journal of Fungi*, 9(4), 453.

White, T. J., Bruns, T., Lee, S. J. W. T., & Taylor, J. (1990). Amplification and direct sequencing of fungal ribosomal RNA genes for phylogenetics. PCR protocols: A guide to methods and applications. *Mycologia*, 18(1), 315–322.

Wu, J., Chen, D., Li, Q., Feng, T., & Xu, J. (2024). Metabolomics-guided discovery of new dimeric xanthones from co-cultures of mangrove endophytic fungi *Phomopsis asparagi* DHS-48 and *Phomopsis* sp. DHS-11. *Marine Drugs*, 22(3), 102.

Wu, X., Zou, K., Liu, X., Fu, S., Zhang, S., Duan, Z., Zhou, J. & Liang, Y. (2022). The novel distribution of intracellular and extracellular flavonoids produced by Aspergillus sp. Gbtc 2, an endophytic fungus from *Ginkgo biloba* root. *Frontiers in Microbiology*, 13, 972294.

Xue, M., Hou, X., Fu, J., Zhang, J., Wang, J., Zhao, Z., & Zhou, L. (2023). Recent advances in search of bioactive secondary metabolites from fungi triggered by chemical epigenetic modifiers. *Journal of Fungi*, 9(2), 172.

Yadav, G., & Meena, M. (2021). Bioprospecting of endophytes in medicinal plants of Thar Desert: An attractive resource for biopharmaceuticals. *Biotechnology Reports*, 30, e00629.

Yan, D. H., Song, X., Li, H., Luo, T., Dou, G., & Strobel, G. (2018). Antifungal activities of volatile secondary metabolites of four *Diaporthe* strains isolated from *Catharanthus roseus*. *Journal of Fungi*, 4(2), 65.

Yang, Y. H., Yang, D. S., Li, G. H., Pu, X. J., Mo, M. H., & Zhao, P. J. (2019). Antibacterial diketopiperazines from an endophytic fungus *Bionectria* sp. Y1085. *The Journal of Antibiotics*, 72(10), 752–758.

Yue, Q., Miller, C. J., White, J. F., & Richardson, M. D. (2000). Isolation and characterization of fungal inhibitors from Epichlo ë festucae. *Journal of Agricultural and Food Chemistry*, 48(10), 4687–4692.

Zhang, X. Y., Hao, H. L., Lau, S. C. K., Wang, H. Y., Han, Y., Dong, L. M., & Huang, R. M. (2019). Biodiversity and antifouling activity of fungi associated with two soft corals from the South China Sea. *Archives of Microbiology*, 201(6), 757–767.

Zhou, H., & Hou, C. L. (2019). Three new species of Diaporthe from China based on morphological characters and DNA sequence data analyses. *Phytotaxa*, 422(2), 157–174.

Zhang, H., Tang, Y., Ruan, C., & Bai, X. (2016). Bioactive secondary metabolites from the endophytic *Aspergillus* genus. *Records of Nature Production*, 10(1), 1–16.

Zhang, H. W., Song, Y. C., & Tan, R. X. (2006). Biology and chemistry of endophytes. *Natural Product Reports,* 23(5), 753–771.

Zhao, Y., Cui, J., Liu, M., & Zhao, L. (2020). Progress on terpenoids with biological activities produced by plant endophytic fungi in China between 2017 and 2019. *Natural Product Communications*, 15(7), 1934578X20937204.

Zhu, A. H., Song, Z. K., Wang, J. F., Guan, H. W., Qu, Z., & Ma, H. X. (2024). Multi-gene phylogenetic analyses reveals heteroxylaria Gen. Nov. and new contributions to Xylariaceae (Ascomycota) from China. *Journal of Fungi*, 10(9), 645.

Zhu, J., Song, L., Shen, S., Fu, W., Zhu, Y., & Liu, L. (2023). Bioactive alkaloids as secondary metabolites from plant endophytic *Aspergillus* genus. *Molecules*, 28(23), 7789.

5 Study of Deep Learning and Bioprospecting for Black Fungus Identification and Treatment

Harsh Lalwani, Yakshit Arora, Samrath Singh Narula, Kamal Rawal, and Navaneet Chaturvedi

5.1 INTRODUCTION

Mucormycosis (MCR), also known as black fungus, in recent times has been seen globally with exceptionally large numbers of cases specifically over the last three decades. The major cancer Centres like MD Anderson and Fred-Hutchinson have seen almost double the major cases of mucormycosis in the last 15 years. Talking about other countries like France, the report shows an increase of 70% between 2007 and 2015 compared to the incidents of 1988-2006. Similarly, another country reports (Switzerland) suggest ten times increase in MCR cases among the patients after 2003. In India, we can clearly conclude the hyper-endemic state of mucormycosis with almost 200,000+ cases annually before COVID-19 pandemic, in which approximately 24% are all invasive of Mould infections (Divakar, 2021; Alqarihi et al., 2023). Another alarming research highlights the reports from the Leading International Fungal Education (LIFE) portal, which estimates approximately 910,000 cases globally after including Indian data, a dramatic increase from their previous estimate of about 10,000 cases worldwide excluding India. India's disproportionate burden of this illness is brought out by this obvious disparity. In addition, according to a computational model, the prevalence of mucormycosis in India is around 140 cases per million people, or between 137,807 and 208,177 cases annually (mean: 171,504), with an attributable mortality of 38.2% (65,500 fatalities). India's significant diabetic population, which is expected to rise from 61. 3 million in 2011 to 101. 2 million by 2030, is directly responsible for this severe burden (Nambiar et al., 2021).

Mucormycosis' causative agent belongs to the order of Mucorales. Globally, the species that dominates is Rhizopus, which is accountable for more than 70%

DOI: 10.1201/9781003633952-6

of all mucormycosis cases. There are also other causative agents like *Mucor* and *Lichtheimia* (formally known as Absidia) species, and these species are collectively accountable for 90% of the cases. Other less common agents are from species *Cunninghamella*, *Apophysomyces*, *Saksenaea*, *Rhizomucor*, *Cokeromyces*, *Actinomucor*, and *Syncephalastrum*, which only accounts for 1%–5% of the total reported cases (Hernández & Buckley, 2023; Alqarihi et al., 2023). This disease affects immunocompromised individuals (whose immune systems are inefficient for any reason). In high-income countries, haematological malignancies are one of the most common conditions causing the immunity reduction in an individual leading to mucormycosis, but in developing countries, uncontrolled diabetes is the leading cause (Skiada et al., 2018). During the COVID-19 pandemic in countries like India as the virus attack was already in the epidemiology of population and then the treatment of which involved the administration of corticosteroids, both of which created a perfect biological environment for the mucormycosis to evolve and spread in a very dynamic way.

To treat this condition, the timely diagnosis was one of the important parameters. Unfortunately, the current diagnostic approaches of mucormycosis face significant limitations. Clinical diagnosis relies on recognizing host factors and clinical manifestations, but this technique is not efficient as it lacks sensitivity and specificity for other fungi species like Aspergillus or fusarium because they have similar clinical signs and prognosis (Sharma & Goel, 2022). For example, tissue death is one of the symptoms of mucormycosis, but it can also be the clinical symptoms of other fungi too. Apart from clinical diagnosis, microscopic and culture diagnosis are also options for mucormycosis but with their own significant drawbacks. Like in microscopy, direct microscopy with specific optical stains can rapidly suggest the presence of infection by mucormycosis but it is time taking and less sensitive especially when the incident rate of mucormycosis patient is on exponential rise in conditions like COVID-19 (Skiada et al., 2018). Detecting black fungus therefore calls for a proper, very fast, and accurate strategy. Because of the epidemiological load, a better understanding of the disease mechanism is needed since it will define the inheritance of the disease to the next generation, as well as the influence several pathophysiological routes might have on expected epidemiological results.

5.2 COMPREHENSIVE PATHOPHYSIOLOGY OF MUCORMYCOSIS

Pathogenesis of mucormycosis results from a complex interaction of fungal virulence agents and host vulnerability systems, where recent research exposes key molecular pathways, which provide its forceful clinical behaviour. The GRP78-CotH Receptor and Ligand Axis dominates host–pathogen interactions occurring at the cellular level. The glucose-regulated protein 78 (GRP78) on endothelial cells interacts with fungal spore coat protein homologues (CotH), Mucorales enters mostly. The predilection for mucormycosis in diabetic ketoacidosis (DKA) patients results from different causes including CotH3/Cot2, a structurally conserved protein, according to this molecular dialogue, found in *Rhizopus arrhizus* that binds to GRP78 highly (Baldin & Ibrahim, 2017; Gebremariam et al., 2014). An additional significant cause could be induction. Structural protein modifications in the event of hyperglycaemia cause endothelial GRP78 upregulation to improve CotH binding (Gebremariam et al., 2014). Mucormycosis's invasion mechanism is yet another significant cause for which CotH-GRP78 engages itself to start the acting cytoskeleton rearrangement through GTPase signalling that allows fungal entry into the subendothelial channels (Baldin & Ibrahim, 2017; Lax et al., 2024). In vivo investigations show the anti-CotH monoclonal antibodies, emphasizing the therapeutic potential of this interaction. Baldin and Ibrahim (2017) and Lax et al. (2024) showed a 92% reduction in fungal load in mouse models.

5.2.1 Iron Acquisition Systems

Mucorales employ sophisticated iron-scavenging strategies critical for pathogenesis, which can be studied in a tabular form in Table 5.1. Notably, serum iron levels >270 μg/dL increase mucormycosis risk 11-fold by upregulating fungal iron permeases (FTR1). The rhizoferrin biosynthesis pathway (involving RFS1–5 genes) is essential for virulence knockout strains, which show 75% reduced mortality in murine models (Larcher et al., 2013; Lax et al., 2024).

5.2.2 Enzymatic Virulence Factors

Tissue-Degrading Enzymes from Mucorales secrete a repertoire of hydrolytic enzymes facilitating tissue invasion. Some of the examples of enzymes are serine proteases (cleave at E-cadherin junction and degrade collagen membranes), Phospholipases (PLB1 Knockout reduces endothelial damage by 68% and generates arachidonic acid precursors for pro-inflammatory eicosanoids), and aspartic proteases, which shows optimal activity at pH 3.5–4.0, which is biocompatible with phagolysosomes and enables the escape from macrophage phagosomes (Lax et al., 2024).

5.2.3 Immune Evasion Strategies

R. arrhizus utilizes multiple strategies to evade the host immune response and withstand oxidative stress. To subvert the complement system, its surface proteins bind Factor H, a regulatory component that reduces C3b deposition by 80%, while its capsular glucuronoxylomannan inhibits membrane attack complex

TABLE 5.1
Study of Different Mechanisms in Different Research Papers for the Clinical Relevance of Mucormycosis

Mechanism	Key Components	Clinical Relevance
Siderophore production	Rhizoferrin (hydroxamate type)	Chelates Fe^{3+} from transferrin (Larcher et al., 2013; Lax et al., 2024)
Feroxamine utilization	Feroxamine B receptors	Exploits deferoxamine therapy (Lax et al., 2024)
Direct haem uptake	Haem oxygenase enzymes	Enables growth in iron-poor environments (Larcher et al., 2013)

(MAC) formation through steric hindrance. In terms of oxidative stress resistance, *R. arrhizus* expresses superoxide dismutases (SOD1–3), which maintain fungal viability within neutrophils—achieving a 73% survival rate compared to 22% in SOD-deficient mutants. Additionally, catalase production allows the fungus to neutralize hydrogen peroxide generated during the respiratory burst, decomposing approximately two million H_2O_2 molecules per second (Parente et al., 2017; Lax et al., 2024). Additionally, this molecular understanding directly informs diagnostic and therapeutic strategies. For instance, PCR targeting CotH genes achieves 90% sensitivity in serum samples, while iron chelators like deferasirox reduce mortality by 40% in clinical trials. Ongoing research explores bispecific antibodies targeting CotH–GRP78 interactions and small-molecule inhibitors of rhizoferrin biosynthesis as novel therapeutics (Gebremariam et al., 2014; Lax et al., 2024). It is very clear till now that the need of pathophysiological studies is one of the crucial steps for the early detection of mucormycosis, but the major problem is the complexity in the pathophysiology, which adds more irrelevant and insignificant data to disrupt and slow down the research process. These problems are actually the motivation for advanced computational systems and architectures that too AI driven to make the diagnostic process that is less time taking and, more importantly, more accurate.

5.3 DEEP LEARNING APPROACHES FOR BLACK FUNGUS IDENTIFICATION

5.3.1 AI-Driven Medical Imaging Analysis

AI is the new age tool in the field of medical diagnosis and analysis with improved accuracy and fast results, which was the exact requirement of our black fungus diagnostic problem. The application of deep learning models is one of the latest improvements being done to identify various radiological patterns associated with mucormycosis. Research indicates that multiple nodules ($\geq$10) presence and pleural effusion view on CT scans are associated with pulmonary mucormycosis, while the reverse halo sign is also another indicator of the infection (Skiada et al., 2018). Latest AI algorithms can detect these subtle patterns more consistently and precisely than human observers, thus enabling earlier diagnosis possible.

5.3.2 Hybrid Learning Methodologies for Disease Detection

In Mucormycosis detection, the analyses of image-based data are very crucial, which is in high-dimensional format and unstructured to directly use. That's why simpler algorithms like logistic regression or basic decision tree are not efficient for this purpose. Thus, for image recognition, the best algorithm choice would be CNN (Convolutional Neural Network), which excels in the extraction of relevant image features like texture, shape, and patterns (Nira & Kumar, 2022). The CNN component excels at hierarchical feature extraction from medical images through its convolutional layers that apply multiple filters to identify patterns relevant to mucormycosis. These layers progressively learn more complex features—from basic edges and textures in early layers to specific pathological patterns in deeper layers (Figure 5.1).

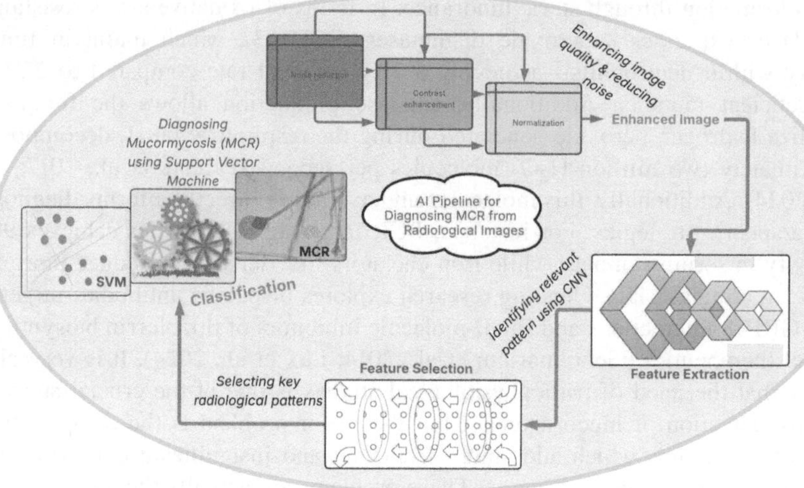

FIGURE 5.1 AI pipeline for diagnosing mucormycosis from radiological images. The process involves image preprocessing to enhance quality, feature extraction using CNN to detect patterns, feature selection of key radiological traits, and final classification using SVM to diagnose mucormycosis.

Using CNN, a high curated image data will be transformed into good feature-based differentiated data that can be further classified using Support Vector Machine (SVM). SVM can manage non-linear boundaries well and often can generalize better with good features without the need of large datasets like deep networks. Medically, we can conclude that CNN will be doing the job of radiologist, examining the images and extracting all the important signals, whereas SVM is a seasoned diagnostician which will take those signals and make a precise diagnosis (Yadav et al., 2025). Henceforth, a hybrid model will be the best possible tool in the detection of mucormycosis. A critical aspect worth elaborating is the feature selection mechanism between the CNN and SVM components (Meister et al., 2021). Implementing feature importance analysis can identify which radiological patterns most strongly correlate with mucormycosis presence. For instance, research indicates that multiple nodules (≥10) and pleural effusion on CT scans strongly associate with pulmonary mucormycosis, while the reverse halo sign serves as another indicator (Pak et al., 2008; Prakash & Chakrabarti, 2019). Sophisticated AI algorithms can detect these subtle patterns more consistently than human observers.

5.4 PERFORMANCE EVALUATION AND CLINICAL IMPLEMENTATION

When this technology was applied in the recent study, very promising results for the future were obtained that needed to be discussed. The accuracy achieved was 99.5% in distinguishing between eyes affected by black fungus and unaffected eyes in the COVID-19 patients. This makes it suitable for use in hospital diagnostic system where a fast screening of patients on the basis of eye image can be done in the case of mucormycosis. However, the real-world application will highly depend on the availability of consistent quality eye images. This achieved accuracy is particularly significant if we also consider the parameter of ocular involvement, which is common in rhinocerebral mucormycosis and early detection of it can prevent vision loss and more severe complications (Karthikeyan et al., 2022). Apart from the fact of accuracy, the cost and time of the whole diagnosis process can also be evolved using this technology as it eliminates the need for costly body scans like CT or MRI, and the automated system can get us the results in seconds. The fact that the clinical deployment of this CNN-SVM architecture will require validation from all different populations and variety of image sources including the regulatory approvals and explainability for medical use can also not to be ignored. This algorithm was designed in a biological stepwise logic that includes image processing, feature selection, and then disease classification. Further, the implementation of this algorithm was done using python tools that were quite common in biological data analysis. Python provides flexibility for large database ecosystem, which will make implementation faster and more reproducible (Kumar et al., 2023). Libraries like TensorFlow will streamline the building and hybridization of both CNN and SVM models. In rural region where CT or MRI scans are still an important thing to be available because of large cost, this model can be perfectly developed with minimum cost.

A flowchart has been generated for clearly understanding and visualizing the CNN-SVM architecture, as shown in Figure 5.1. The flowchart starts with image inputs into the system for the diagnosis of mucormycosis, and these images are obtained from various sources like Clinical photography, radiological investigations (CT, MRI), and histopathological investigations. The clinical image analysis can be used for the mucormycosis detection due to its morphological signs on the patients; for example, in histopathological studies, it will show characteristic features like ribbon-like hyphae (Alexander et al., 2021; Nielsen et al., 2023). Some common features include black lesion on the palate, facial swelling, and preorbital oedema. The affected organs can vary from sinuses, lungs, GI tract, or soft tissues with distinctive visual characteristics (Nielsen et al., 2023). Then comes the Image processing part, which is one of the important crucial steps for any image analysis as the obtained images are not always structured or are with unwanted artifacts. The process steps that refine these images include Noise reduction, contrast enhancement, Normalization, and resizing. One important thing that is not included in the flowchart is segmentation. Segmentation is done to isolate the region of interest that may contain the potential mucormycosis-related features. In image processing pipelines, the segmentation is founded as the very crucial stage in the epidemiological survey (Nira & Kumar, 2022). Segmentation includes thresholding (isolating foreground from background), region-based segmentation (grouping similar pixels), edge detection (identifying boundaries), and deep learning-based segmentation. The next crucial step is the feature extraction, which is done to extract the notable features to study from the region of interest, and this uses pooling of layers, convolutional layers, and activation functions. As the required data is registered in the architecture, it is time for SVM for the decision-making process. SVM creates an optimal boundary that separates mucormycosis positive from negative by implementing kernel functions on the output of CNN layers, and it does this by maximization, support vectors, and regularization.

5.4.1 Clinical Manifestations of Mucormycosis

Mucormycosis is a serious fungal infection characterized by its aggressive and invasive nature, particularly affecting immunocompromised individuals. Clinically, it can present in various forms, including rhino-orbital-cerebral, pulmonary, cutaneous, gastrointestinal, and disseminated types (Hernández & Buckley, 2023). The most common manifestation is rhino-orbital-cerebral mucormycosis, which often begins with nasal congestion, facial swelling, and black lesions on the nasal bridge or upper inside of the mouth. As the infection progresses, it may lead to vision loss, neurological deficits, or even death if not treated promptly. Early diagnosis and rapid intervention are crucial due to the infection's rapid progression and high mortality rate (Table 5.2).

TABLE 5.2
Representation of Different Clinical Manifestations in Terms of Its Symptoms, Populations, Mortality Rate, and Diagnostic Challenges in Different Cited Studies

Clinical Form	Key Symptoms & Diagnostic Features	Typical Patient Populations	Mortality Rate	Diagnostic Challenges
Rhino-orbital-cerebral	• One-sided facial swelling • Black nasal/oral lesions • Vision changes • Sinus congestion • Headache	• Poorly controlled diabetes (A1C >8%) • Organ transplant recipients • Post-COVID patients	~30%–50%	Mimics bacterial sinusitis; requires invasive biopsy for confirmation
Pulmonary	• Fever • Cough • Chest pain • Shortness of breath • Haemoptysis (rare)	• Haematologic malignancies • Stem cell transplant recipients • Prolonged neutropenia	Up to 76%	Overlaps with aspergillosis; requires bronchoscopy or lung biopsy
Cutaneous	• Blisters/ulcers • Blackened skin • Localized pain/swelling • Necrotic wound edges	• Trauma/burn victims • IV drugs' users • Surgical patients • Immunocompetent individuals	10%–30% (localized)	Often misdiagnosed as bacterial cellulitis; requires deep tissue sampling

(*Continued*)

TABLE 5.2 (*Continued*)
Representation of Different Clinical Manifestations in Terms of Its Symptoms, Populations, Mortality Rate, and Diagnostic Challenges in Different Cited Studies

Clinical Form	Key Symptoms & Diagnostic Features	Typical Patient Populations	Mortality Rate	Diagnostic Challenges
Gastrointestinal	• Abdominal pain • Nausea/ vomiting • GI bleeding • Peritoneal signs	• Premature infants • Malnourished patients • Iron overload conditions	~85%	Rare presentation; requires endoscopy or surgical exploration
Disseminated	• Multi-organ failure • Mental status changes • Coma • Cutaneous involvement in 60% of cases	• Severe immunosuppression • Untreated localized infections • Haematologic malignancies	~70%	Symptoms masked by primary illness; often diagnosed post-mortem

5.5 CURRENT CLINICAL MANAGEMENT OF MUCORMYCOSIS

Treatment of mucormycosis is very challenging given the destructive nature of the disease and the usual delay in diagnosis. The current standard of care combines antifungal medication with surgical debridement using a multimodal approach. To get rid of dead tissue, increase drug penetration, and lower fungal load, surgical intervention is still required. Radical debridement frequently involves significant tissue excision, which can result in significant morbidity. The primary ingredient of antifungal treatment is amphotericin B, particularly its lipid formulations. It has become the first line of medical treatment (Hernández & Buckley, 2023) because its nephrotoxicity is reduced in comparison to traditional medications. Triazoles such as posaconazole and isavuconazole have been shown to be successful for patients unable to tolerate amphotericin B or as a stepping-down therapy. Isavuconazole has grown in popularity because of its bioavailability, outstanding safety record, and demonstrated effectiveness in recent clinical trials (Dogra et al., 2022). Despite these therapies, mortality rates remain between 40% and 80%, depending on the underlying condition, the severity of the infection, and the intervention time (Motamedi et al., 2022). The CNNSVM's hybrid architecture presents one possibility for the automated diagnosis of mucormycosis. This design successfully identifies this potentially fatal fungal infection by combining the powerful feature extraction capabilities of CNN with the classifying strengths of SVM. The model follows a disciplined process from input image capture to preprocessing, feature extraction, and at last categorization.

5.5.1 Bioprospecting for Novel Antifungal Compounds

The next big pharmaceutical and biological problem human civilization may see is the increase in antibiotic resistance in humans; as we are using antibiotics on ourselves from almost 5 generations, our body has started developing resistance to it. This establishes a need of continuous new drug/bio-active compound research for different diseases including mucormycosis. This establishes the need of bioprospecting that is systemic exploration of biodiversity or bioactive compounds that can be used for the generation of new antifungal molecule. In recent study, an endophytic fungi is isolated from *Bauhinia guianensis*, a traditional Amazonian medicinal plant, having significant antimicrobial potential 5 out of 17 fungal isolates; the methanolic extract of *Exserohilum rostratum* has shown potent activity against *Escherichia coli* and *Bacillus subtilis* (Pinheiro et al., 2016). The active compound that was identified is termed as monocerin, a polyketide. Antifungal, insecticidal, and even anti-malarial properties have been observed in fungal species like *Fusarium* and *Microdochium* (Figure 5.2).

5.5.2 Natural Sources of Antimicrobial Agents

Nature has been our best chemical reservoir till now. Plant like *B. guianensis* produces bioactive compounds or phytochemicals, such as flavonoids and glycosides, and host endophytes capable of synthesizing novel antimicrobial drugs (Pinheiro et al., 2016; Chaturvedi et al., 2024. These microbial organisms can mimic the therapeutic characteristics of their host plant. For example, endophytes from plants like *Urospermum picroides* and *Erica arborea* have the ability to yield bio-active compounds because of the presence of strong antimicrobial and antifungal properties. Research have seen overall 17 endophytic fungi from the species *B. guianensis* and *E. rostratum*, which can produce a metabolic extract with good bio-activity against most bacteria (Tiwari & Bae, 2022; Arora et al., 2025).

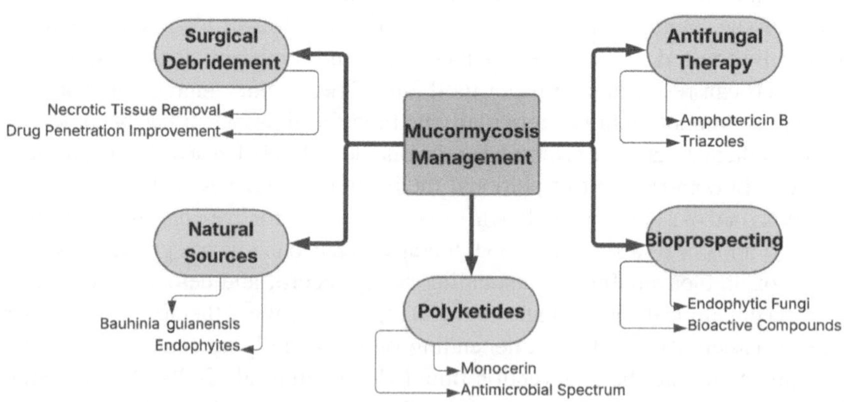

FIGURE 5.2 Integrated approaches for mucormycosis management. Representation of key strategies for managing mucormycosis, including surgical debridement, antifungal therapy, exploration of natural sources and polyketides, and bioprospecting for novel bioactive compounds. These approaches aim to enhance treatment efficacy and support drug development.

5.5.3 AI-Assisted Screening Methodologies

The combination of AI with bioprospecting is developing as a revolutionary tactic in the hunt for new antifungal molecules. AI algorithms have the accuracy to forecast bioactivity depending on chemical structures, therefore allowing the discovery of potential candidates from large chemical compound libraries and the optimization of lead compounds for better efficacy and safety. Modern AI-driven drug screening techniques comprise ligand-based methods, structure-based virtual screening, quantum mechanical computations, and multi-task learning frameworks (Wainwright et al., 2022; Ocana et al., 2025).

Deep learning models like AtomNet in structure-based virtual screening use 3D convolutional networks to examine molecular structures and properly estimate binding affinity with fungal target proteins, outperforming more conventional docking methods. Based on the idea that structurally comparable chemicals usually show comparable biological activity, ligand-based method models such as Mol2Vec and graph neural networks convert molecular structures into embeddings that grasp both chemical and functional characteristics. Additionally, with rapid prediction of electronic properties and chemical reactivity of possible antifungal compounds, AI helps quantum mechanical simulations while maintaining retentive accuracy and considerably lowering computational time (Malhotra et al., 2025a; Rehman et al., 2024). Likewise, multitask learning patterns offer a more inclusive evaluation of therapeutic potential by consenting the prediction of some characteristics, which include antifungal activity, toxicity, bioavailability, and metabolism.

These computer techniques meaningfully decrease the time and resources, essential for first screening, therefore enabling researchers to emphasize their experimental studies on the maximum potent compounds. Virtual screening using molecular docking simulations ranks chemicals with the highest binding affinity and possible inhibitory activity by examining how they might engage with target fungus (Hashemi et al., 2024).

5.6 AI-DRIVEN DRUG REPURPOSING FOR MUCORMYCOSIS TREATMENT

5.6.1 Deep Learning Frameworks for Drug Repurposing

Deep learning architecture can be integrated with our casual inference process by using computational frameworks that are capable of emulating random clinical trials data. Therefore, identification of new novel drug can be done very cost effectively by computational frameworks, and the generation of random clinical trials data can be done by longitudinal analysis and observation from electronic health records (EHRs) including insurance claims data. This also solves the major challenge in healthcare analysis, which deals with real-world factors which contains non-structured random data (Yang et al., 2022; Dharmani et al., 2024). Computational methods have the ability to leverage their long short-term memory (LSTM) to capture temporal dependency or small observable patterns, which can predict and analyse the change of patient's health over time. Though this can generate control selection bias from the computer, this is further corrected by inverse probability of treatment weighting (IPTW).

5.6.2 Methodologies for Therapeutic Effect Prediction

A variety of AI-driven methods are being used to precisely forecast therapeutic outcomes against mucormycosis, each of which provides unique perspectives at the molecular, clinical, and systems biology levels. As previously said, EHR analysis makes use of longitudinal patient data in order to identify temporal trends in treatment outcomes and provide real-world data for the efficacy of drugs. AI-accelerated molecular docking and dynamics simulations model interactions between drugs and fungal targets, evaluating binding affinities and the stability of complexes over time in order to determine therapeutic potential. Gene expression analysis employs machine learning to find substances that reverse transcriptional signatures connected to infection, indicating that they may be able to combat the course of the illness (Tanwar et al., 2024). Network-based drug repurposing models biological processes as networks, where graph neural networks assist in identifying drugs that can target crucial nodes or pathways associated with the disease. Transfer learning methods apply models trained on similar fungal infections to mucormycosis, utilizing established data to enhance predictions for this uncommon illness. Ensemble methods combine outputs from various models, such as clinical data, transcriptomic studies, and structure-based screens, to produce more precise and complete predictions in order to increase robustness (Li et al., 2024).

Approaches such as recurrent neural networks (RNNs) with LSTM cells, have become useful, which provide a dynamic viewpoint of therapeutic effect over time (Godoy et al., 2012). Additionally, in generative models, this can even be accomplished without relying on clinical trial data, which create new guidelines for hypothesis formation and medicine repurposing. This method generates a high-throughput computational pipeline that, given a disease cohort, extracts possible repurposing candidate drugs and associated user and non-user sub-cohorts (Lee et al., 2025).

5.7 CASE STUDIES AND VALIDATION APPROACHES

One of the significant variables for assessing correctness and relevance in the biological system for medication repurposing is the deep learning's credibility. Additionally, data-driven trial-emulating approach that may be utilized more can be demonstrated by means of a large-scale case study centred on coronary artery disease (CAD). Insurance claim archives of more than three million patients (Wan et al., 2025) have been systematically assessed, with over 5,000 frequently arranged medications. The similar approach possesses high translational value for rare or new infections like mucormycosis. Some approved medications showed mechanistic plausibility and safety, therefore pointing towards repurposed potential (Wang et al., 2023).

5.7.1 AI Diagnosis with Targeted Therapeutics

A holistic approach to deal mucormycosis is the combination of particular treatment techniques with AI-driven diagnostic systems. Deep learning models might be positively associated to the choice of therapies, which include this combined approach that enables the development of tailored treatment plans by conventional antifungal treatments and repurposed drugs discovered by AI analysis (Hasan Pour, 2025).

5.7.2 Implementation Challenges in Clinical Settings

The introduction of AI-based solutions turns across some application issues. These problems include the need for ample clinical validation of AI models before their use, integration of the models with current hospital information systems, compliance with regulatory requirements for AI-enabled medical decision-making, and education of healthcare professionals to properly interpret and utilize AI outputs (Maleki Varnosfaderani & Forouzanfar, 2024). These issues would mandate for the development of a whole outline that includes technologists, health professionals, regulatory organizations, and public health experts.

5.7.3 Cost-Effectiveness Analysis

Enhanced early identification of treatment failures would permit rapid intervention and accuracy, which would reduce unnecessary medicine, and better general patient consequences lower healthcare costs (Khanna et al., 2022). Beside these specific local occurrences, a cost-effectiveness analysis should reflect these factors of mucormycosis and the specific healthcare environment.

5.8 CHALLENGES AND FUTURE DIRECTIONS

5.8.1 Technical Limitations of Current AI Models

A major weakness is the absence of AI models through many patient groups, imaging equipment, and clinical scenarios. Models may be difficult to apply in that data from one community or healthcare system might not provide the best results when utilized somewhere else (Maleki Varnosfaderani & Forouzanfar, 2024).

5.8.2 Ethical Considerations in AI-Based Healthcare

The use of AI in mucormycosis management increases ethical problems. Concerns were about the potential for algorithmic bias, especially if training data does not properly signify the variety of patient populations (Harishbhai et al., 2024). Biased algorithms lead to dissimilar performance across diverse demographic groups, which thus impair current inequalities in healthcare access and quality (Figure 5.3). Future investigations should focus on the creation of international registries and future studies for the effective decrease in worldwide burden of mucormycosis diagnosis and risk evaluation.

5.8.3 Future Research Priorities

Another future consideration can be the creation and stardardization for quick molecular diagnostics for species-level identification (Spellberg et al., 2005; Kontoyiannis et al., 2012; Garre, 2022). We must continually expand the molecular toolbox to a wider variety of Mucorales species to enable functional studies and validation of drug targets (Garre, 2022; Tahiri et al., 2023). We perform randomized clinical trials to assess novel antifungal drugs and combination treatments, with outcomes that

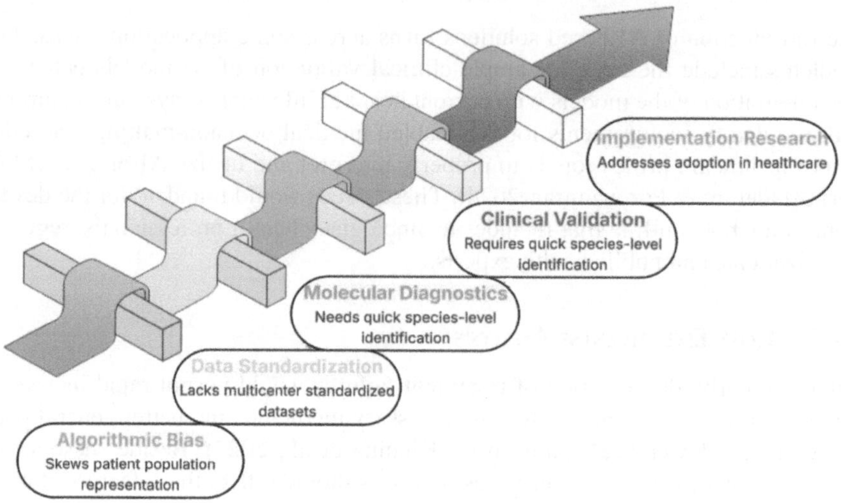

FIGURE 5.3 Tree diagram illustrates major ineffective AI application in mucormycosis management due to technical, ethical, and research limitations.

capture the natural progression of the disease (Kontoyiannis et al., 2012; Suganya et al., 2019). Second, and importantly, on the basis of present knowledge, a number of high-priority research directions have been identified to further the diagnosis and treatment of mucormycosis. First, what is urgently needed is the creation of multicentre, standardized datasets to train and validate the AI models with higher accuracy and generalizability. Platforms that integrate imaging analysis with molecular diagnosis of Mucorales species would also enhance diagnostic accuracy. Simultaneously, the identification of natural products with antifungal activity, driven by AI-assisted bioactivity prediction, is a compelling future direction for new therapeutic identification. Optimizing drug repurposing strategies to focus on mucormycosis specifically may expedite treatment alternatives (Garre, 2022). In addition, future clinical trials are needed to determine the actual in-clinic effect of AI-supported diagnosis and treatment on patient outcomes. Finally, practical issues in the adoption of these technologies in several healthcare environments must be addressed by implementation research. Furthermore, promising for studies on mucormycosis is the inclusion of AI and deep learning techniques in vaccine creation. Using predictive models, immunogenic epitope identification from Mucorales genomes may be fastly done, enabling the in silico generation of peptide-based vaccinations. Using extra specificity and effectiveness, these techniques can dramatically speed up vaccine development pathways (Malhotra et al., 2025b; Mishra et al., 2024).

5.9 CONCLUSION

Mucormycosis is still an infectious disease with fatal potential especially in individuals whose immune systems are compromised, even if its case mortality rate is high and diagnostics and treatment obstacles exist. Over the last ten years, the COVID-19 epidemic, rising incidence of diabetes, and more use of immunosuppressive medication have all contributed to the worldwide rising rate of mucormycosis. The evolution of rapid, molecular diagnostics, perhaps with the use of cell-free DNA analysis and CRISPR-Cas technology, can possibly lead to earlier and more precise detection, which is essential for enhancing patient outcomes (Garre, 2022). Future studies must establish these methods in the clinical setting and incorporate them into established diagnostic protocols.

The site and severity of infection play a large role in the prognosis of mucormycosis, though mortality remains too high, especially in pulmonary and disseminated forms (Suganya et al., 2019). Combination regimens (e.g., amphotericin B with azoles or echinocandins) are under investigation to enhance efficacy, though high-dose liposomal amphotericin B remains the drug of choice for monotherapy. New antifungal drugs and immunotherapies that act on ligands on the fungus (e.g., mucoricin and CotH proteins) are at an early stage of development and meet significant unmet needs. In addition, drug repurposing technology, AI-assisted virtual screening, and molecular docking have accelerated the identification of antifungal as well as novel ones. By lowering the time and cost of traditional medication development, we aid to save both time and cost (Yadav et al., 2024). The imperative to develop and execute these novel strategies is underscored by the global rise in mucormycosis, especially during the COVID-19 pandemic. Nevertheless, achieving their full potential requires overcoming significant technical, clinical, and ethical challenges through concerted, multidisciplinary efforts. Additionally, future studies must focus on cross-validation of these methodologies across a range of clinical environments, providing equitable access to resultant technologies, and ongoing refinement of methods to enhance performance. By matching the pattern recognition capabilities of computer science with the chemical diversity of natural products, researchers can develop more effective methods of addressing mucormycosis and reducing the risk of death for infected patients.

ACKNOWLEDGEMENT

The authors gratefully acknowledge Amity University, Noida for providing the necessary resources and academic environment that facilitated the successful completion of this chapter.

REFERENCES

Alexander, B. D., Lamoth, F., Heussel, C. P., Prokop, C. S., Desai, S. R., Morrissey, C. O., & Baddley, J. W. (2021). Guidance on imaging for invasive pulmonary Aspergillosis and Mucormycosis: From the Imaging Working Group for the Revision and Update of the Consensus Definitions of Fungal Disease from the EORTC/MSGERC. *Clinical Infectious Diseases*, 72(Supplement_2), S79–S88. https://doi.org/10.1093/CID/CIAA1855

Alqarihi, A., Kontoyiannis, D. P., & Ibrahim, A. S. (2023). Mucormycosis in 2023: An update on pathogenesis and management. *Frontiers in Cellular and Infection Microbiology*, 13, 1254919. https://doi.org/10.3389/FCIMB.2023.1254919

Arora, Y., Lalwani, H., Kumar, A., Mathur, P., Chaturvedi, N., & Mishra, V. (2025). AI-Driven antimicrobial peptides for drug development. In *Methods in Microbiology* (pp. 385–406). Elsevier. Aakansha Shrivastava and Vaibhav Mishra https://doi.org/10.1016/bs.mim.2024.08.003

Baldin, C., & Ibrahim, A. S. (2017). Molecular mechanisms of mucormycosis—The bitter and the sweet. *PLoS Pathogens*, 13(8), e1006408. https://doi.org/10.1371/JOURNAL.PPAT.1006408

Chaturvedi, N., Yadav, M. K., & Sharma, M. (2024). Applications of artificial intelligence and machine learning in microbial diagnostics and identification. In Srivastava, A., & Mishra, V. (Eds.), *Methods in Microbiology* (pp. 213–230). Elsevier. https://doi.org/10.1016/bs.mim.2024.05.013

Dharmani, K., Sinha, A., & Chaturvedi, N*. (2024). AI-driven microbial medicine development. In Srivastava, A., & Mishra, V. (Eds.), *Methods in Microbiology* (pp. 101–124). Elsevier. https://doi.org/10.1016/bs.mim.2024.05.008

Divakar, P. K. (2021). Fungal taxa responsible for mucormycosis/"Black Fungus" among COVID-19 patients in India. *Journal of Fungi*, 7(8), 641. https://doi.org/10.3390/jof7080641

Dogra, S., Arora, A., Aggarwal, A., Passi, G., Sharma, A., Singh, G., & Barnwal, R. P. (2022). Mucormycosis amid COVID-19 crisis: Pathogenesis, diagnosis, and novel treatment strategies to combat the spread. *Frontiers in Microbiology*, 12, 794176. https://doi.org/10.3389/fmicb.2021.794176

Garre, V. (2022). Recent advances and future directions in the understanding of mucormycosis. *Frontiers in Cellular and Infection Microbiology*, 12, 850581. https://doi.org/10.3389/fcimb.2022.850581

Gebremariam, T., Liu, M., Luo, G., Bruno, V., Phan, Q. T., Waring, A. J., Edwards, J. E., Filler, S. G., Yeaman, M. R., & Ibrahim, A. S. (2014). CotH3 mediates fungal invasion of host cells during mucormycosis. *Journal of Clinical Investigation*, 124(1), 237–250. https://doi.org/10.1172/JCI71349

Godoy, M., Viswanathan, C., Marchiori, E., Truong, M. T., Benveniste, M. F., Rossi, S., & Marom, E. M. (2012). The reversed halo sign: Update and differential diagnosis. *The British Journal of Radiology*, *85*(1017), 1226. https://doi.org/10.1259/BJR/54532316

Harishbhai Tilala, M., Kumar Chenchala, P., Choppadandi, A., Kaur, J., Naguri, S., Saoji, R., & Devaguptapu, B. (2024). Ethical considerations in the use of artificial intelligence and machine learning in health care: A comprehensive review. *Cureus*, 16(6), e62443. https://doi.org/10.7759/cureus.62443

Hasan Pour, B. (2025). Superficial fungal infections and artificial intelligence: A review on current advances and opportunities: REVISION. *Mycoses*, 68(1), e70007. https://doi.org/10.1111/myc.70007

Hashemi, S., Vosough, P., Taghizadeh, S., & Savardashtaki, A. (2024). Therapeutic peptide development revolutionized: Harnessing the power of artificial intelligence for drug discovery. *Heliyon*, 10(22), e40265. https://doi.org/10.1016/j.heliyon.2024.e40265

Hernández, J. L., & Buckley, C. J. (2023). Mucormycosis. *StatPearls*. https://www.ncbi.nlm.nih.gov/books/NBK544364/

Karthikeyan, S., Ramkumar, G., Aravindkumar, S., Tamilselvi, M., Ramesh, S., & Ranjith, A. (2022). A novel deep learning-based black fungus disease identification using modified hybrid learning methodology. *Contrast Media & Molecular Imaging*, 2022(1), 4352730. https://doi.org/10.1155/2022/4352730

Khanna, N. N., Maindarkar, M. A., Viswanathan, V., Fernandes, J. F. E., Paul, S., Bhagawati, M., Ahluwalia, P., Ruzsa, Z., Sharma, A., Kolluri, R., Singh, I. M., Laird, J. R., Fatemi, M., Alizad, A., Saba, L., Agarwal, V., Sharma, A., Teji, J. S., Al-Maini, M., Rathore, V., … Suri, J. S. (2022). Economics of artificial intelligence in healthcare: Diagnosis vs. treatment. *Healthcare*, 10(12), 2493. https://doi.org/10.3390/healthcare10122493.

Kontoyiannis, D. P., Lewis, R. E., Lotholary, O., Spellberg, B., Petrikkos, G., Roillides, E., Ibrahim, A., & Walsh, T. J. (2012). Future directions in mucormycosis research. *Clinical Infectious Diseases: An Official Publication of the Infectious Diseases Society of America*, 54(Suppl 1), S79. https://doi.org/10.1093/CID/CIR886

Kumar, Y., Koul, A., Singla, R., & Ijaz, M. F. (2023). Artificial intelligence in disease diagnosis: A systematic literature review, synthesizing framework and future research agenda. *Journal of Ambient Intelligence and Humanized Computing*, 14(7), 8459–8486. https://doi.org/10.1007/s12652-021-03612-z

Larcher, G., Dias, M., Razafimandimby, B., Bomal, D., & Bouchara, J. P. (2013). Siderophore production by pathogenic mucorales and uptake of deferoxamine B. *Mycopathologia*, 176(5–6), 319–328. https://doi.org/10.1007/S11046-013-9693-5.

Lax, C., Nicolás, F. E., Navarro, E., & Garre, V. (2024). Molecular mechanisms that govern infection and antifungal resistance in Mucorales. *Microbiology and Molecular Biology Reviews*, 88(1). https://doi.org/10.1128/mmbr.00188-22

Lee, J., Ma, S., Serban, N., & Yang, S. (2025). Accurate treatment effect estimation using inverse probability of treatment weighting with deep learning. *JAMIA Open*, 8(2), ooaf032. https://doi.org/10.1093/jamiaopen/ooaf032

Li, W., Ma, W., Yang, M., et al. (2024). Drug repurposing based on the DTD-GNN graph neural network: Revealing the relationships among drugs, targets and diseases. *BMC Genomics*, 25, 584. https://doi.org/10.1186/s12864-024-10499-5

Maleki Varnosfaderani, S., & Forouzanfar, M. (2024). The role of AI in hospitals and clinics: Transforming healthcare in the 21st century. Bioengineering, 11(4), 337. https://doi.org/10.3390/bioengineering11040337

Malhotra, Y., John, J., Yadav, D., Sharma, D., Vanshika, Rawal, K., … Chaturvedi, N.* (2025a). Advancements in protein structure prediction: A comparative overview of AlphaFold and its derivatives. *Computers in Biology and Medicine*, 188(109842), 109842. https://doi.org/10.1016/j.compbiomed.2025.109842

Malhotra, Y., Yadav, D., Chaturvedi, N., Gujar, A., John, R., & Ahmad, K. (2025b). AI in infectious disease diagnosis and vaccine development. In Srivastava, A., & Mishra, V. (Eds.), *Methods in Microbiology* (pp. 191–218). Academic Press. https://doi.org/10.1016/bs.mim.2024.12.005

Meister, S., Wermes, M., Stüve, J., & Groves, R. M. (2021). Cross-evaluation of a parallel operating SVM – CNN classifier for reliable internal decision-making processes in composite inspection. *Journal of Manufacturing Systems*, 60, 620–639. https://doi.org/10.1016/j.jmsy.2021.07.022

Mishra, V., Mishra, S. K., Srivastava, A., Dubey, C. K., Dharmani, K., & Chaturvedi, N. (2024). Leveraging artificial intelligence (AI) and machine learning (ML) for enhanced drug discovery and development from microbes. In Srivastava, A., & Mishra, V. (Eds.), *Methods in Microbiology* (pp. 125–140). Elsevier. https://doi.org/10.1016/bs.mim.2024.05.005

Motamedi, M., Golmohammadi, Z., Yazdanpanah, S., Saneian, S. M., & Shafiekhani, M. (2022). Epidemiology, clinical features, therapeutic interventions and outcomes of mucormycosis in Shiraz: An 8-year retrospective case study with comparison between children and adults. *Scientific Reports*, 12(1), 1–10. https://doi.org/10.1038/S41598-022-21611-8

Nambiar, M., Varma, S. R., & Damdoum, M. (2021). Post-Covid alliance-mucormycosis, a fatal sequel to the pandemic in India. *Saudi Journal of Biological Sciences*, 28(11), 6461–6464. https://doi.org/10.1016/j.sjbs.2021.07.004

Nielsen, M. C., Cerqueira, F. M., Kavuri, S. B., Raymond, C. M., Muneeb, A., Kudlicki, A. S., Tariq, S., Liu, M., Routh, A. L., Qiu, S., & Ren, P. (2023). Diverse clinical manifestations and challenges of mucormycosis: insights from serial cases. *Open Forum Infectious Diseases*, *10*(11). https://doi.org/10.1093/OFID/OFAD527

Nira, & Kumar, H. (2022). Epidemiological mucormycosis treatment and diagnosis challenges using the adaptive properties of computer vision techniques based approach: A review. *Multimedia Tools and Applications*, 81(10), 14217–14245. https://doi.org/10.1007/S11042-022-12450-W

Ocana, A., Pandiella, A., Privat, C., Bravo, I., Luengo-Oroz, M., Amir, E., & Gyorffy, B. (2025). Integrating artificial intelligence in drug discovery and early drug development: A transformative approach. *Biomarker Research*, 13(1), 45. https://doi.org/10.1186/s40364-025-00758-2

Pak, J., Tucci, V. T., Vincent, A. L., Sandin, R. L., & Greene, J. N. (2008). Mucormycosis in immunochallenged patients. *Journal of Emergencies, Trauma and Shock*, 1(2), 106. https://doi.org/10.4103/0974-2700.42203

Parente, R., Clark, S. J., Inforzato, A., & Day, A. J. (2017). Complement factor H in host defense and immune evasion. *Cellular and Molecular Life Sciences: CMLS*, 74(9), 1605–1624. https://doi.org/10.1007/s00018-016-2418-4

Pinheiro, E. A. A., Pina, J. R. S., Feitosa, A. O., Carvalho, J. M., Borges, F. C., Marinho, P. S. B., & Marinho, A. M. R. (2016). Bioprospecting of antimicrobial activity of extracts of endophytic fungi from *Bauhinia guianensis*. *Revista Argentina de Microbiología*, 49(1), 3–6. https://doi.org/10.1016/J.RAM.2016.08.005

Prakash, H., & Chakrabarti, A. (2019). Global epidemiology of mucormycosis. *Journal of Fungi*, 5(1), 26. https://doi.org/10.3390/JOF5010026

Rehman, A. U., Li, M., Wu, B., Ali, Y., Rasheed, S., Shaheen, S., … Zhang, J. (2024). Role of artificial intelligence in revolutionizing drug discovery. *Fundamental Research*. https://doi.org/10.1016/j.fmre.2024.04.021

Sharma, A., & Goel, A. (2022). Mucormycosis: risk factors, diagnosis, treatments, and challenges during COVID-19 pandemic. *Folia Microbiologica*, 67(3), 363–387. https://doi.org/10.1007/s12223-021-00934-5

Skiada, A., Lass-Floerl, C., Klimko, N., Ibrahim, A., Roilides, E., & Petrikkos, G. (2018). Challenges in the diagnosis and treatment of mucormycosis. *Medical Mycology*, 56(Suppl 1), S93. https://doi.org/10.1093/MMY/MYX101

Spellberg, B., Edwards, J., & Ibrahim, A. (2005). Novel perspectives on mucormycosis: Pathophysiology, presentation, and management. *Clinical Microbiology Reviews*, 18(3), 556. https://doi.org/10.1128/CMR.18.3.556-569.2005

Suganya, R., Malathi, N., Karthikeyan, V., & Janagaraj, V. D. (2019). Mucormycosis: A brief review. *Journal of Pure and Applied Microbiology*, 13(1), 161–165. https://doi.org/10.22207/JPAM.13.1.16

Tahiri, G., Lax, C., Cánovas-Márquez, J. T., Carrillo-Marín, P., Sanchis, M., Navarro, E., Garre, V., & Nicolás, F. E. (2023). Mucorales and mucormycosis: Recent insights and future prospects. *Journal of Fungi*, 9(3), 335. https://doi.org/10.3390/JOF9030335

Tanwar, M., Singh, A., Singh, T. P., Sharma, S., & Sharma, P. (2024). Comprehensive review on the virulence factors and therapeutic strategies with the aid of artificial intelligence against mucormycosis. *ACS Infectious Diseases*, 10(5), 1431–1457. https://doi.org/10.1021/acsinfecdis.4c00082

Tiwari, P., & Bae, H. (2022). Endophytic fungi: Key insights, emerging prospects, and challenges in natural product drug discovery. *Microorganisms*, 10(2), 360. https://doi.org/10.3390/microorganisms10020360

Wainwright, C. L., Teixeira, M. M., Adelson, D. L., Buenz, E. J., David, B., Glaser, K. B., … Wolfender, J.-L. (2022). Future directions for the discovery of natural product-derived immunomodulating drugs: An IUPHAR positional review. *Pharmacological Research: The Official Journal of the Italian Pharmacological Society*, 177(106076), 106076. https://doi.org/10.1016/j.phrs.2022.106076

Wan, Z., Sun, X., Li, Y., Chu, T., Hao, X., Cao, Y., & Zhang, P. (2025). Applications of artificial intelligence in drug repurposing. *Advanced Science*, 12(14), e2411325. https://doi.org/10.1002/advs.202411325

Wang, S. V., Schneeweiss, S., RCT-DUPLICATE Initiative, Franklin, J. M., Desai, R. J., Feldman, W., … Quinto, K. (2023). Emulation of randomized clinical trials with nonrandomized database analyses: Results of 32 clinical trials. *JAMA: The Journal of the American Medical Association*, 329(16), 1376–1385. doi:10.1001/jama.2023.4221.

Yadav, M. K., Dahiya, V., Tripathi, M. K., Chaturvedi, N., Rashmi, M., Ghosh, A., & Raj, V. S. (2024). Unleashing the future: The revolutionary role of machine learning and artificial intelligence in drug discovery. *European Journal of Pharmacology*, 985, 177103. https://doi.org/10.1016/j.ejphar.2024.177103

Yadav, M. K., Tripathi, M. K., Chaturvedi, N., Nath, A., & Kumar, U. (2025). Chapter 12 - Deep learning in predictive medicine: Current state of the art. In K. Raza (Ed.), *Deep Learning in Genetics and Genomics* (pp. 261–291). Elsevier. https://doi.org/10.1016/B978-0-443-27574-6.00008-4

Yang, F., Zhang, Q., Ji, X., Zhang, Y., Li, W., Peng, S., & Xue, F. (2022). Machine learning applications in drug repurposing. *Interdisciplinary Sciences, Computational Life Sciences*, 14(1), 15–21. https://doi.org/10.1007/s12539-021-00487-8

Part 2

Comprehending Pathobiology of Fungus

6 Genetic and Molecular Mechanisms Driving Antifungal Resistance in Emerging Fungal Species

Shumaiza Asif, Ashik Francis, Emmanuel Ijabani, and Anil Thakur

6.1 INTRODUCTION

Fungal pathogens, once relegated to opportunistic infections in severely immunocompromised hosts, are now increasingly recognized as major agents of morbidity and mortality across diverse patient populations. The increasing use of immunosuppression therapies, invasive procedures, broad-spectrum antibiotics, and ageing populations has led to invasive fungal infections (IFIs) becoming a critical, yet under-addressed component of global infectious disease burden (Bongomin et al., 2017; Brown et al., 2012). Conservative estimates suggest that fungal infections are responsible for over 1.5 million deaths annually, which are attributable to IFIs—more than those caused by tuberculosis or malaria—yet public health efforts and antifungal drug development remain severely lacking (Hawksworth & Lucking, 2017). Over the past decade multidrug-resistant (MDR) fungus has been increased rapidly such as Candida as Candida auris, Candida glabrata (now Nakaseomyces glabratus), Aspergillus fumigatus, Cryptococcus neoformans, Fusarium spp., and Trichosporon asahii (Perfect, 2017). Pathogens such as *Candida auris*, *Candida glabrata* (now *Nakaseomyces glabratus*), *Aspergillus fumigatus*, *Cryptococcus neoformans*, *Fusarium* spp., and *Trichosporon asahii* are no longer rare curiosities; they represent a growing clinical and public health emergency (Pfaller & Diekema, 2010; Wiederhold, 2017).

Among these, *C. auris* has garnered particular notoriety. First identified in 2009, it has since caused outbreaks across every inhabited continent, often with simultaneous resistance to all three major antifungal classes: azoles, echinocandins, and polyenes (Chowdhary et al., 2017). Alarmingly, *C. auris* exhibits intrinsic multidrug resistance, prolonged environmental persistence, and diagnostic challenges due to frequent misidentification by conventional methods (Jeffery-Smith et al., 2018). *C. glabrata*, while part of the normal microbiota, has emerged as a major threat due to its inherent low susceptibility to azoles and rapid acquisition of echinocandin resistance through hotspot mutations in *FKS* genes (Healey & Perlin, 2018). In parallel, *A. fumigatus* has become a focal point for environmental resistance selection. The widespread

DOI: 10.1201/9781003633952-8

use of triazole fungicides in agriculture has led to azole-resistant *A. fumigatus* strains disseminating globally — threatening not just immunocompromised patients but also those with chronic respiratory diseases (Verweij et al., 2016). In contrast, *C. neoformans*, an encapsulated basidiomycete yeast, continues to cause cryptococcal meningitis in patients with advanced HIV/AIDS, particularly in low-resource settings. While echinocandins are ineffective against *Cryptococcus*, adaptive mechanisms such as capsule enlargement, titan cell formation, and altered membrane lipid composition allow evasion of both host and drug defences (Zaragoza et al., 2010). The spectrum of antifungal drug resistance can be conceptualized through three overlapping categories:

- *Intrinsic resistance*, where certain fungal species are naturally refractory to specific antifungal classes (e.g., *C. glabrata* and *Candida krusei* to fluconazole, *Fusarium* spp. to amphotericin B);
- *Acquired resistance*, which evolves via point mutations, gene amplification, efflux pump upregulation, or chromosomal rearrangements under antifungal pressure; and
- *Clinical resistance*, a broader term capturing treatment failure due to subtherapeutic drug levels, bioavailability issues, or host immune dysfunction, despite laboratory susceptibility (White et al., 1998).

Mechanistically, antifungal resistance arises through a confluence of genetic mutations, epigenetic modulation, transcriptional rewiring, biofilm formation, and phenotypic plasticity. Advances in genomics have uncovered new complex regulatory networks and pathways by which fungi adapt to antifungal stress, including ergosterol pathway remodelling, efflux pump overexpression, chromatin remodelling, and heteroresistance phenomena (Cowen et al., 2001; Robbins et al., 2017). Epigenetic mechanisms, such as histone modification, aneuploidy, and non-coding RNA-mediated regulation, add another layer of rapid adaptability—often enabling transient, reversible resistance even in the absence of fixed genetic changes (Iracane et al., 2021; Rabaan et al., 2023). This chapter offers a comprehensive, mechanistic exploration of antifungal resistance in emerging human fungal pathogens, focusing on:

- Classical and novel antifungal drug targets;
- genetic and epigenetic reprogramming of resistance pathways;
- efflux pump systems and their regulation;
- morphological adaptations, such as yeast–hyphae transitions or capsule remodelling;
- and the contribution of biofilms and persister cells to drug recalcitrance.

In framing the issue from both molecular and evolutionary perspectives, we aim to underscore the urgent need for integrated solutions: improved diagnostics, global surveillance, innovative therapeutic strategies, and a reimagined pipeline of antifungals that target non-canonical, essential pathways in fungal physiology. Importantly, we also highlight the species-specific nuances in resistance strategies across key pathogens, which must inform both laboratory research and clinical management.

6.2 ANTIFUNGAL THERAPEUTICS: MECHANISMS OF ACTION AND EMERGING MOLECULAR TARGETS

6.2.1 Antifungal Drug Targets

The antifungal pharmacopeia, although limited compared to antibacterial agents, targets several conserved fungal cellular processes crucial for survival and virulence. Classic drug classes have primarily focused on pathways or structures unique or essential to fungi, such as ergosterol biosynthesis, fungal cell wall synthesis, and nucleic acid metabolism (Perfect, 2017; Whaley et al., 2016). However, the shared eukaryotic ancestry of fungi poses a considerable therapeutic challenge: these pathogens share many cellular pathways with their human hosts, constraining the therapeutic window and demanding exceptional selectivity. Consequently, resistance development—whether intrinsic, acquired, or clinical—often exploits this overlap, either through mutations that alter drug-target affinity or by compensatory pathway activation (Cowen et al., 2014; Robbins et al., 2016).

6.2.2 Molecular Mechanisms Underpinning Classic Antifungal Drug Classes

Ergosterol Biosynthesis Inhibitors (*Azoles*): Azoles remain the cornerstone of antifungal therapy due to their oral bioavailability and broad-spectrum activity. Their primary mode of action involves inhibition of lanosterol 14α-demethylase (*ERG11*), a cytochrome P450 enzyme pivotal for ergosterol synthesis, a sterol indispensable for fungal membrane integrity and function (Emami et al., 2017). Azole binding disrupts ergosterol production, leading to the accumulation of toxic sterol intermediates and membrane destabilization. However, mutations within *ERG11*, overexpression of efflux pumps (*e.g.*, *CDR1, MDR1*), and alterations in membrane composition have been extensively documented mechanisms for azole resistance, particularly in *C. glabrata* and *C. auris* (Berkow & Lockhart, 2017; Chowdhary et al., 2017; Healey et al., 2018).

β-1,3-Glucan Synthase Inhibitors (*Echinocandins*): Echinocandins target the fungal cell wall, a structure absent in humans, by inhibiting the β-1,3-glucan synthase complex—specifically the Fks1 subunit responsible for polymerizing glucan chains vital for cell wall strength and shape (Walker et al., 2010). This mechanism confers fungicidal activity against *Candida* species and fungistatic effects against *Aspergillus*. Resistance arises predominantly via hotspot mutations in *FKS* genes, reducing drug binding affinity without compromising enzyme function. The rapid emergence of echinocandin resistance in *C. glabrata* underscores the clinical significance of target site alterations and the need for vigilant susceptibility testing (Pham et al., 2014).

Membrane Disruptors (*Polyenes*): Polyenes, such as amphotericin B, bind directly to ergosterol in fungal membranes, creating transmembrane pores that cause ion leakage and cell death (Ahmady et al., 2024; Ghannoum & Rice, 1999). Their broad-spectrum efficacy and low resistance rates reflect the critical structural role of ergosterol. Nevertheless, toxicity and infusion-related side effects limit their clinical use. Resistance, though rare, can involve changes in ergosterol content or composition, effectively reducing polyene binding sites, as observed in some *Candida* and *Aspergillus* isolates (Mesa-Arango et al., 2012).

Nucleic Acid Analogues (*Flucytosine*): Flucytosine (5-FC) is converted intracellularly into 5-fluorouracil, which disrupts fungal DNA and RNA synthesis by inhibiting thymidylate synthase and incorporating into RNA (Vermes et al., 2000). Rapid resistance through mutations in cytosine deaminase or permease genes often occurs when used as monotherapy, necessitating combination with amphotericin B or azoles. Despite this limitation, 5-FC remains essential for cryptococcal meningitis treatment due to its CNS penetration and synergistic potential (Perfect et al., 2010).

6.2.3 Next-Generation Antifungal Agents and Novel Molecular Targets

In light of increasing resistance, toxic liabilities, and a paucity of druggable targets, recent antifungal innovation has pivoted towards first-in-class compounds and fungal-specific processes. Unlike traditional approaches, these novel strategies aim to bypass cross-resistance, mitigate toxicity, and expand the spectrum of action, especially against MDR pathogens like *C. auris*, *Fusarium*, and *Scedosporium*. Table 6.1 summarizes the key next-generation antifungals under clinical or preclinical development, along with their mechanisms and spectrum of activity.

These compounds circumvent traditional ergosterol or glucan targets, offering new avenues where legacy drugs fail. Table 6.2 outlines some of the most promising molecular targets explored in next-generation antifungal strategies, including fungal-specific pathways, stress adaptation mechanisms, and virulence regulators.

TABLE 6.1
Next-Generation Antifungal Agents and Their Molecular Targets

Drug	Target/Mechanism	Active against	Development Status
Olorofim (F901318)	Inhibits dihydroorotate dehydrogenase (pyrimidine biosynthesis)	*Aspergillus* spp., *Scedosporium* spp.	Phase III (du Pre et al., 2020)
Fosmanogepix (APX001)	Inhibits Gwt1, essential for GPI anchor biosynthesis	Broad spectrum, incl. *Candida auris*	FDA approved as Manogepix (2023) (Trzoss et al., 2019)
Ibrexafungerp (SCY-078)	Non-echinocandin glucan synthase inhibitor	*Candida* spp., incl. Echinocandin-resistant	FDA-approved (Quindos et al., 2022)
Rezafungin	Long-acting echinocandin analogue	*Candida* spp., *Aspergillus* spp.	FDA-approved for Candidemia (Sofjan et al., 2018)
ATI-2307	Selectively disrupts mitochondrial membrane potential	*Cryptococcus*, *Candida auris*, *Aspergillus fumigatus*	Preclinical (Mitsuyama et al., 2008)

TABLE 6.2
Novel Molecular Targets in Antifungal Drug Discovery

Target/Pathway	Biological Role in Fungi	Therapeutic Potential
Gwt1	Required for GPI-anchor attachment to cell wall proteins	Targeted by fosmanogepix; highly selective (Gebremariam et al., 2020)
Dihydroorotate dehydrogenase (DHODH)	Rate-limiting enzyme in pyrimidine biosynthesis	Targeted by olorofim; no human cross-reactivity (Vanbiervliet et al., 2024)
Inositol phosphorylceramide (IPC) synthase	Central to fungal sphingolipid biosynthesis	Not present in humans; ideal selectivity (Zhong et al., 2000)
Histone deacetylases (HDACs)	Regulate fungal stress responses and morphogenesis	Epigenetic modulators under study (Garnaud et al., 2016)
Calcineurin–Hsp90 axis	Crucial for stress adaptation, antifungal tolerance, and biofilms	Synergistic inhibition with azoles or echinocandins (Cowen et al., 2009)whose efficacy is compromised by host toxicity, a limited activity spectrum, or the emergence of drug resistance. We previously established that the molecular chaperone Hsp90 enables the emergence and maintenance of fungal drug resistance. For the most prevalent fungal pathogen of humans, Candida albicans, Hsp90 mediates resistance to azoles, which inhibit ergosterol biosynthesis and are the most widely deployed antifungals in the clinic. For the emerging opportunistic pathogen Aspergillus terreus, Hsp90 is required for basal resistance to echinocandins, which inhibit β(1, 3
Transcriptional regulators (e.g., TAC1, MRR1, and UPC2)	Govern expression of efflux pumps and ergosterol genes	Modulating these could prevent resistance emergence (Hill et al., 2013)

These targets represent a paradigm shift towards "anti-virulence" therapies that impair pathogenicity without imposing direct fungicidal pressure—an approach expected to reduce resistance evolution. The antifungal armamentarium, though effective in many settings, is clearly under siege from rising resistance and toxicity constraints. The past decade has witnessed a promising resurgence in antifungal research, with next-generation agents targeting fungal-specific enzymes, mitochondrial function, and stress adaptation mechanisms. Unlike earlier efforts that mimicked antibacterial paradigms, current innovations embrace fungal biology's unique vulnerabilities—from epigenetics to cell wall assembly. Sustained investment in these novel strategies, paired with stewardship and diagnostic refinement, will be critical to future-proofing antifungal therapy.

6.3 MOLECULAR BASIS OF ANTIFUNGAL RESISTANCE IN EMERGING FUNGAL PATHOGENS

6.3.1 Overview

The rapid emergence and global spread of MDR fungal pathogens pose a significant threat to public health, especially in immunocompromised patients. Species such as *C. auris*, *A. fumigatus*, *Fusarium* spp., and *Scedosporium* spp. exhibit a spectrum of resistance mechanisms complicating clinical management and resulting in increased morbidity and mortality (Chowdhary et al., 2017). Unlike classical fungal pathogens, these emerging species demonstrate both intrinsic and acquired resistance mechanisms shaped by environmental and clinical antifungal pressures (Perfect, 2017; Robbins et al., 2016). Given the limited arsenal of antifungal drugs and the eukaryotic similarities between fungi and humans, resistance evolution is particularly problematic. This section provides an in-depth analysis of molecular resistance mechanisms, emphasizing genetic mutations, drug efflux systems, transcriptional regulation, and compensatory biochemical pathways that enable survival under antifungal stress. Although biofilm-mediated resistance significantly contributes to clinical persistence, it is discussed in a separate section.

6.3.2 Azole Resistance Mechanism

Azole antifungals target the cytochrome P450 enzyme lanosterol 14α-demethylase (*ERG11* in yeasts, *Cyp51A/B* in moulds), disrupting ergosterol synthesis critical for fungal membrane integrity and function (Rosam et al., 2020). Resistance to azoles is among the most extensively studied and clinically relevant mechanisms in emerging fungal pathogens.

6.3.3 Target Site Mutations

In *C. auris*, point mutations in *ERG11*, notably Y132F and K143R, have been identified as major contributors to azole resistance (Chowdhary et al., 2017; Healey & Perlin, 2018; Jangir et al., 2023). These mutations alter the azole binding site, reducing drug affinity and leading to elevated minimum inhibitory concentrations (MICs). Whole-genome sequencing of clinical isolates confirms clade-specific patterns of these mutations, suggesting independent evolutionary events under antifungal selection pressure (Berkow & Lockhart, 2017). Similarly, azole resistance in *A. fumigatus* is primarily driven by mutations in the *Cyp51A* gene. Tandem repeats in the promoter region combined with coding sequence mutations—specifically TR34/L98H and TR46/Y121F/T289A—have been linked to high-level azole resistance worldwide (Verweij et al., 2016). These mutations alter enzyme conformation and expression, decreasing azole binding and increasing ergosterol biosynthesis despite drug presence (Kumar et al., 2024a).

In *Fusarium* spp. and *Scedosporium* spp., mechanisms are less well characterized, but clinical isolates exhibit decreased azole susceptibility associated with point mutations in *Cyp51* paralogues and potential overexpression (Vermeulen et al., 2022).

6.3.4 Efflux Pumps and Transcriptional Regulation

Efflux-mediated drug extrusion is a conserved and potent resistance mechanism among emerging fungi. In *C. auris* and other *Candida* species, ATP-binding cassette (ABC) transporters such as CDR1 and major facilitator superfamily (MFS) transporters like *MDR1* actively export azole drugs, reducing intracellular concentrations below fungicidal levels (Healey & Perlin, 2018; Prasad & Rawal, 2014). Transcriptional regulators such as *TAC1* (regulating CDR pumps), *MRR1* (regulating MDR pumps), and *UPC2* (regulating ergosterol biosynthesis genes) orchestrate efflux pump expression and ergosterol pathway components, enabling dynamic resistance modulation in response to antifungal exposure (Lee et al., 2021; Schubert et al., 2011). Notably, gain-of-function mutations in these transcription factors have been documented in clinical isolates of *C. glabrata* and *C. auris*, underpinning multidrug resistance phenotypes (Healey et al., 2018).

In *A. fumigatus*, while efflux pumps contribute to azole resistance, their role is less dominant than target enzyme mutations. However, the ABC transporter *Cdr1B* has been implicated in resistance, particularly in environmental isolates exposed to agricultural azoles (Fraczek et al., 2013).

6.3.5 Compensatory and Alternative Pathways

Fungi can also develop resistance through compensatory alterations in ergosterol biosynthesis and membrane composition. For instance, mutations or downregulation of *ERG3* (C-5 sterol desaturase) in *C. albicans* and possibly *C. auris* can bypass toxic sterol accumulation caused by azole inhibition, mitigating drug effects. Although direct evidence in *C. auris* remains limited, transcriptional profiling suggests similar mechanisms may operate. Alternative sterol pathways or increased synthesis of sterol intermediates may also confer a survival advantage under azole pressure, though these pathways are poorly understood and warrant further investigation (Berkow & Lockhart, 2017; Song et al., 2025).

6.3.6 Echinocandin Resistance: Target Site Mutations and Adaptations

Echinocandins, as noncompetitive inhibitors of β-1,3-glucan synthase, disrupt fungal cell wall biosynthesis, primarily targeting the enzyme complex encoded by *FKS* gene family members (*FKS1, FKS2*, and less frequently *FKS3* in some species). Resistance to echinocandins is almost exclusively linked to point mutations in two highly conserved "hotspot" regions within *FKS* genes, which alter the conformation of the glucan synthase catalytic subunit and reduce drug binding affinity without severely impairing enzyme function (Fraczek et al., 2013; Healey et al., 2018)

In *C. glabrata*, a species notorious for multidrug resistance, mutations such as S629P and f659del in *FKS2* and *FKS1* mutations (e.g., S629P) confer elevated MICs to caspofungin, micafungin, and anidulafungin (Arendrup & Perlin, 2014). These mutations have been correlated with increased echinocandin treatment failures in clinical settings, representing a significant public health concern. *C. auris* presents a parallel, yet alarming, pattern. Molecular surveillance reveals that a substantial proportion of clinical *C. auris* isolates harbour mutations in *FKS1*, particularly S639F/P substitutions within hotspot 1, resulting in reduced susceptibility or resistance to all echinocandins (Chowdhary et al., 2018; Kordalewska et al., 2018). The rapid emergence and clonal spread of these resistant strains underscore the limited antifungal arsenal available.

In moulds such as *A. fumigatus*, echinocandin resistance mechanisms remain less elucidated. While FKS-homologous mutations are rare or not well documented, echinocandin resistance appears to be associated with compensatory increases in chitin synthesis and cell wall remodelling mediated by signalling pathways like the cell wall integrity (CWI) MAPK cascade (Arendrup & Perlin, 2014; Wiederhold, 2018). This phenotype leads to a paradoxical "paradoxical effect" or "Eagle effect," where fungal growth resumes at higher echinocandin concentrations, complicating therapeutic decisions (Walker et al., 2008).

Diagnostic challenges arise because conventional susceptibility testing often fails to detect resistance at early stages due to slow growth or heteroresistance, while molecular assays targeting FKS hotspot mutations, though promising, lack standardized protocols across clinical laboratories (Healey & Perlin, 2018). The evolving landscape mandates the incorporation of rapid genotypic tests and close surveillance to mitigate clinical failures and resistance spread.

6.3.7 Polyene Resistance and Membrane Remodelling

Polyenes, principally amphotericin B, kill fungi by binding ergosterol in the plasma membrane, forming ion-permeable channels that disrupt membrane integrity, leading to leakage of intracellular contents and cell death (Kristanc et al., 2019). Polyene resistance, while historically uncommon due to the critical role of ergosterol in fungal viability, has been increasingly documented in certain emerging pathogens, demanding a nuanced understanding of underlying molecular adaptations. Resistance mechanisms primarily involve mutations or regulatory changes in ergosterol biosynthesis pathway genes such as *ERG3*, *ERG6*, and *ERG11*, resulting in quantitative and qualitative alterations in membrane sterol composition (Kanafani & Perfect, 2008; Mesa-Arango et al., 2012). For instance, *ERG3* loss-of-function mutations yield accumulation of sterol intermediates (e.g., fecosterol) that exhibit reduced amphotericin B binding affinity, thereby diminishing drug efficacy without completely abolishing ergosterol biosynthesis, which would be lethal (Tanwar et al., 2024).

In *Candida lusitaniae* and emerging *C. auris* isolates, documented polyene resistance correlates with such altered sterol profiles (Czajka et al., 2023). In moulds such as *Aspergillus terreus*, intrinsic polyene resistance is linked to unique sterol content and biophysical membrane characteristics that limit amphotericin B binding and pore formation (Revankar & Sutton, 2010). At the biophysical level, changes in membrane fluidity, lipid raft distribution, and ergosterol accessibility modulate polyene insertion and channel stability, reducing fungicidal activity (Bieberich, 2018). Such membrane remodelling can be induced by environmental stressors, highlighting the plasticity of fungal membrane composition as an adaptive strategy.

Clinically, polyene resistance complicates the treatment of invasive fungal infections, especially in the context of MDR pathogens where therapeutic alternatives are scarce (Perfect, 2017). These findings underscore the need for diagnostic methods that can detect sterol composition changes and pharmacodynamic models that incorporate membrane biophysics for predicting amphotericin B efficacy.

6.3.8 Stress Response Networks, Epigenetics Regulation, and Resistance Modulation

The complexity of antifungal resistance transcends genetic mutations, encompassing dynamic stress response networks and epigenetic regulation that collectively enable fungi to withstand antifungal assault and environmental stresses. Central to this network is the molecular chaperone Hsp90, which stabilizes multiple client proteins critical for fungal cell signaling and survival, including the calcium/calmodulin-dependent phosphatase calcineurin. The Hsp90–calcineurin axis orchestrates stress responses that promote survival under drug pressure by maintaining cell wall integrity, regulating morphogenetic transitions, and modulating antifungal tolerance (Cowen & Lindquist, 2005; Singh et al., 2009). Pharmacological inhibition of Hsp90 or calcineurin dramatically sensitizes resistant strains to echinocandins and azoles, providing a promising combinatorial therapeutic strategy (Robbins et al., 2016).

Epigenetic modifications—such as histone acetylation, methylation, and DNA methylation—affect chromatin architecture and gene expression, facilitating phenotypic plasticity without altering the underlying genome. In *C. albicans*, epigenetically mediated phenotypic switching (white–opaque transition) is well characterized and influences virulence and antifungal susceptibility (Braun & Johnson, 2000). Though less studied in emerging pathogens like *C. auris*, preliminary evidence suggests that similar epigenetic mechanisms may enable rapid adaptation and transient tolerance (Zhang et al., 2025). Furthermore, transcriptional regulators, including zinc cluster proteins like *TAC1* and *MRR1*, drive the overexpression of multidrug efflux pumps (*CDR1*, *MDR1*), while *UPC2* regulates ergosterol biosynthesis genes, collectively contributing to resistance phenotypes (Morschhauser, 2016). These regulators integrate environmental cues with cellular metabolism, underscoring resistance as a multifaceted, regulated phenotype rather than a fixed trait. Targeting these stress response pathways and epigenetic modifiers could curtail resistance evolution and enhance antifungal efficacy. However, comprehensive mechanistic studies, especially in non-albicans emerging fungi, remain sparse, representing a critical knowledge gap (Table 6.3).

TABLE 6.3
Molecular Basis of Established Resistance Mechanism as Seen in Various Emerging Fungal Species

Resistance Mechanism	Fungal Species/ Pathogens	Molecular/Genetic Basis	Phenotypic Effect	Clinical/ Diagnostic Implications
Echinocandin Resistance	*Candida glabrata, Candida auris*	Point mutations in *FKS1* and *FKS2* hotspot regions (e.g., S639F in *Candida auris*) (Doorley et al., 2025; Szymankiewicz et al., 2021).	Elevated MICs; reduced drug binding	Treatment failure risk; molecular tests needed
	Aspergillus fumigatus	Cell wall remodelling, increased chitin synthesis (Satish & Perlin, 2019).	Paradoxical growth at high echinocandin doses	Diagnostic challenges; phenotypic resistance hard to detect
Polyene resistance	*Candida auris, Candida lusitaniae*	Mutations in *ERG3* and *ERG6*, altered ergosterol content (Eliaš et al., 2024).	Reduced amphotericin B binding	Reduced drug efficacy; diagnostic sterol profiling recommended
	Aspergillus terreus	Unique sterol composition and membrane biophysics (Posch et al., 2018).	Intrinsic resistance	Limits amphotericin B use in invasive aspergillosis
Stress Response & Regulatory	Various emerging fungi	Hsp90-calcineurin signalling, epigenetic modifications, transcription factors (*TAC1*, *MRR1*, and *UPC2*) (Iyer et al., 2022; Segéňová et al., 2025).	Antifungal tolerance, phenotypic plasticity	Potential targets for adjunct therapy; complex regulation complicates resistance management

6.4 EPIGENETIC MECHANISMS AND THEIR IMPACT ON DRUG RESISTANCE

Antifungal resistance is increasingly understood not as a static consequence of genetic mutations alone, but as a dynamic process shaped by epigenetic regulation, enabling fungal pathogens to swiftly adapt to environmental pressures, including antifungal drugs. Epigenetics refers to heritable changes in gene expression that occur without alterations in the DNA sequence, chiefly through chromatin remodelling, histone modifications, DNA methylation (although less prominent in fungi), and non-coding RNAs. This layer of regulation orchestrates rapid phenotypic shifts such as transient resistance, heteroresistance, and tolerance—phenomena frequently observed in emerging MDR species like *C. auris, C. albicans, C. neoformans*, and *A. fumigatus*.

6.4.1 Histone Modifications and Chromatin Dynamics

Post-translational modifications (PTMs) of histone proteins—including methylation, acetylation, phosphorylation, ubiquitination—govern nucleosome structure and accessibility, thus controlling transcriptional activity of resistance-associated genes.

- *Histone acetylation and deacetylation*: Acetylation of histone H3 and H4 tails by histone acetyltransferases (HATs) neutralizes positive charges, loosening DNA–histone interactions and facilitating transcription. In *C. albicans*, increased histone H3 acetylation at promoter regions of *CDR1* (an ABC transporter) and *ERG11* (lanosterol demethylase) is linked to azole resistance. Conversely, histone deacetylases (HDACs), such as Hda1 and Rpd3, repress these genes by removing acetyl groups, thus HDAC inhibitors like trichostatin A (TSA) sensitize resistant strains by reversing transcriptional repression (O'Kane et al., 2020; Zhang et al., 2025).
- *Histone methylation*: Methylation of histone H3 lysine residues (e.g., h3k4me3 and h3k36me3) is associated with active transcription, while H3K9 and H3K27 methylation generally mediate repression. *C. neoformans* exhibits H3K4 trimethylation enrichment at efflux pump and ergosterol biosynthesis genes under fluconazole exposure, promoting drug resistance. Intriguingly, H3K9 methylation marks can create facultative heterochromatin domains that silence antifungal target genes, contributing to reversible drug tolerance (Rai & Rai, 2024).
- *Crosstalk and combinatorial histone codes*: Resistance phenotypes often involve interplay between acetylation and methylation marks. For instance, in *C. albicans*, mutations in the Set1 histone methyltransferase reduce H3K4 methylation, dampening expression of multidrug resistance genes and increasing azole susceptibility, illustrating how combinatorial PTMs tune the epigenetic landscape (Baker et al., 2022).
- *Chromatin remodelling complexes*: ATP-dependent remodellers like SWI/SNF, RSC, and ISWI dynamically reposition nucleosomes to regulate access to promoter regions. In *C. glabrata*, the SWI/SNF complex is required for the activation of PDR1-regulated ABC transporters under azole stress; loss of SWI/SNF subunits decreases pump expression and restores azole susceptibility (Kumar et al., 2024b). Similarly, in *A. fumigatus*, chromatin remodellers influence transcriptional responses to azoles by modulating promoter accessibility of cyp51a and other resistance-associated genes (Kuhbacher et al., 2022).

6.4.2 Chromosomal Instability, Aneuploidy, and Copy Number Variation

Fungal genomes demonstrate remarkable plasticity via aneuploidy and copy number variation (CNV), serving as epigenetic-like mechanisms enabling rapid resistance evolution without point mutations.

- *Aneuploidy as a reversible resistance mechanism:* In *C. albicans*, trisomy of chromosome 5 is repeatedly selected during fluconazole treatment. This chromosome harbours *ERG11* and the transcription factor *TAC1*, whose

amplification increases azole target and efflux pump expression (Anderson et al., 2017). This gain is reversible, allowing cells to shed extra chromosomes when drug pressure wanes, minimizing fitness costs.
- *Segmental duplications and isochromosomes:* Formation of isochromosome 5L, duplicating the left arm of chromosome 5, also confers azole resistance. Similar segmental duplications have been detected in *C. auris*, involving chromosomal regions with *FKS1* and *ERG* genes, indicating parallel resistance strategies (Narayanan et al., 2022).
- *Heteroresistance and subpopulation dynamics:* Aneuploidy contributes to heteroresistance, where only a subpopulation within a fungal culture exhibits resistance. This transient genomic variation complicates diagnostics and treatment because the resistant subpopulation can expand rapidly under antifungal exposure (Su et al., 2025).

6.4.3 Therapeutic Implications and Future Directions

Epigenetic plasticity complicates conventional antifungal strategies by masking resistance and allowing rapid adaptive shifts. However, it also unveils novel therapeutic vulnerabilities:

- *Epigenetic modifiers as adjunctive therapies*: HDAC inhibitors like vorinostat and TSA potentiate azole efficacy by disrupting chromatin states that maintain resistance gene expression. Similarly, inhibitors targeting methyltransferases and chromatin remodellers are under preclinical evaluation.
- *Diagnostics*: Epigenetic markers may serve as biomarkers for early detection of resistance states, enabling personalized antifungal regimens.
- *Integrated multi-omics approaches*: Combining epigenomic mapping (chip-seq for histone marks), transcriptomics, and genome sequencing is vital to decode the multilayered resistance networks in clinical isolates.

The interplay between genetic mutations and epigenetic regulation in fungi is a sophisticated evolutionary strategy, allowing rapid, reversible adaptation to antifungal pressure while preserving genomic integrity. Deciphering these mechanisms is critical to developing durable antifungal therapies against MDR pathogens like *C. auris* and *A. fumigatus.*

6.5 EFFLUX PUMPS: KEY MEDIATORS OF ANTIFUNGAL DRUG RESISTANCE

6.5.1 Overview of Efflux Pump Systems in Fungi

Efflux pumps were initially known for their crucial role in nutrient uptake in organisms. However, research have unveiled their broader role in drug resistance particularly in pathogenic bacteria and fungi. These transporters are involved in the active transportation of unwanted molecules out of the cell such as drugs, heavy metals, and organic compounds. The overexpression of efflux pumps significantly contributes to antifungal resistance in *Candida*. Two major families involved are ABC transporters and MFS pumps. Genes such as *CDR1* and *CDR2* in *C. albicans*, along with MDR1 in *C. glabrata*, enhance drug efflux capabilities, thereby reducing intracellular drug concentrations (Rogers & Barker, 2002). This phenomenon is well pronounced in resistance towards mainly the azoles. In contrast, efflux-mediated resistance mechanisms are not involved in adaptive resistance to echinocandins or polyenes, as they are poor pump substrates (Cannon et al., 2009; Lee et al., 2021). Fungal cells possess two major families of drug efflux transporters that transport antifungal agents and other xenobiotics out of the cell; the ABC transporters, which utilize the energy of ATP hydrolysis, and the MFS transporters, which generally use the proton motive force to drive efflux. Both families contribute to multidrug resistance. In *C. albicans*, for example, azole-resistant isolates typically overexpress the ABC pump Cdr1p (and its paralog Cdr2p) and/or the MFS pump Mdr1p, resulting in rapid efflux of the drug and reduced intracellular accumulation (Prasad & Rawal, 2014). Structurally, fungal ABC transporters have two nucleotide-binding domains and two transmembrane domains, whereas MFS pumps are single-polypeptide secondary transporters.

In general, both ABC and MFS efflux pumps are embedded in the plasma membrane and reduce drug concentration by actively exporting drugs out of the cell, and elevated expression or activity of either of the pumps lowers intracellular drug levels, thereby mediating antifungal tolerance and permitting fungal survival under antifungal therapy.

6.5.2 Major Efflux Pumps Implicated in Antifungal Resistance

C. albicans and non-*albicans* Candida alongside other emerging fungal pathogens utilize different classes of drug transporters to efflux drugs out of their cells. In *C. albicans*, the best characterized efflux pumps are the two ABC transporters Cdr1p and Cdr2p, as well as the MFS transporter Mdr1p. Overexpression of Cdr1p/Cdr2p or Mdr1p is a hallmark of clinical azole resistance. The upregulation of *CDR1* and CDR2 (ABC pumps) is frequently observed in azole-resistant isolates and is a leading cause of multidrug resistance (Jain et al., 2023). Similarly, gain-of-function activation of MDR1 (an MFS efflux pump) underlies many cases of fluconazole resistance (Lohberger et al., 2014). *C. glabrata* has been shown to harbour orthologous ABC transporters (cgcdr1, cgcdr2) under the control of the *PDR1* regulator. *PDR1* gain-of-function mutations drive overexpression of CgCDR pumps and high-level azole resistance. Other species of non-*albicans Candida* like *Candida parapsilosis* and *Candida tropicalis* similarly express MDR-type pumps that contribute to resistance. Notably, the emerging MDR *C. auris* has *CDR1* and *MDR1* homologues that are often upregulated in resistant isolates (efflux-mediated resistance) (Duggan & Usher, 2023). In *Cryptococcus* species, three ABC pumps, namely, Afr1, Afr2, and Mdr1, have been identified in both *C. neoformans* and *Cryptococcus gattii*, and all the three pumps can export multiple azole drugs (Basso et al., 2015). Baso and colleagues also showed that Afr1 particularly plays a dominant role in fluconazole efflux. In their work, they have demonstrated that *C. neoformans* cells overexpressing *AFR1, AFR2*, or *MDR1* have much higher azole MICs than controls, and deletion of *AFR1* resulted in a dramatic increase in drug susceptibility.

6.5.3 Regulation of Efflux Pumps Expression

The expression of efflux pump genes is tightly regulated by specific transcription factors. These regulations often occur through gain-of-function mutations in resistant clinical strains. The ABC transporter genes such as Cdr1 and Cdr2 in *C. albicans* are mainly regulated by Tac1, whereas others like Cdr4 of *C. glabrata* is regulated by Pdr1p. In C. *albicans* and also non-*albicans Candida*, the Zn(II)2Cys6 transcription factor Tac1 activates the ABC transporter genes CDR1 and *CDR2* (in *C. albicans*). Gain-of-function mutations in TAC1 have been shown to induce constitutive overexpression of Cdr1p/Cdr2p (Li et al., 2024; Lohberger et al., 2014). Similarly, the MFS efflux gene *MDR1* is regulated by Mrr1p. Mrr1 is also a Zn(II)2Cys6 transcription factor, which is also activated through gain-of-function mutations. These activating mutations in MRR1 have been shown to result in permanent upregulation of Mdr1p (Schneider & Morschhauser, 2015). In a recent report by Franconi et al. (2025), mutations in *C. parapsilosis* Mrr1 (cpMrr1) are responsible for fluconazole resistance.

The third important factor in the expression of efflux pumps in fungi is Upc2p, which is primarily known to regulate sterol biosynthesis genes and also impact azole resistance, as mentioned earlier (Li et al., 2024; Lohberger et al., 2014; Song et al., 2025; Vasicek et al., 2014). Based on recent reports by Li and colleagues on the role of Upc2 in *C. auris*, Upc2 has been shown to upregulate Mdr1 directly or indirectly via activation of Mrr1, a key regulator of Mdr1 transporters. Additionally, while Upc2 hyperactivation has been shown to significantly increase TAC1b expression, the expression of *CDR1* remains unchanged. This suggests that its role in the upregulation of *CDR1* is minimal.

In general, *TAC1* controls *CDR1/CDR2*, *MRR1* controls *MDR1*, and *UPC2* controls *ERG11*, and gain-of-function alleles of these factors are commonly found in resistant clinical strains. For example, *C. albicans* with hyperactive *TAC1* alleles show high *CDR1/2* expression even without drug, and those with *MRR1* mutations overexpress *MDR1*. Another notable feature of efflux pump regulators is their possession of the xenobiotic-binding domain (usually in the middle homology region, e.g., in Tac1). This domain allows them to sense and respond to drugs or other chemicals (Jain et al., 2023). Thus, exposure to azoles or other xenobiotics can activate transcription factors and induce *CDR1/2* and/or *MDR1* expression. Mrr1 has been shown to mediate both basal and inducible *MDR1* expression. Other fungicidal chemicals like benomyl or peroxide have also been shown to trigger MDR1 induction via Mrr1 (Duggan & Usher, 2023).

6.5.4 Detection and Quantification of Efflux Activity for Identification of Drug-Resistant Fungi

Efflux pump activity is not restricted to the removal of drugs; a wide range of other xenobiotics including some dyes used in the laboratory can serve as substrates for these transporters. Dye assays are widely used in the determination of efflux activity. Examples of the most commonly used dyes are Nile red and Rhodamine 6G. Nile red is a lipophilic dye that is usually pumped out of the cell by Cdr1/Cdr2 and Mdr1p, whereas Rhodamine 6G is a classical ABC pump substrate. Therefore, the accumulation of dyes in cells inversely correlates to their efflux activity. The process normally involves incubating cells with dye while de-energized in order to allow dye uptake and accumulation and then re-energizing them by introducing glucose. Then, fluorescent levels in the cells are quantified using spectrofluorometry, flow cytometry, or fluorescence microscopy. A decrease in fluorescence indicates efflux activity, and more dye retention reflects less efflux activity. These dyes can also be used to determine the efficiency of inhibitors that block these transporters. One example of such a study was performed by Singh et al. (2018), where they used R6G and Nile red to validate their claim of the fungicidal nature of geraniol. Their results confirmed its effect on cacdrlp, as evidenced by Nile red and Rhodamine 6G accumulation in the cells following treatment with geraniol. Quantification of efflux activity can also be performed at the gene level through quantitative PCR (qPCR) and RNA sequencing. Elevated mRNA levels of the transporters in resistant isolates relative to susceptible controls are commonly detected by qPCR. Genome-wide RNAseq can also identify upregulation of multiple transporters in resistant strains. These methods complement functional assays, confirming which pump genes are overexpressed in a given isolate (Figure 6.1).

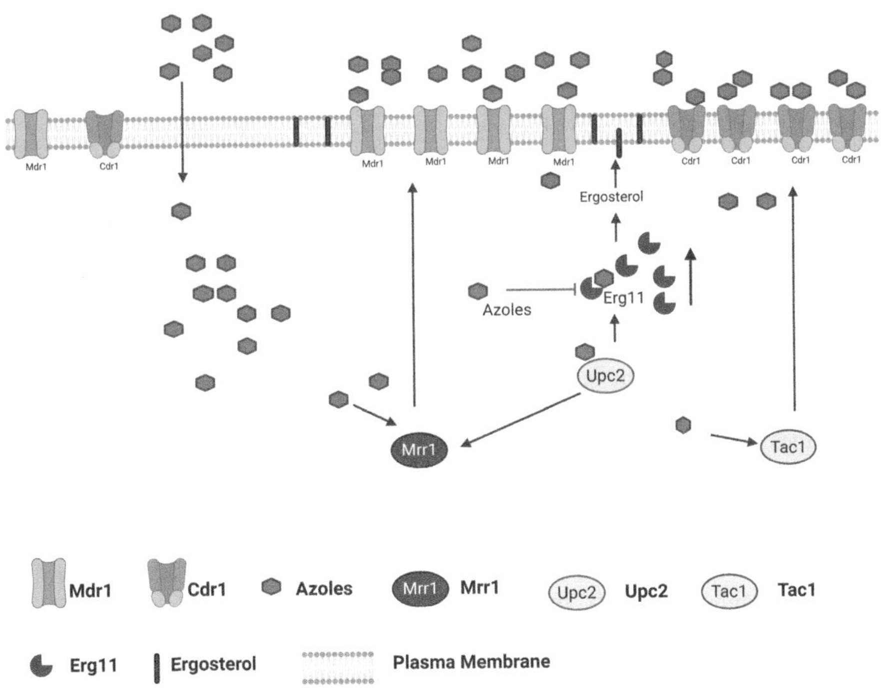

FIGURE 6.1 Key regulators of efflux pumps activity under drug conditions; specifically, the azoles. The azoles are sensed in the cellular environment by Mrr1 which upregulates expression of Mdr1. Mrr1 can be activated directly by azoles or indirectly by Upc2 as seen in C. auris. Upc2 (a key regulator of ergosterol biosynthesis) senses depletion of ergosterol and upregulates the expression of Erg11. Tac1 senses azoles and upregulates the expression of Cdr1. Both transporters extrude azoles out of the cell thereby conferring resistance.

6.6 ADAPTIVE MEMBRANE DYNAMICS IN ANTIFUNGAL RESISTANCE

6.6.1 Brief Overview of Fungal Membrane and Function

The plasma membrane in living systems is quite critical as it plays crucial role in the cell's integrity and survival. The membranes' major function includes maintaining all homeostasis, signal transmission, facilitating material exchange, and providing resistance to environmental stimuli such as osmotic pressure, temperature fluctuations, and drug resistance, among others. It also provides organized matrix for enzymes pathways involved in the biosynthesis of cell components (Stewart, 2017). Just like other eukaryotes, all fungi possess a plasma membrane composed of a lipid bilayer enriched with lipids including glycerophospholipids, sphingolipids, and sterols with embedded proteins (Sant et al., 2016). The phospholipids include phosphatidyl inositol, phosphatidyl serine, phosphatidyl choline, and phosphatidyl ethanolamine. The glycerophospholipids constitute about 55%–75% of total lipids, sphingolipids constitute 7%–16% and the sterols, and majorly, ergosterol constitutes about 30%–40% of the fungal membranes (Mehta et al., 2023). The major sterol component is ergosterol. Ergosterol maintains the integrity and fluidity of the fungal membranes. The containment of ergosterol by fungal membrane serves as a unique feature that makes them distinguishable from other cells. This makes important targets for antifungal drugs. The protein component of the fungal plasma membrane is in approximately equal proportion with the lipid component, and most of them are assumed to be functional not structural. Fungal membrane proteins are responsible for diverse cellular activities; notable examples are those involved in regulation of solute transport across the membrane, signalling and communication, and cell wall synthesis (Stewart, 2017).

6.6.2 Membrane Components That Serve as Targets for Antifungals

The fungal cell membrane is a crucial target for several classes of antifungal agents, especially due to its unique lipid composition that distinguishes it from mammalian membranes. The most widely used antifungals targeting the fungal membrane are the polyenes (such as amphotericin B), azoles (such as fluconazole), and allylamines (such as terbinafine) (Bhattacharya et al., 2018). These agents primarily disrupt the structure and function of the fungal membrane, leading to cell death. The central lipid component that serves as a primary target is ergosterol, an essential sterol that maintains membrane fluidity, integrity, and function in fungi (Mehta et al., 2023).

Ergosterol biosynthesis follows a complex pathway involving a cascade of 25 enzymes, and it is divided into three different parts starting from the mevalonate pathway, the late pathway, and the alternate pathway (Bhattacharya et al., 2018). Membrane targeting antifungals have their targets along the late pathway of ergosterol biosynthesis (see Figure 6.2). Polyenes, such as amphotericin B, bind directly to ergosterol, creating pores that lead to ion leakage and cell death (Kanafani & Perfect, 2008). Sphingolipids are a component of the plasma membrane that often interacts with ergosterol, influencing the effectivity of amphotericin binding to ergosterol.

This way sphingolipids modulate the drug targets in the membrane, increasing the susceptibility of the pathogen towards the drug (Kalra et al., 2024; Madaan & Bari, 2023). Azoles, including fluconazole and itraconazole, inhibit the enzyme lanosterol 14α-demethylase (Erg11), a key enzyme in the ergosterol biosynthetic pathway, resulting in depletion of ergosterol and accumulation of toxic sterol intermediates (Ghannoum & Rice, 1999). The allylamines, most notably terbinafine, target squalene epoxidase (Erg1), another important enzyme in the ergosterol biosynthesis pathway, leading to the buildup of squalene and subsequent membrane dysfunction (Lepesheva & Waterman, 2007). See Figure 6.2.

In addition to lipids, some fungal membrane proteins may also be implicated as antifungal targets, though they are less extensively exploited. For instance, H^+-ATPases involved in maintaining electrochemical gradients have been proposed as potential antifungal targets due to their essential roles in fungal viability. Disruption of these membrane components alters the architecture of the fungal membrane, affecting permeability, membrane potential, and the proper localization of proteins, all of which compromise fungal survival (Cowen & Steinbach, 2008). Ultimately, antifungal agents that disrupt membrane integrity can have cascading

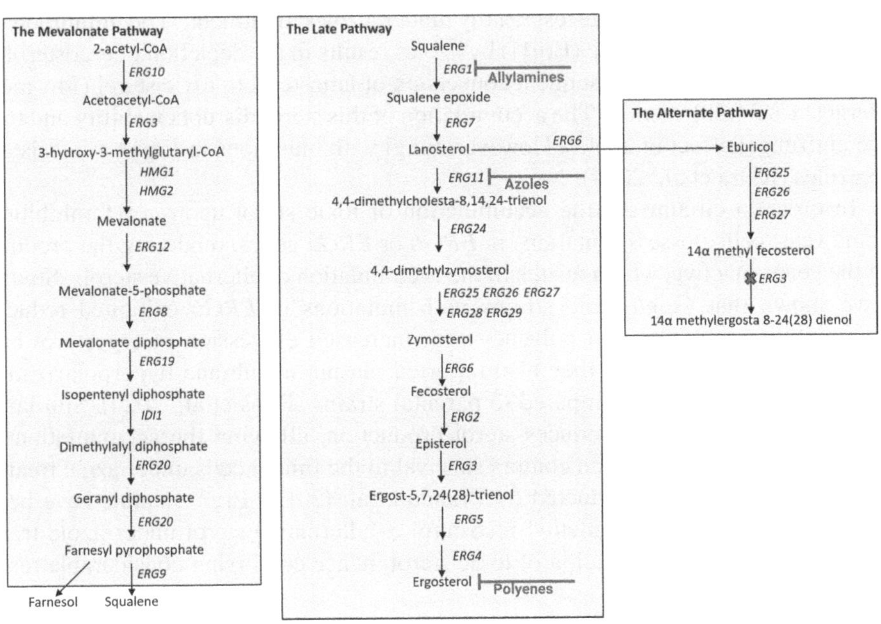

FIGURE 6.2 Membrane antifungals and their targets in the ergosterol biosynthesis pathway. Basically, the allylamines inhibit squalene epoxidase (Erg1), the azoles targets lanosterol 14α-demethylase. Polyenes directly binds to ergosterol resulting in membrane damage and leakage. The alternate pathway is induced upon inhibition of Erg11 which leads to production of the toxic sterol (14α methylergosta 8-24(28) dienol). However, Erg3 mutant strains bypass synthesis of toxic sterols and use and incorporate the alternative sterol (14α methyl fecosterol in their membrane thereby aiding their survival (Bhattacharya et al., 2018; Song et al., 2025).

effects on cellular homeostasis and metabolism, underscoring the significance of membrane components as critical targets.

6.6.3 Mechanisms of Membrane Remodelling in Response to Antifungals

Fungi have evolved sophisticated mechanisms to adapt to antifungal-induced membrane stress through membrane remodelling and transcriptional reprogramming. Additionally, fungi can modify membrane lipid composition by incorporating saturated fatty acids or alternative sterols, thereby altering membrane fluidity and reducing drug susceptibility (Barker et al., 2004).

The study of C-24 sterol methyl transferase in the filamentous fungi *Mucor lusitanicus* has identified at least four alternative pathways for sterol biosynthesis. This would render most ergosterol-targeting antifungals ineffective against this fungus. In contrast to filamentous fungi, the primary sterol biosynthetic pathway in yeast is ergosterol biosynthesis. So far, there is no experimental evidence of alternative ergosterol production pathways in yeast, but alternative pathway can be activated under certain conditions including during drug treatment (Bauer et al., 2023; Eliaš et al., 2024). While ergosterol is the main sterol of the fungal cell membranes, other sterols can also accumulate especially under azoles treatment. The inhibition of lanosterol 14-α demethylase (Erg11) by azoles results in the depletion of ergosterol in the cell membrane and subsequent conversion of lanosterol to toxic sterol (14α-methylergosta-8,24(28)-dienol). The accumulation of this sterol disrupts viability and thus the antifungal effect of azoles. However, fungi with mutations in Erg11 are resistant to azoles (Kalra et al., 2024).

In order to circumvent the accumulation of toxic sterol upon erg11 inhibition, some yeast cells possess mutations in *ERG6* or *ERG3* genes, rendering the products of the genes inactive, which results in the accumulation of alternative sterols. Studies have shown that *C. glabrata* strains with mutations in *ERG6* exhibited reduced susceptibility to azoles and polyenes with increased expression of ergosterol biosynthetic genes. However, they also reported plasma membrane hyperpolarization and decrease in fluidity compared to parental strains (Eliaš et al., 2024). Similarly, the loss of Erg3 activity reduces sterol production, allowing the accumulation of 14α-methyl fecosterol, which confers survival to the fungal cells under azole treated conditions. In a study conducted by Branco et al. (2017), Erg3 mutants have been shown to accumulate 14α-methyl fecosterol as alternative sterol under azole treatment and prevent the formation of toxic sterol, hence conferring considerable resistance to azoles.

6.6.4 Genetic and Regulatory Pathways Governing Membrane Adaptation

The genetic regulation of membrane adaptation and homeostasis in fungi is a complex process involving many transcription factors. A primary response to ergosterol depletion caused by azoles or other membrane-targeting agents involves the upregulation of ERG genes involved in sterol biosynthesis, often regulated by transcription factors such as Upc2p in *C. albicans* (MacPherson et al., 2005). Upc2 is a zinc cluster factor, which senses changes in sterol levels. When sterol level depletes due to sterol inhibiting drugs, Upc2 is activated. This activation leads to the upregulation of genes in sterol biosynthesis and uptake, enabling fungus to compensate the sterol deficiency caused by the antifungal, hence contributing to drug resistance.

A number of researches have demonstrated that Upc2 is central to regulation of ergosterol biosynthesis and azole resistance in *C. albicans* and *C. glabrata* (Vasicek et al., 2014; Vu et al., 2021). Basically, Upc2 upregulates Erg11; however, studies have also indicated that it plays an important role in the regulation of other genes involved in azole resistance including the MFS or the ABC transporters. In a recent study, Li et al. (2024) reported hyperactivation of Upc2 in *C. auris* in the presence of azoles, resulting in increased levels of Mdr1 via direct upregulation or indirectly by controlling Mrr1. Additionally, they have also reported its role in the upregulation of Tac1b, resulting in positive regulation of Cdr1.

While Ucp2 has been considered as the primary regulator of ergosterol biosynthesis genes, another key regulator, Adr1, has also been identified in *C. albicans*. Adr1 has been known to regulate fatty acid, ethanol, and glycerol metabolism in other yeasts, most notably *Saccharomyces cerevisiae*; however, recent reports have suggested that it can independently activate the transcription of ergosterol biosynthesis genes. Song et al. (2025) suggested that while Upc2 activation drives fluconazole resistance, its upregulation of ergosterol biosynthesis genes heavily depends on Adr1. And the loss of Upc2 has less impact on strains with activated Adr1. However, Adr1 expression has been shown to be regulated by Upc2, whereas Adr1 does not directly affect Upc2 expression, placing Upc2 as the master regulator of ergosterol biosynthesis genes.

Another transcription factor, Rpn4, has been identified to contribute to azole tolerance in *C. albicans* and *C. glabrata* where reports have shown that loss in Rpn4 results in increase in sensitivity to azoles and decrease in transcription levels of ergosterol biosynthesis genes including Upc2 (Yau et al., 2023). However, Yau and colleagues also reported that Rpn4 regulatory role differs in *C. albicans* and *C. glabrata* wherein it directly binds to the Erg11p promoter and required only for fluconazole-induced gene expression in *C. glabrata*. Whereas no direct link has been identified between Rpn4 and Erg11 promoter in *C. albicans*, suggesting that its activation of Erg11 genes in *C. albicans* occurs indirectly, most likely, by modulating Upc2 expression.

Recent reports have unveiled that in *A. fumigatus*, a 34bp promoter region of Erg11/Cyp51 can integrate signals from multiple transcription factors through dynamic interactions, offering new insights into antifungal resistance mechanism (Song et al., 2025). *A. fumigatus* has been identified to employ srba to regulate ergosterol biosynthesis where it binds to sterol response elements (SRE) within the *ERG11/CYP51A* promoter, which in turns results in direct activation of *ERG3*, *ERG24A*, *ERG25A* together with the core genes *ERG11/CYP51A* (Chung et al., 2014; Paul et al., 2022). Hagiwara et al. (2017) have also identified another transcription factor atrr in *A. fumigatus*, which is a zinc finger transcription factor that also binds to a second site within the 34bp promoter region of ER1/CYP51A and in turn enhances the expression *ERG11/CYP51A* genes (Figure 6.3).

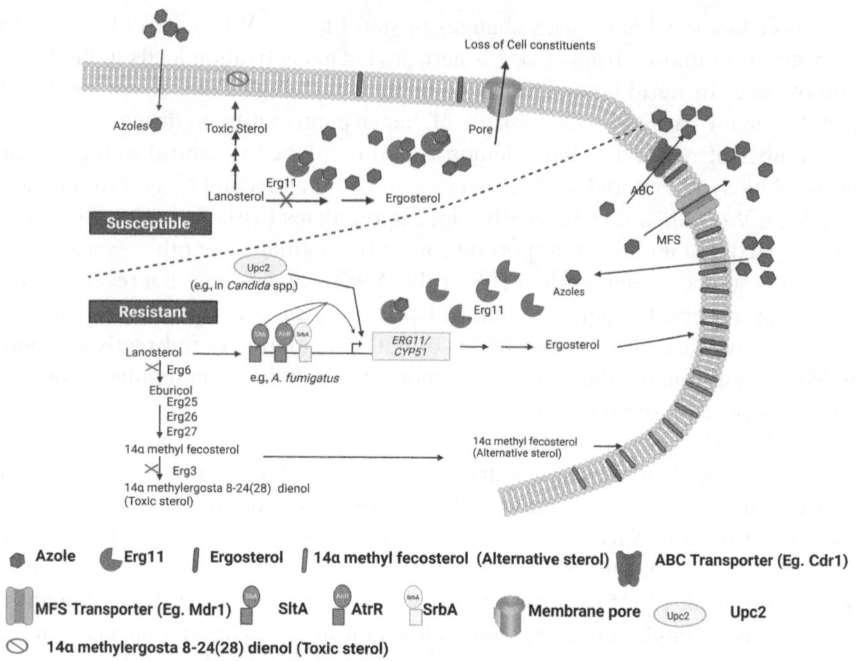

FIGURE 6.3 Membrane remodelling and adaptation in response to antifungals. Membrane targeting antifungals such as azoles passively enters the cell. In susceptible cells, they inhibit Erg11 resulting in ergosterol depletion in the membrane, pore formation and subsequent loss of cellular contents. Resistant cells activate Upc2 which upregulates Erg11, and at the same time deactivates Erg6 and Erg3 to prevent formation of toxic sterols. In moulds specifically, A. fumigatus, SltA, AtrR and SrbA upregulates CYP51A. When Erg11 is depleted, resistant cells also deactivate Erg3 to prevent synthesis of toxic sterol (14α methylergosta 8-24(28) dienol) and incorporate an alternative sterol (14α methyl fecosterol) in their membrane.

6.7 BIOFILM FORMATION AND ITS CONTRIBUTION TO DRUG RESISTANCE

Fungal biofilms are composed of fungal cells and an extracellular matrix. Fungal cells come together to form communities covered in a matrix. The extracellular matrix consists of molecules such as carbohydrates, lipids, proteins, and nucleic acids, aiding in adhesion, aggregation, and defence. The infections spread through both abiotic and biotic surfaces. The majority of fungal infections are nosocomial, starting by adhering to medical appliances such as catheter, prosthesis, and feeding tubes. Biotic surfaces include enamel, CNS infection, the genitourinary tract, and non-healing wounds. *Candida* species occupy different areas of the body with *C. albicans* and *C. glabrata* present in the oral cavity. *C. auris* mainly colonizes the skin surfaces, recovered from fingertips, palms, and rooms of the infected person. Other fungal pathogens like spore-forming *Aspergillus* and *Fusarium* spp. have been found to be present in the head and neck region. *C. neoformans* spores once inhaled develop respiratory problems from where they can disseminate to the brain causing

Cryptococcus meningitis (Ramage et al., 2025). The study by Mukherjee et al. (2003) explores into how biofilm formation contributes to the early stages of antifungal resistance in *Candida* species. The antifungal resistance is majorly achieved by the upregulation of efflux pumps and the interference in penetration of drugs via the extracellular matrix (Mukherjee et al., 2003).

6.7.1 Steps in Biofilm Formation

The fungal biofilm formation in general unfolds in four distinct steps, including colonization, proliferation, maturation, and dispersion.

- *Colonization*: It refers to the attachment of the fungus to a suitable substratum. This phase begins with adhesion. The surface can be both biotic or abiotic. This adhesion is carried by the surface adhesin-like molecules, which facilitate the binding of the fungal cell to surfaces. Few examples include ALS (agglutinin like sequence) in *C. albicans*. This family has eight proteins identified, whose orthologues have been present in different *Candida* species such as *C. tropicalis* and *C. auris* (Sundstrom, 2002). ALS3 is extensively studied in *C. albicans*, which is primarily involved in the attachment (Liu & Filler, 2011). In *C. glabrata*, EPA (epithelial adhesin) gene family mediates the adhesion phase (Timmermans et al., 2018). In *A. fumigatus*, initiation begins when the conidia adhere to the substratum, followed by the swelling of conidia. About 30 genes are involved in adhesion specific phase. One of the important proteins is Cal A, known to bind to host laminin that is involved in adhesion. Another class includes hydrophobins, which mediate the adherence by surface interactions; for example, the usage of collagen coated coverslips was shown to mimic the human skin better by forming aggressive biofilms (Biswas et al., 2023; Dague et al., 2008; Liu et al., 2016; Upadhyay et al., 2009). In *C. neoformans*, yeast cells adhere to the surface with the help of capsular component known as GXM, glucuronoxylomannan. It binds to host ECM proteins to mediate the attachment (Barbosa et al., 2006).
- *Colony formation/proliferation*: Once the cells adhere to the specific substratum, they divide and attach to their nearby cells forming microcolonies. The cells proliferate, leading to increase in biomass. SCF1 is surface colonization factor 1 identified in *C. auris*, which facilitates the adhesion by their cationic hydrophobic regions onto the surface (Santana et al., 2023). ALS5 promotes adherence to the surface by forming aggregates like beta structures helpful in the cell-cell attachment (Golan et al., 2022). This phase is marked by the morphological transformation of yeast cells into hyphae in few species such as *C. albicans* and *A. fumigatus*. In *C. albicans*, *HWP* (hyphal wall protein), a hyphal-specific adhesin, is responsible for binding of hyphal cells to the host cells. *C. neoformans* uses its capsular polysaccharides to proliferate. At this stage, the cells start to encapsulate in the extracellular matrix moving towards a mature biofilm stage (Ajesh & Sreejith, 2012).

- *Maturation phase*: This phase involves complexity and formation of three-dimensional biofilm. It is characterized by the compaction of cells wrapped in an extracellular matrix (Iñigo & Del Pozo, 2018). The matrix formation is mediated by extracellular vesicles regulated by the endosomal sorting complexes required for transport (ESCRT) complex. EVs are bilayered structures transporting DNA, polysaccharides, proteins, and lipids across the cell (Heredia & Andes, 2021). The ECM composition varies in different fungal species. Proteins, polysaccharides, and nucleic acids are main components. For example, in *A. fumigatus*, ECM components include eDNA, polyols, proteins, lipids, and exopolysaccharides such as glucans, galactomannan, and galactosaminogalactan. However, in *C. albicans*, glucans and mannans harbour the most of the polysaccharides. Whereas, hexoamines (chitin) and lipids are most abundant in *C. tropicalis* ECM (Al-Fattani & Douglas, 2006; Baillie & Douglas, 2000; Reichhardt et al., 2015).
- *Dispersal*: It is the last stage of biofilm formation, where once the matured biofilm is being formed, the cells utilize the nutrients and result in nutrient deprivation. In this stress condition, the cells from a matrix enclosed biofilm escape to establish at new niches in the host. This results in an increase in pathogenicity of the pathogen as it disseminates to different parts of the body (Uppuluri et al., 2010).

6.8 GENETIC REGULATION OF FUNGAL BIOFILM

It is a complex, orderly process at molecular level, with a number of signalling processes driving fungal biofilm formation. Six transcriptional regulators related to *Candida* spp. have been identified namely EFG1, TEC1, NDT80, BCR1, ROB1, and BRG1. During the adhesion phase, *BCR1* is mainly involved, and Als1, Als3, and Hwp1 are under the control of *BCR1*. Mutant *bcr1* formed thin biofilm and had decreased the expression of Als1, Als3, and Hwp1. *TEC1* controls the expression of *BCR1*, which fine-tunes the downstream adhesion genes. *TEC1* is regulated by *EFG1*, which is involved in filamentous growth. The maturation phase of biofilm is controlled by *NDT80, ROB1, and BRG1* (Araújo et al., 2017).

In *C. auris*, SCF1, a surface colonization factor, is an important adhesin responsible for adherence to abiotic and biotic surfaces. It mediates the adhesion via cationic residues exposed on the surface. The study shows that SCF1 aids not only in adhesion but also in biofilm formation and virulence in mice catheter infection model. Another important adhesin, ALS4, showed that an increase in copy number is related to enhanced adhesion and biofilm formation (Bing et al., 2023; Santana et al., 2023). Recently, the study shows that Ume6 hyperactivation is responsible for adhering to surfaces as well as biofilm formation. Hyperactivation of Ume6 regulates the expression of ALS*4498*, HGC1, and *SCF1* involved in biofilm formation (Louvet et al., 2024).

In *A. fumigatus*, orthologues of *C. albicans* biofilm regulators Flo8, Ada2, and EFG1 are GAG regulators. MedA and StuA regulate the expression of genes involved in GAG synthesis vital for adhesion and biofilm formation. Another important transcription factor SomA has binding sites in genes responsible for GAG synthesis *agd3* and *ega3* along with its regulators *MedA* and *StuA*. Further, SomA also has binding sites in the upstream region of cell wall stress genes for chitin synthases and β-1,3-glucan synthase. SomA interacts with ptab, a LIM-binding protein to regulate the biofilm formation. In whole, SomA not only maintains the cell wall architecture but it is also involved in the development of biofilm. Spt20, a subunit of the SAGA (Spt-Ada-Gcn5-acetyltransferase) complex, is identified as a regulator of biofilm. Mutants of *spt20* were completely defective in conidiation and had severe defects in biofilm formation (Chen et al., 2020; Gravelat et al., 2010; Lin et al., 2015).

6.9 QUORUM SENSING AND INTERCELLULAR SIGNALLING REGULATE BIOFILM MATURATION

As cell density increases, cells communicate with each other by the secretion of small molecules that regulate virulence and biofilm formation in fungal species. This is known as quorum sensing. Important quorum-sensing molecules include farnesol and tyrosol. Quorum sensing during biofilm formation is extensively studied in *C. albicans*.

Farnesol is a quorum-sensing molecule regulating the virulence properties of *C. albicans*. Farnesol is involved in modulating the morphology, efflux pump expressions, apoptosis, and biofilm formation. Farnesol inhibits the yeast to hyphal transition and represses the genes involved in filamentation. The yeast to hyphal transition is governed by the inhibition of Ras1-Cdc35-PKA signalling pathway, and also farnesol blocks the ergosterol synthesis by causing alterations in the cell membrane. Farnesol inhibits Cyr1, adenylyl cyclase, downregulating Nrg1 expression and attenuating the cAMP/PKA signalling. Farnesol inhibits UBR1, a ubiquitin ligase, which causes degradation of CUP9, relieving Sok1, and Sok1 activates NRG1, which is a negative regulator of morphological transition (Lu et al., 2014). Farnesol is also reported to cause a decrease in antifungal resistance against *Candida* cells. Chen et al. (2018) showed an increase in susceptibility of antifungals towards *Candida* biofilm by regulating the expression of CYR1 and PDE2 genes. Katragkou et al. (2015) showed the synergism of micafungin and farnesol towards *Candida* biofilm. Due to the amphiphilic nature of farnesol, it can easily integrate into the cell membrane, influencing its integrity and permeability. This causes the cellular polarity disruption in the case of *C. parapsilosis* (Kovács & Majoros, 2020).

Tyrosol is another quorum-sensing molecule that impacts the biofilm formation during early and intermediate stage of biofilm formation in *C. albicans*. It accelerates the yeast hyphal transition during the initiation phase of biofilm development. Tyrosol alone or in synergism with amphotericin inhibits biofilm formation in *C. albicans* and *C. tropicalis*, whereas combination of tyrosol and azoles did not show any effect on mature biofilm (Cordeiro et al., 2015).

Quorum-sensing-like peptide 1 (QSP1) is reported in *C. neoformans*. Pro-Qsp1 is cleaved by (predicted serine protease) Psp1 into functional Qsp1 extracellularly, which is taken up by cells via OPT1 transporter. Qsp1 is known to regulate a variety of virulence factors, cell wall integrity, and secreted aspartyl protease activity. Qsp1 also regulates the binding of 3 transcription factors, Cqs2, Liv3, and Nrg1, modulating the expression of Qsp1-regulated genes. Mutants of Qsp1 showed reduced inhalation of pathogen by mice and slower uptake by macrophages revealing that it is indeed an important virulence factor. Also, mutants of Psp1 failed to cleave proqsp1 into functional Qsp1 (Summers et al., 2020; Tian et al., 2018).

6.10 ANTIFUNGAL DRUG RESISTANCE DURING BIOFILM FORMATION

Fungal biofilms present a serious problem to healthcare due to its resistance to the available antifungals, as compared to the planktonic infections. Antifungal resistance is a multifaceted phenomenon arising due to a number of factors including increased expression of drug pumps, extracellular matrix, and the presence of persister cells in the biofilm.

1. *Extracellular matrix*: The biofilm ECM sequesters antifungals. In *Candida* spp., the ECM composition includes glycoproteins (55%), carbohydrates (25%), lipids (15%), and nucleic acid (5%). Glucan and mannan form a protective complex layer, preventing the drugs penetration. Increased expression of *FKS1* genes encoding glucan synthesis is reported to contribute to antifungal drug resistance. The matrix also binds to azole classes of drugs, sequestering them and decreasing their concentration (Nett et al., 2010). DNAse treatment caused the degradation of eDNA in the biofilm, leading to increased sensitivity of biofilms to antifungals (Martins et al., 2010). In addition to ECM production, *Candida* spp. also release extracellular vesicles during the biofilm formation. These are bilayered structures transporting components of ECM extracellularly. Mutants deficient in EVs showed greater susceptibility to fluconazole in a catheter infection model, as compared to wild type during biofilm (Zarnowski et al., 2021). GAG, galactosaminogalactan, is a major contributor of ECM in *Aspergillus* spp. Reports show that the degradation of GAGs by hydrolases leads to severe impairment of drug tolerance during 9 hours of biofilm development. In addition to hydrolases, DNase treatment enhanced the biofilm susceptibility to antifungals (Liu et al., 2022).
2. *Efflux pumps*: Overexpression of efflux pumps is one of the major driving forces of antifungal resistance in biofilm. There are 2 efflux pump classes in *C. albicans*: ABC and MFS. Upregulation of efflux pumps occurs due to mutations in the promoter of genes encoding Cdr1, Cdr2, or Mdr1. They expel out the azole class of antifungals. During biofilm development, from the start of the adherence phase to the maturation, the efflux pumps remain upregulated even when antifungals are not present (Kaur & Nobile, 2023; Mukherjee et al., 2003). In *A. fumigatus*, Mdr4, an efflux pump of ABC transporter, transcript levels were significantly increased during biofilm, which showed an increase in antifungal resistance (Rajendran et al., 2011).

   Persister cells: These cells are described as metabolically inactive cells formed during the biofilm development. Few of the cells still remain viable even after the antifungal treatment, and as a result, they become resistant to high doses of antifungals. This small population of cells is thus drug resistant, and it has the ability to survive ROS accumulation induced by drugs. The cells either overexpress the superoxide dismutase or enhance the expression of stress responsive genes such as HSP90 or HSP21 to protect the cells from oxidative stress. The increased expression of alkyl hydroperoxide

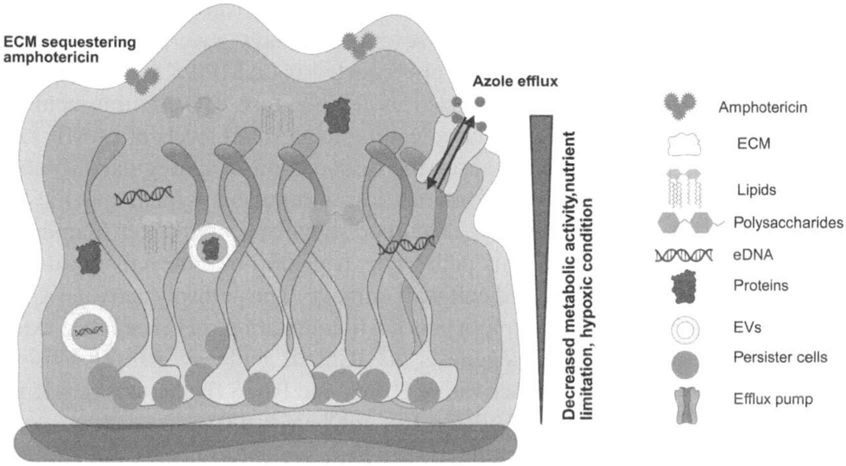

FIGURE 6.4 Schematic representing drug resistance mechanisms within a fungal biofilm. As the cell density towards the substratum increases, there is a decrease in metabolic activity and limitation of nutrients at the bottom. ECM covers the dense fungal cells, preventing the penetration of drugs. Efflux pumps in the ECM expel out the fluconazole. Extracellular vesicles transport the cargo (lipids, proteins, and eDNA) to the outside.

reductase 1 (AHP1) in persister cells in response to high concentration of AmB is another mechanism of regulating the oxidative stress. Persister cells utilize alternative carbon sources such as glycogen and trehalose. Persister cells show a downregulation of glycolytic pathway, suggesting that the cells adapt to the absence of glucose (Bink et al., 2011; Diezmann et al., 2015; Wuyts et al., 2018) (Figure 6.4).

6.11 THE ROLE OF MORPHOGENETIC ADAPTATION IN FUNGAL DRUG RESISTANCE

Morphogenetic transitions refer to different changes in shape, size, and structure of cell. This is triggered by the environmental surroundings and different signalling pathways. Different morphological forms are important to adapt to a particular niche in the host and survive. Moreover, it is an important adaptation seen across the fungal pathogens to survive and persist in the host. This section will uncover the different morphological transitions in *Candida* spp., *Cryptococcu*s, and *Aspergillus* spp.

6.12 MORPHOLOGICAL TRANSITIONS IN *CANDIDA* SPECIES

Candida spp. undergo yeast to hyphal transition, specifically more pronounced in *C. albicans* and *C. tropicalis*. *C. glabrata* and *C. parapsilosis* undergo yeast to pseudohyphae formation, whereas *C. auris* rarely undergoes filamentation but forms clumps of cells known as aggregates (Chow et al., 2021). Filamentation was observed only when *C. auris* cells passed through the mammalian body (Yue et al., 2018). Hyphal forms are branched elongated cells with a defined septum having a large surface area. Pseudohyphal cells remain attached together after division to form a chain (Sudbery et al., 2004). There are several pathways governing morphogenesis in *Candida* spp. One of the extensively reported is cAMP/PKA pathway, where cells in response to external stimuli such as temperature, CO_2, amino acids, and glucose activate Cyr1, an adenyl cyclase, leading to elevated levels of cAMP. cAMP in turn activates PKA, and PKA causes the dissociation of its regulatory subunits Bcy1 from catalytic subunit Tpk1. Tpk1 once free regulates hyphal-specific genes such as Efg1, Flo8, Ume6, and Tec1. Another signalling regulating filamentation includes MAP kinase pathway. This pathway is triggered by external stimulus such as osmotic stress, low nitrogen, and cell wall damage. The pathway activates Cph1, a downstream transcription factor important for filamentation. pH is another external stimulus, which regulates morphogenesis via Rim101 pathway. Rim21 senses the change in pH in external milieu, activating Rim8, and Rim8 further triggers Rim101 initiating the response (Chow et al., 2021; Lin & Chen, 2018).

However, there are negative regulators as well, which do not permit the yeast–hyphal transition. Tup1 and Nrg1 are transcription factors negatively regulating the hyphae formation. Farnesol-mediated hyphal state inhibition is discussed in quorum sensing section. Azoles, an essential class of antifungals, target the ergosterol biosynthetic pathway, one of the critical components of the fungal cell membrane, and are reported to inhibit the hyphal formation in *C. albicans*. This occurs because azoles target lanosterol 14α-demethylase involved in sterol biosynthesis, leading to the accumulation of the sterol biosynthetic intermediate, farnesyl pyrophosphate, indirectly increasing the synthesis of farnesol (Bandara et al., 2012). Farnesol further prevents yeast–hyphal transition via Cup9 degradation. Further, a secondary network involving Eed1 and Ume6 acting downstream of Efg1is inhibited by Tup1 and Nrg1. Eed1 is activated by Efg1, and Δ/Δefg1 slightly rescued the hyphal phenotype in Eed1 overexpressed cells. Ume6 acts downstream of Eed1. Ume6 is repressed by Nrg1-Tup1 complex (Braun et al., 2001).

Aggregation is another unique morphology specific to *C. auris*. Aggregation refers to the cell–cell attachment forming clumps or multicellular aggregates. The aggregation is governed by cell surface adhesins, extracellular matrix, or the cytokinetic defect. Als4112 is a cell surface adhesin involved in cell–cell interaction. Mutants of Als4112 caused the disruption of clumps (Wang et al., 2024). Cell separation defects can occur due to mutations in genes that are responsible for cell division, cytokinesis, cell wall integrity, and cellular polarization. Bing et al. (2024) showed that mutations in essential genes such as CHS1, BNI1, LRG, and CAS4 contribute to the multicellular aggregative morphotype in *C. auris*. The defect in cell separation can arise due to the failure of septum disruption. Malavia-Jones and his associates found that ECM was able to induce aggregation at 37°C. In contrast, there was no aggregative phenotype at 30°C with no production of ECM. The immune system poses a challenge in clearing up the aggregates as macrophages do not readily engulf them. This poor clearance of aggregates by macrophages may provide an advantage to the pathogen for immune evasion. Bing et al. (2024) and Malavia-Jones et al. (2023) showed that the phagocytosis of aggregates by macrophages was readily reduced as compared to the single yeast cells, which were easily engulfed by macrophages.

6.13 *CRYPTOCOCCUS NEOFORMANS* UNDERGOES NUMEROUS MORPHOLOGICAL TRANSITIONS

The outer cell architecture of the pathogen includes cell membrane covered with a cell wall, and the cell wall lies beneath the capsule. The capsule is the main factor involved in virulence. It is composed of glucuronoxylomannan (GXM) and glucuronoxylomannogalactan (gxmgal). *Cryptococcus* has a remarkable characteristic of undergoing a number of morphological changes in response to a specific trigger. The transitions include yeast–hyphal, pseudohyphal, titan cell formation, and microcells.

The transition between yeast to hyphal form is related to mating and signalling pathways. Hyphal formation takes place through mating, either by opposite mating cells (a-α) or same mating type cells (a-a). However, hyphal transition can occur independent of mating and is driven by ZNF2 transcription factor. Mutant znf2 locked the cells in yeast form. Thy hyphal form exhibited attenuated virulence due to the problem in disseminating to different parts of the body. Low nutrients and temperature promote the hyphal form in *C. neoformans* (Lin et al., 2019).

The pseudohyphal formation is triggered in response to nitrogen starvation where ammonium is the sole nitrogen source. AMT1 and AMT2, ammonium permeases, are required for pseudohyphal growth. Double mutants of ammonium permease rendered the cells in yeast form. Whereas wild type or single mutants exhibited the pseudohyphal cells. The RAM (Regulation of Ace2p activity and cellular Morphogenesis) pathway is central to pseudohyphal transition. RAM pathway includes 2 important kinases, Cbk1 and Kic1, and accessory proteins, Mob2, Tao3, and Sog2. Mutation in any of the RAM components leads to the formation of pseudohyphae due to a cytokinetic defect where daughter cells fail to separate from the parent cells (Chadwick et al., 2022; Magditch et al., 2012).

6.14 TITAN CELL FORMATION

C. neoformans has a unique morphological feature where the cells undergo enlargement, called as titan formation, so as to evade the immune system and disseminate to different parts of the body. Usually, cell size is 4–6 μm, but it changes to more than 10 μm, and the cell is said to undergo titan formation. The cells have thickened cell wall covered with an enlarged capsule. It is cAMP/PKA regulated pathway dependent on adenyl cyclase CAC1 as well as mating dependent (Dyląg et al., 2020). Titan cell is induced only during lung infection in mammalian host and also by amoeba lipid membrane components. Of the known seven GPCRs, only GPR5 is central to titan formation. Δgpr5 does not undergo cell enlargement as compared to wild type. Gpr5 regulates through Gpa1 and Ste3. Rim101 is downstream of Gpa1, known to play an important role in titan formation. Mutants of rim101 showed severed defects in titan formation. The titan formation decreases the susceptibility of oxidative stress caused by fluconazole treatment (Crabtree et al., 2012; Stempinski et al., 2023) (Figure 6.5).

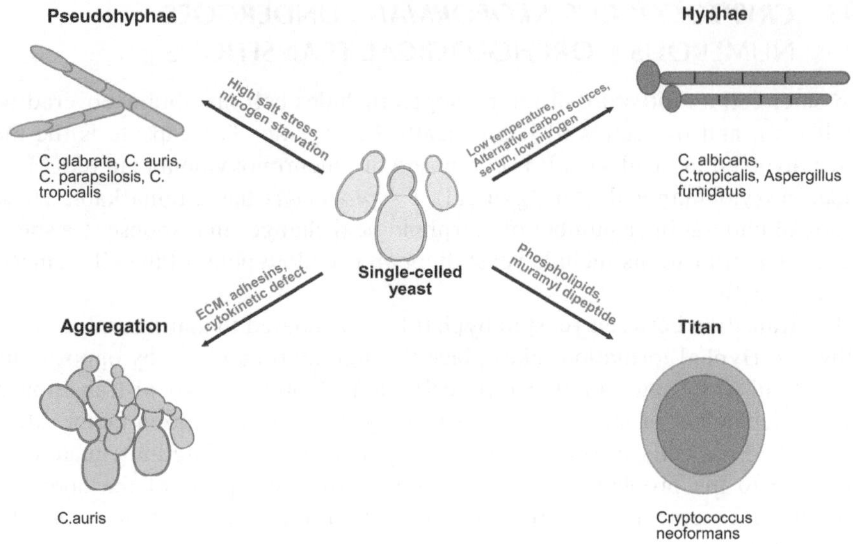

FIGURE 6.5 Morphological transitions in fungal species. The fungal cells attain different morphologies under various stress conditions. The cells form hyphae, pseudohyphae, aggregates, or large encapsulated cell (titan)

6.15 CONCLUSION AND FUTURE DIRECTIONS

The global acceleration of antifungal resistance, particularly among emerging and MDR fungal species, calls for a recalibration of current research and clinical paradigms. While substantial progress has been made in characterizing classical resistance mechanisms, from efflux pump overexpression to ergosterol pathway mutations, new insights underscore a far more complex, dynamic, and species-specific resistance landscape. The convergence of genetic mutations, epigenetic plasticity, and environmental adaptation strategies has made fungal pathogens remarkably agile in evading antifungal pressures.

A pressing imperative is the expansion of genomic surveillance efforts, not only in clinical settings but also across environmental reservoirs, where selective pressure from agricultural fungicides may drive resistance evolution in pathogens like *A. fumigatus* and *C. auris*. High-throughput sequencing and phylogenomic tools can reveal clade-specific resistance determinants, track transmission chains, and identify emerging threats before they reach outbreak potential. In this line, it is equally important for the integration of multi-omics approaches—genomics, transcriptomics, proteomics, and lipidomics—to construct a systems-level understanding of resistance evolution. Multi-layered datasets can help uncover context-dependent vulnerabilities, compensatory networks, and epistatic interactions that cannot be captured by single-layer analyses. For example, transcriptomic rewiring of efflux regulators, proteomic shifts in membrane constituents, and lipidomic remodelling of ergosterol pathways all contribute to phenotypic resistance in ways that are not always predictable from genotypes alone.

Moving beyond traditional targets, novel antifungal compounds that exploit non-classical pathways hold significant promise. Therapeutic strategies targeting heat shock proteins (e.g., Hsp90), the calcineurin axis, sphingolipid biosynthesis, or histone-modifying enzymes not only circumvent existing resistance mechanisms but also open the door to synergistic combination therapies. These agents may suppress stress tolerance or morphogenesis—key hallmarks of fungal virulence—without exerting direct fungicidal pressure, potentially reducing the risk of resistance development. Moreover, robust antifungal stewardship programmes are indispensable. Delayed diagnoses, inappropriate drug selection, and empiric broad-spectrum therapy all contribute to resistance selection. Improving diagnostic resolution—through rapid molecular assays, biomarker-based stratification, and resistance gene panels—will be key to precision-guided therapy. Clinical microbiology must transition from passive identification to real-time resistance prediction.

Finally, governments and policymakers must become active stakeholders in the antifungal resistance agenda. Antifungal resistance, much like antimicrobial resistance (AMR), is as much a political and economic challenge as it is a biological one. Yet it remains conspicuously absent from most national AMR action plans, funding frameworks, and global health security agendas. Public health campaigns focused on awareness, coupled with funding incentives for antifungal drug development, are urgently needed. International collaborations between research institutes, public health authorities, and the agricultural sector will be critical in establishing unified guidelines and policies. The framing of antifungal resistance as a silent pandemic—with implications as far-reaching as antibiotic resistance—may catalyse long-overdue action. At the international level, multilateral organizations such as WHO, FAO, and OIE must incorporate fungal pathogens into global AMR initiatives, ensuring that fungal resistance is monitored alongside bacterial and viral threats. Coordinated data-sharing platforms, antifungal resistance alert systems, and cross-border surveillance projects must be established, especially given the pandemic potential of pathogens like *C. auris*.

To safeguard the future of antifungal therapeutics, an integrated approach is essential, one that fuses scientific innovation with policy foresight, global coordination with local action, and molecular insight with ecological awareness. Resistance is not an inevitable outcome; it is a preventable crisis, but only if science, stewardship, and statecraft act in concert.

REFERENCES

Ahmady, L., Gothwal, M., Mukkoli, M. M., & Bari, V. K. (2024). Antifungal drug resistance in Candida: A special emphasis on amphotericin B. *Apmis*, 132(5), 291–316. https://doi.org/10.1111/apm.13389

Ajesh, K., & Sreejith, K. (2012). *Cryptococcus laurentii* biofilms: Structure, development and antifungal drug resistance. *Mycopathologia*, 174(5–6), 409–419. https://doi.org/10.1007/s11046-012-9575-2

Al-Fattani, M. A., & Douglas, L. J. (2006). Biofilm matrix of *Candida albicans* and *Candida tropicalis*: Chemical composition and role in drug resistance. *J Med Microbiol*, 55(8), 999–1008. https://doi.org/https://doi.org/10.1099/jmm.0.46569-0

Anderson, M. Z., Saha, A., Haseeb, A., & Bennett, R. J. (2017). A chromosome 4 trisomy contributes to increased fluconazole resistance in a clinical isolate of *Candida albicans*. *Microbiology*, 163(6), 856–865. https://doi.org/10.1099/mic.0.000478

Araújo, D., Henriques, M., & Silva, S. (2017). Portrait of Candida species biofilm regulatory network genes. *Trends Microbiol*, 25(1), 62–75. https://doi.org/10.1016/j.tim.2016.09.004

Arendrup, M. C., & Perlin, D. S. (2014). Echinocandin resistance: An emerging clinical problem? *Curr Opin Infect Dis*, 27(6), 484–492. https://doi.org/10.1097/QCO.0000000000000111

Baillie, G. S., & Douglas, L. J. (2000). Matrix polymers of Candida biofilms and their possible role in biofilm resistance to antifungal agents. *J Antimicrob Chemother*, 46(3), 397–403. https://doi.org/10.1093/jac/46.3.397

Baker, K. M., Hoda, S., Saha, D., Gregor, J. B., Georgescu, L., Serratore, N. D., Zhang, Y., Cheng, L., Lanman, N. A., & Briggs, S. D. (2022). The Set1 histone H3K4 methyltransferase contributes to azole susceptibility in a species-specific manner by differentially altering the expression of drug efflux pumps and the ergosterol gene pathway. *Antimicrob Agents Chemother*, *66*(5), e0225021. https://doi.org/10.1128/aac.02250-21

Bandara, H. M., Lam, O. L., Jin, L. J., & Samaranayake, L. (2012). Microbial chemical signaling: A current perspective. *Crit Rev Microbiol*, 38(3), 217–249. https://doi.org/10.3109/1040841x.2011.652065

Barbosa, F. M., Fonseca, F. L., Holandino, C., Alviano, C. S., Nimrichter, L., & Rodrigues, M. L. (2006). Glucuronoxylomannan-mediated interaction of *Cryptococcus neoformans* with human alveolar cells results in fungal internalization and host cell damage. *Microb Infect*, 8(2), 493–502. https://doi.org/https://doi.org/10.1016/j.micinf.2005.07.027

Barker, K. S., Crisp, S., Wiederhold, N., Lewis, R. E., Bareither, B., Eckstein, J., Barbuch, R., Bard, M., & Rogers, P. D. (2004). Genome-wide expression profiling reveals genes associated with amphotericin B and fluconazole resistance in experimentally induced antifungal resistant isolates of *Candida albicans*. *J Antimicrob Chemother*, 54(2), 376–385. https://doi.org/10.1093/jac/dkh336

Basso, L. R., Jr., Gast, C. E., Bruzual, I., & Wong, B. (2015). Identification and properties of plasma membrane azole efflux pumps from the pathogenic fungi *Cryptococcus gattii* and *Cryptococcus neoformans*. *J Antimicrob Chemother*, 70(5), 1396–1407. https://doi.org/10.1093/jac/dku554

Bauer, K., Rafael, B., Vágó, B., Kiss-Vetráb, S., Molnár, A., Szebenyi, C., Varga, M., Szekeres, A., Vágvölgyi, C., Papp, T., & Nagy, G. (2023). Characterization of the sterol 24-C-methyltransferase genes reveals a network of alternative sterol biosynthetic pathways in *Mucor lusitanicus*. *Microbiol Spectr*, 11(3), e0031523. https://doi.org/10.1128/spectrum.00315-23

Berkow, E. L., & Lockhart, S. R. (2017). Fluconazole resistance in Candida species: A current perspective. *Infect Drug Resist*, 10, 237–245. https://doi.org/10.2147/IDR.S118892

Bhattacharya, S., Esquivel, B. D., & White, T. C. (2018). Overexpression or deletion of ergosterol biosynthesis genes alters doubling time, response to stress agents, and drug susceptibility in *Saccharomyces cerevisiae*. *mBio*, *9*(4). https://doi.org/10.1128/mBio.01291-18

Bieberich, E. (2018). Sphingolipids and lipid rafts: Novel concepts and methods of analysis. *Chem Phys Lipids*, 216, 114–131. https://doi.org/10.1016/j.chemphyslip.2018.08.003

Bing, J., Guan, Z., Zheng, T., Ennis, C. L., Nobile, C. J., Chen, C., Chu, H., & Huang, G. (2024). Rapid evolution of an adaptive multicellular morphology of *Candida auris* during systemic infection. *Nature Communications*, 15(1), 2381. https://doi.org/10.1038/s41467-024-46786-8

Bing, J., Guan, Z., Zheng, T., Zhang, Z., Fan, S., Ennis, C. L., Nobile, C. J., & Huang, G. (2023). Clinical isolates of *Candida auris* with enhanced adherence and biofilm formation due to genomic amplification of ALS4. *PLoS Pathog*, *19*(3), e1011239. https://doi.org/10.1371/journal.ppat.1011239

Bink, A., Vandenbosch, D., Coenye, T., Nelis, H., Cammue, B. P., & Thevissen, K. (2011). Superoxide dismutases are involved in *Candida albicans* biofilm persistence against miconazole. *Antimicrob Agents Chemother*, 55(9), 4033–4037. https://doi.org/10.1128/aac.00280-11

Biswas, B., Rana, A., Gupta, N., Gupta, I., Puria, R., & Thakur, A. (2023). A novel robust method mimicking human substratum to dissect the heterogeneity of *Candida auris* biofilm formation. *Microbiology Spectrum*, 11(4), e00892-23. https://doi.org/10.1128/spectrum.00892-23

Bongomin, F., Gago, S., Oladele, R. O., & Denning, D. W. (2017). Global and multi-national prevalence of fungal diseases-estimate precision. *J Fungi (Basel)*, *3*(4). https://doi.org/10.3390/jof3040057

Branco, J., Ola, M., Silva, R. M., Fonseca, E., Gomes, N. C., Martins-Cruz, C., Silva, A. P., Silva-Dias, A., Pina-Vaz, C., Erraught, C., Brennan, L., Rodrigues, A. G., Butler, G., & Miranda, I. M. (2017). Impact of ERG3 mutations and expression of ergosterol genes controlled by UPC2 and NDT80 in *Candida parapsilosis* azole resistance. *Clin Microbiol Infect*, 23(8), 575.e571–575.e578. https://doi.org/10.1016/j.cmi.2017.02.002

Braun, B. R., & Johnson, A. D. (2000). TUP1, CPH1 and EFG1 make independent contributions to filamentation in *Candida albicans*. *Genetics*, 155(1), 57–67. https://doi.org/10.1093/genetics/155.1.57

Braun, B. R., Kadosh, D., & Johnson, A. D. (2001). NRG1, a repressor of filamentous growth in *C. albicans*, is down-regulated during filament induction. *EMBO J*, 20(17), 4753–4761. https://doi.org/10.1093/emboj/20.17.4753

Brown, G. D., Denning, D. W., Gow, N. A., Levitz, S. M., Netea, M. G., & White, T. C. (2012). Hidden killers: Human fungal infections. *Sci Transl Med*, 4(165), 165rv113. https://doi.org/10.1126/scitranslmed.3004404

Cannon, R. D., Lamping, E., Holmes, A. R., Niimi, K., Baret, P. V., Keniya, M. V., Tanabe, K., Niimi, M., Goffeau, A., & Monk, B. C. (2009). Efflux-mediated antifungal drug resistance. *Clin Microbiol Rev*, 22(2), 291–321. https://doi.org/10.1128/CMR.00051-08

Chadwick, B. J., Pham, T., Xie, X., Ristow, L. C., Krysan, D. J., & Lin, X. (2022). The RAM signaling pathway links morphology, thermotolerance, and CO2 tolerance in the global fungal pathogen *Cryptococcus neoformans*. *eLife*, *11*, e82563. https://doi.org/10.7554/eLife.82563

Chen, S., Xia, J., Li, C., Zuo, L., & Wei, X. (2018). The possible molecular mechanisms of farnesol on the antifungal resistance of *C. albicans* biofilms: The regulation of CYR1 and PDE2. *BMC Microbiol*, 18(1), 203. https://doi.org/10.1186/s12866-018-1344-z

Chen, Y., Le Mauff, F., Wang, Y., Lu, R., Sheppard Donald, C., Lu, L., & Zhang, S. (2020). The transcription factor SomA synchronously regulates biofilm formation and cell wall homeostasis in *Aspergillus fumigatus*. *mBio*, *11*(6). https://doi.org/10.1128/mbio.02329-20

Chow, E. W. L., Pang, L. M., & Wang, Y. (2021). From Jekyll to Hyde: The yeast-hyphal transition of *Candida albicans*. *Pathogens*, *10*(7), 859. https://doi.org/10.3390/pathogens10070859

Chowdhary, A., Prakash, A., Sharma, C., Kordalewska, M., Kumar, A., Sarma, S., Tarai, B., Singh, A., Upadhyaya, G., Upadhyay, S., Yadav, P., Singh, P. K., Khillan, V., Sachdeva, N., Perlin, D. S., & Meis, J. F. (2018). A multicentre study of antifungal susceptibility patterns among 350 *Candida auris* isolates (2009–17) in India: Role of the ERG11 and FKS1 genes in azole and echinocandin resistance. *J Antimicrob Chemother*, 73(4), 891–899. https://doi.org/10.1093/jac/dkx480

Chowdhary, A., Sharma, C., & Meis, J. F. (2017). *Candida auris*: A rapidly emerging cause of hospital-acquired multidrug-resistant fungal infections globally. *PLoS Pathog*, *13*(5), e1006290. https://doi.org/10.1371/journal.ppat.1006290

Chung, D., Barker, B. M., Carey, C. C., Merriman, B., Werner, E. R., Lechner, B. E., Dhingra, S., Cheng, C., Xu, W., Blosser, S. J., Morohashi, K., Mazurie, A., Mitchell, T. K., Haas, H., Mitchell, A. P., & Cramer, R. A. (2014). ChIP-seq and in vivo transcriptome analyses of the *Aspergillus fumigatus* SREBP SrbA reveals a new regulator of the fungal hypoxia response and virulence. *PLoS Pathog*, *10*(11), e1004487. https://doi.org/10.1371/journal.ppat.1004487

Cordeiro, R. D. A., Teixeira, C. E. C., Brilhante, R. S. N., Castelo-Branco, D. S. C. M., Alencar, L. P., de Oliveira, J. S., Monteiro, A. J., Bandeira, T. J. P. G., Sidrim, J. J. C., Moreira, J. L. B., & Rocha, M. F. G. (2015). Exogenous tyrosol inhibits planktonic cells and biofilms of Candida species and enhances their susceptibility to antifungals. *FEMS Yeast Res.*, *15*(4). https://doi.org/10.1093/femsyr/fov012

Cowen, L. E., Singh, S. D., Köhler, J. R., Collins, C., Zaas, A. K., Schell, W. A., Aziz, H., Mylonakis, E., Perfect, J. R., Whitesell, L., & Lindquist, S. (2009). Harnessing Hsp90 function as a powerful, broadly effective therapeutic strategy for fungal infectious disease. *Proc. Natl. Acad. Sci. USA*, *106*(8), 2818–2823. https://doi.org/10.1073/pnas.0813394106

Cowen, L. E., Kohn, L. M., & Anderson, J. B. (2001). Divergence in fitness and evolution of drug resistance in experimental populations of *Candida albicans*. *J Bacteriol*, 183(10), 2971–2978. https://doi.org/10.1128/JB.183.10.2971-2978.2001

Cowen, L. E., & Lindquist, S. (2005). Hsp90 potentiates the rapid evolution of new traits: Drug resistance in diverse fungi. *Science*, 309(5744), 2185–2189. https://doi.org/10.1126/science.1118370

Cowen, L. E., Sanglard, D., Howard, S. J., Rogers, P. D., & Perlin, D. S. (2014). Mechanisms of antifungal drug resistance. *Cold Spring Harb Perspect Med*, *5*(7), a019752. https://doi.org/10.1101/cshperspect.a019752

Cowen, L. E., & Steinbach, W. J. (2008). Stress, drugs, and evolution: The role of cellular signaling in fungal drug resistance. *Eukaryotic Cell*, 7(5), 747–764. https://doi.org/10.1128/ec.00041-08

Crabtree, J. N., Okagaki, L. H., Wiesner, D. L., Strain, A. K., Nielsen, J. N., & Nielsen, K. (2012). Titan cell production enhances the virulence of *Cryptococcus neoformans*. *Infect Immun*, 80(11), 3776–3785. https://doi.org/10.1128/iai.00507-12

Czajka, K. M., Venkataraman, K., Brabant-Kirwan, D., Santi, S. A., Verschoor, C., Appanna, V. D., Singh, R., Saunders, D. P., & Tharmalingam, S. (2023). Molecular mechanisms associated with antifungal resistance in pathogenic Candida species. *Cells*, *12*(22). https://doi.org/10.3390/cells12222655

Dague, E., Alsteens, D., Latgé, J. P., & Dufrêne, Y. F. (2008). High-resolution cell surface dynamics of germinating *Aspergillus fumigatus* conidia. *Biophys J*, 94(2), 656–660. https://doi.org/10.1529/biophysj.107.116491

Diezmann, S., Leach, M. D., & Cowen, L. E. (2015). Functional divergence of Hsp90 genetic interactions in biofilm and planktonic cellular states. *PLoS ONE*, *10*(9), e0137947. https://doi.org/10.1371/journal.pone.0137947

Doorley, L. A., Meza-Perez, V., Jones, S. J., & Rybak, J. M. (2025). A candidozyma (Candida) auris–Optimized episomal plasmid–induced Cas9-editing system reveals the direct impact of the S639F-encoding FKS1 mutation. *J Infect Dis*. https://doi.org/10.1093/infdis/jiaf285

Duggan, S., & Usher, J. (2023). *Candida glabrata*: A powerhouse of resistance. *PLoS Pathog*, *19*(10), e1011651. https://doi.org/10.1371/journal.ppat.1011651

Dyląg, M., Colon-Reyes, R. J., & Kozubowski, L. (2020). Titan cell formation is unique to Cryptococcus species complex. *Virulence*, 11(1), 719–729. https://doi.org/10.1080/21505594.2020.1772657

Eliaš, D., Tóth Hervay, N., & Gbelská, Y. (2024). Ergosterol biosynthesis and regulation impact the antifungal resistance and virulence of Candida spp. *Stresses*, 4(4), 641–662. https://www.mdpi.com/2673-7140/4/4/41

Emami, S., Tavangar, P., & Keighobadi, M. (2017). An overview of azoles targeting sterol 14alpha-demethylase for antileishmanial therapy. *Eur J Med Chem*, 135, 241–259. https://doi.org/10.1016/j.ejmech.2017.04.044

Fraczek, M. G., Bromley, M., Buied, A., Moore, C. B., Rajendran, R., Rautemaa, R., Ramage, G., Denning, D. W., & Bowyer, P. (2013). The cdr1B efflux transporter is associated with non-cyp51a-mediated itraconazole resistance in *Aspergillus fumigatus*. *J Antimicrob Chemother*, 68(7), 1486–1496. https://doi.org/10.1093/jac/dkt075

Franconi, I., Poma, N., Rizzato, C., Maltinti, L., Falcone, M., Tavanti, A., & Lupetti, A. (2025). The S862C amino acid change in CpMrr1 confers fluconazole resistance in *Candida parapsilosis*. *JAC Antimicrob Resist*, *7*(3), dlaf051. https://doi.org/10.1093/jacamr/dlaf051

Garnaud, C., Champleboux, M., Maubon, D., Cornet, M., & Govin, J. (2016). Histone deacetylases and their inhibition in Candida species. *Front Microbiol*, *7*, 1238. https://doi.org/10.3389/fmicb.2016.01238

Gebremariam, T., Alkhazraji, S., Alqarihi, A., Wiederhold, N. P., Shaw, K. J., Patterson, T. F., Filler, S. G., & Ibrahim, A. S. (2020). Fosmanogepix (APX001) is effective in the treatment of pulmonary murine mucormycosis due to *Rhizopus arrhizus*. *Antimicrob Agents Chemother*, *64*(6). https://doi.org/10.1128/AAC.00178-20

Ghannoum, M. A., & Rice, L. B. (1999). Antifungal agents: Mode of action, mechanisms of resistance, and correlation of these mechanisms with bacterial resistance. *Clin Microbiol Rev*, 12(4), 501–517. https://doi.org/10.1128/CMR.12.4.501

Golan, N., Schwartz-Perov, S., Landau, M., & Lipke, P. N. (2022). Structure and conservation of amyloid spines from the *Candida albicans* Als5 adhesin. *Front Mol Biosci*, 9, 926959. https://doi.org/10.3389/fmolb.2022.926959

Gravelat, F. N., Ejzykowicz, D. E., Chiang, L. Y., Chabot, J. C., Urb, M., Macdonald, K. D., Al-Bader, N., Filler, S. G., & Sheppard, D. C. (2010). *Aspergillus fumigatus* MedA governs adherence, host cell interactions and virulence. *Cell Microbiol*, 12(4), 473–488. https://doi.org/10.1111/j.1462-5822.2009.01408.x

Hagiwara, D., Miura, D., Shimizu, K., Paul, S., Ohba, A., Gonoi, T., Watanabe, A., Kamei, K., Shintani, T., Moye-Rowley, W. S., Kawamoto, S., & Gomi, K. (2017). A novel Zn2-Cys6 transcription factor AtrR plays a key role in an azole resistance mechanism of *Aspergillus fumigatus* by Co-regulating cyp51A and cdr1B expressions. *PLoS Pathog*, *13*(1), e1006096. https://doi.org/10.1371/journal.ppat.1006096

Hawksworth, D. L., & Lucking, R. (2017). Fungal diversity revisited: 2.2 to 3.8 million species. *Microbiol Spectr*, *5*(4). https://doi.org/10.1128/microbiolspec.FUNK-0052-2016

Healey, K. R., Kordalewska, M., Jimenez Ortigosa, C., Singh, A., Berrio, I., Chowdhary, A., & Perlin, D. S. (2018). Limited ERG11 mutations identified in isolates of *Candida auris* directly contribute to reduced azole susceptibility. *Antimicrob Agents Chemother*, *62*(10). https://doi.org/10.1128/AAC.01427-18

Healey, K. R., & Perlin, D. S. (2018). Fungal resistance to echinocandins and the MDR phenomenon in *Candida glabrata*. *J Fungi (Basel)*, *4*(3). https://doi.org/10.3390/jof4030105

Heredia, M., & Andes, D. (2021). Contributions of Extracellular vesicles to fungal biofilm pathogenesis. *Curr Top Microbiol Immunol*, 432, 67–79. https://doi.org/10.1007/978-3-030-83391-6_7

Hill, J. A., Ammar, R., Torti, D., Nislow, C., & Cowen, L. E. (2013). Genetic and genomic architecture of the evolution of resistance to antifungal drug combinations. *PLoS Genet*, 9(4), e1003390. https://doi.org/10.1371/journal.pgen.1003390

Iñigo, M., & Del Pozo, J. L. (2018). Fungal biofilms: From bench to bedside. *Rev Esp Quimioter*, 31(Suppl 1), 35–38.

Iracane, E., Vega-Estevez, S., & Buscaino, A. (2021). On and off: Epigenetic regulation of *C. albicans* morphological switches. *Pathogens*, *10*(11). https://doi.org/10.3390/pathogens10111463

Iyer, K. R., Robbins, N., & Cowen, L. E. (2022). The role of *Candida albicans* stress response pathways in antifungal tolerance and resistance. *iScience*, *25*(3). https://doi.org/10.1016/j.isci.2022.103953

Jain, T., Mishra, P., Kumar, S., Panda, G., & Banerjee, D. (2023). Molecular dissection studies of TAC1, a transcription activator of Candida drug resistance genes of the human pathogenic fungus *Candida albicans*. *Front Microbiol*, 14, 994873. https://doi.org/10.3389/fmicb.2023.994873

Jangir, P., Kalra, S., Tanwar, S., & Bari, V. K. (2023). Azole resistance in *Candida auris*: Mechanisms and combinatorial therapy. *Apmis*, 131(8), 442–462. https://doi.org/10.1111/apm.13336

Jeffery-Smith, A., Taori, S. K., Schelenz, S., Jeffery, K., Johnson, E. M., Borman, A., Candida auris Incident Management, T., Manuel, R., & Brown, C. S. (2018). Candida auris: A review of the literature. *Clin Microbiol Rev*, 31(1). https://doi.org/10.1128/CMR.00029-17

Kalra, S., Tanwar, S., & Bari, V. K. (2024). Insights into the role of sphingolipids in antifungal drug resistance. *Fung Biol Rev*, 47, 100342. https://doi.org/10.1016/j.fbr.2023.100342

Kanafani, Z. A., & Perfect, J. R. (2008). Antimicrobial resistance: Resistance to antifungal agents: Mechanisms and clinical impact. *Clin Infect Dis*, 46(1), 120–128. https://doi.org/10.1086/524071

Katragkou, A., McCarthy, M., Alexander, E. L., Antachopoulos, C., Meletiadis, J., Jabra-Rizk, M. A., Petraitis, V., Roilides, E., & Walsh, T. J. (2015). In vitro interactions between farnesol and fluconazole, amphotericin B or micafungin against *Candida albicans* biofilms. *J Antimicrob Chemother*, 70(2), 470–478. https://doi.org/10.1093/jac/dku374

Kaur, J., & Nobile, C. J. (2023). Antifungal drug-resistance mechanisms in Candida biofilms. *Curr Opin Microbiol*, 71, 102237. https://doi.org/10.1016/j.mib.2022.102237

Kordalewska, M., Lee, A., Park, S., Berrio, I., Chowdhary, A., Zhao, Y., & Perlin, D. S. (2018). Understanding echinocandin resistance in the emerging pathogen *Candida auris*. *Antimicrob Agents Chemother*, *62*(6). https://doi.org/10.1128/AAC.00238-18

Kovács, R., & Majoros, L. (2020). Fungal quorum-sensing molecules: A review of their antifungal effect against Candida biofilms. *J Fungi*, *6*(3). https://doi.org/10.3390/jof6030099

Kristanc, L., Bozic, B., Jokhadar, S. Z., Dolenc, M. S., & Gomiscek, G. (2019). The pore-forming action of polyenes: From model membranes to living organisms. *Biochim Biophys Acta Biomembr*, 1861(2), 418–430. https://doi.org/10.1016/j.bbamem.2018.11.006

Kuhbacher, A., Peiffer, M., Hortschansky, P., Merschak, P., Bromley, M. J., Haas, H., Brakhage, A. A., & Gsaller, F. (2022). Azole resistance-associated regulatory motifs within the promoter of cyp51A in *Aspergillus fumigatus*. *Microbiol Spectr*, 10(3), e0120922. https://doi.org/10.1128/spectrum.01209-22

Kumar, A., Francis, A., Hans, S., & Thakur, A. (2024a). Unravelling drug resistance in candida species: Genetic, biofilm, transcriptional, and epigenetic perspectives. In S. Hameed & P. Vijayaraghavan (Eds.), *Recent Advances in Human Fungal Diseases: Progress and Prospects* (pp. 203–237). Springer Nature Singapore. https://doi.org/10.1007/978-981-97-4909-6_8

Kumar, K., Pareek, A., & Kaur, R. (2024b). SWI/SNF complex-mediated chromatin remodeling in *Candida glabrata* promotes immune evasion. iScience, 27(4), 109607. https://doi.org/10.1016/j.isci.2024.109607

Lee, Y., Puumala, E., Robbins, N., & Cowen, L. E. (2021). Antifungal drug resistance: Molecular mechanisms in *Candida albicans* and beyond. *Chem Rev*, 121(6), 3390–3411. https://doi.org/10.1021/acs.chemrev.0c00199

Lepesheva, G. I., & Waterman, M. R. (2007). Sterol 14alpha-demethylase cytochrome P450 (CYP51), a P450 in all biological kingdoms. *Biochim Biophys Acta*, 1770(3), 467–477. https://doi.org/10.1016/j.bbagen.2006.07.018

Li, J., Aubry, L., Brandalise, D., Coste, A. T., Sanglard, D., & Lamoth, F. (2024). Upc2-mediated mechanisms of azole resistance in *Candida auris*. *Microbiol Spectr*, 12(2), e0352623. https://doi.org/10.1128/spectrum.03526-23

Lin, C.-J., & Chen, Y.-L. (2018). Conserved and divergent functions of the cAMP/PKA signaling pathway in *Candida albicans* and *Candida tropicalis*. *J Fungi*, 4(2), 68. https://www.mdpi.com/2309-608X/4/2/68

Lin, C. J., Sasse, C., Gerke, J., Valerius, O., Irmer, H., Frauendorf, H., Heinekamp, T., Straßburger, M., Tran, V. T., Herzog, B., Braus-Stromeyer, S. A., & Braus, G. H. (2015). Transcription factor SomA is required for adhesion, development and virulence of the human pathogen *Aspergillus fumigatus*. *PLoS Pathog*, *11*(11), e1005205. https://doi.org/10.1371/journal.ppat.1005205

Lin, J., Zhao, Y., Ferraro, A. R., Yang, E., Lewis, Z. A., & Lin, X. (2019). Transcription factor Znf2 coordinates with the chromatin remodeling SWI/SNF complex to regulate cryptococcal cellular differentiation. *Commun Biol*, 2(1), 412. https://doi.org/10.1038/s42003-019-0665-2

Liu, Y., & Filler, S. G. (2011). *Candida albicans* Als3, a multifunctional adhesin and invasin. *Eukaryot Cell*, 10(2), 168–173. https://doi.org/10.1128/ec.00279-10

Liu, S., Le Mauff, F., Sheppard, D. C., & Zhang, S. (2022). Filamentous fungal biofilms: Conserved and unique aspects of extracellular matrix composition, mechanisms of drug resistance and regulatory networks in *Aspergillus fumigatus*. *npj Biofilms Microb*, 8(1), 83. https://doi.org/10.1038/s41522-022-00347-3

Liu, H., Lee, M. J., Solis, N. V., Phan, Q. T., Swidergall, M., Ralph, B., Ibrahim, A. S., Sheppard, D. C., & Filler, S. G. (2016). *Aspergillus fumigatus* CalA binds to integrin α(5) β(1) and mediates host cell invasion. *Nat Microbiol*, 2, 16211. https://doi.org/10.1038/nmicrobiol.2016.211

Lohberger, A., Coste, A. T., & Sanglard, D. (2014). Distinct roles of *Candida albicans* drug resistance transcription factors TAC1, MRR1, and UPC2 in virulence. *Eukaryot Cell*, 13(1), 127–142. https://doi.org/10.1128/EC.00245-13

Louvet, M., Li, J., Brandalise, D., Bachmann, D., Sala de Oyanguren, F., Labes, D., Jacquier, N., Genoud, C., Mucciolo, A., Coste, A. T., Sanglard, D., & Lamoth, F. (2024). Ume6-dependent pathways of morphogenesis and biofilm formation in *Candida auris*. *Microbiol Spectr*, 12(11), e0153124. https://doi.org/10.1128/spectrum.01531-24

Lu, Y., Su, C., Unoje, O., & Liu, H. (2014). Quorum sensing controls hyphal initiation in *Candida albicans* through Ubr1-mediated protein degradation. *Proc Natl Acad Sci*, 111(5), 1975–1980. https://doi.org/10.1073/pnas.1318690111

MacPherson, S., Akache, B., Weber, S., De Deken, X., Raymond, M., & Turcotte, B. (2005). *Candida albicans* zinc cluster protein Upc2p confers resistance to antifungal drugs and is an activator of ergosterol biosynthetic genes. *Antimicrob Agents Chemother*, 49(5), 1745–1752. https://doi.org/10.1128/aac.49.5.1745-1752.2005

Madaan, K., & Bari, V. K. (2023). Emerging role of sphingolipids in amphotericin B drug resistance. *Microb Drug Resist*, 29(8), 319–332. https://doi.org/10.1089/mdr.2022.0353

Magditch, D. A., Liu, T.-B., Xue, C., & Idnurm, A. (2012). DNA mutations mediate microevolution between host-adapted forms of the pathogenic fungus *Cryptococcus neoformans*. *PLOS Pathog*, 8(10), e1002936. https://doi.org/10.1371/journal.ppat.1002936

Malavia-Jones, D., Farrer, R. A., Stappers, M. H. T., Edmondson, M. B., Borman, A. M., Johnson, E. M., Lipke, P. N., & Gow, N. A. R. (2023). Strain and temperature dependent aggregation of *Candida auris* is attenuated by inhibition of surface amyloid proteins. *Cell Surf*, 10, 100110. https://doi.org/10.1016/j.tcsw.2023.100110

Martins, M., Uppuluri, P., Thomas, D. P., Cleary, I. A., Henriques, M., Lopez-Ribot, J. L., & Oliveira, R. (2010). Presence of extracellular DNA in the *Candida albicans* biofilm matrix and its contribution to biofilms. *Mycopathologia*, 169(5), 323–331. https://doi.org/10.1007/s11046-009-9264-y

Mehta, D., Saini, V., & Bajaj, A. (2023). Recent developments in membrane targeting antifungal agents to mitigate antifungal resistance. *RSC Med Chem*, 14(9), 1603–1628. https://doi.org/10.1039/D3MD00151B

Mesa-Arango, A. C., Scorzoni, L., & Zaragoza, O. (2012). It only takes one to do many jobs: Amphotericin B as antifungal and immunomodulatory drug. *Front Microbiol*, 3, 286. https://doi.org/10.3389/fmicb.2012.00286

Mitsuyama, J., Nomura, N., Hashimoto, K., Yamada, E., Nishikawa, H., Kaeriyama, M., Kimura, A., Todo, Y., & Narita, H. (2008). In vitro and in vivo antifungal activities of T-2307, a novel arylamidine. *Antimicrob Agents Chemother*, 52(4), 1318–1324. https://doi.org/10.1128/AAC.01159-07

Morschhauser, J. (2016). The development of fluconazole resistance in *Candida albicans* - An example of microevolution of a fungal pathogen. *J Microbiol*, 54(3), 192–201. https://doi.org/10.1007/s12275-016-5628-4

Mukherjee, P. K., Chandra, J., Kuhn, D. M., & Ghannoum, M. A. (2003). Mechanism of fluconazole resistance in *Candida albicans* biofilms: Phase-specific role of efflux pumps and membrane sterols. *Infect Immun*, 71(8), 4333–4340. https://doi.org/10.1128/iai.71.8.4333-4340.2003

Narayanan, A., Kumar, P., Chauhan, A., Kumar, M., Yadav, K., Banerjee, A., Sharma, R. D., Rudramurthy, S. M., Chakrabarti, A., Sanyal, K., & Prasad, R. (2022). Directed evolution detects supernumerary centric chromosomes conferring resistance to azoles in *Candida auris*. *mBio*, *13*(6), e0305222. https://doi.org/10.1128/mbio.03052-22

Nett, J. E., Sanchez, H., Cain, M. T., & Andes, D. R. (2010). Genetic basis of Candida biofilm resistance due to drug-sequestering matrix glucan. *J Infect Dis*, 202(1), 171–175. https://doi.org/10.1086/651200

O'Kane, C. J., Weild, R., & Hyland, E. M. (2020). Chromatin structure and drug resistance in Candida spp. *J Fungi*, *6*(3). https://doi.org/10.3390/jof6030121

Paul, S., Verweij, P. E., Melchers, W. J. G., & Moye-Rowley, W. S. (2022). Differential functions of individual transcription factor binding sites in the tandem repeats found in clinically relevant cyp51A promoters in *Aspergillus fumigatus*. *mBio*, *13*(3), e0070222. https://doi.org/10.1128/mbio.00702-22

Perfect, J. R. (2017). The antifungal pipeline: A reality check. *Nat Rev Drug Discov*, 16(9), 603–616. https://doi.org/10.1038/nrd.2017.46

Perfect, J. R., Dismukes, W. E., Dromer, F., Goldman, D. L., Graybill, J. R., Hamill, R. J., Harrison, T. S., Larsen, R. A., Lortholary, O., Nguyen, M. H., Pappas, P. G., Powderly, W. G., Singh, N., Sobel, J. D., & Sorrell, T. C. (2010). Clinical practice guidelines for the management of cryptococcal disease: 2010 update by the Infectious Diseases Society of America. *Clin Infect Dis*, 50(3), 291–322. https://doi.org/10.1086/649858

Pfaller, M. A., & Diekema, D. J. (2010). Epidemiology of invasive mycoses in North America. *Crit Rev Microbiol*, 36(1), 1–53. https://doi.org/10.3109/10408410903241444

Pham, C. D., Iqbal, N., Bolden, C. B., Kuykendall, R. J., Harrison, L. H., Farley, M. M., Schaffner, W., Beldavs, Z. G., Chiller, T. M., Park, B. J., Cleveland, A. A., & Lockhart, S. R. (2014). Role of FKS mutations in *Candida glabrata*: MIC values, echinocandin resistance, and multidrug resistance. *Antimicrob Agents Chemother*, 58(8), 4690–4696. https://doi.org/10.1128/AAC.03255-14

Posch, W., Blatzer, M., Wilflingseder, D., & Lass-Flörl, C. (2018). *Aspergillus terreus*: Novel lessons learned on amphotericin B resistance. *Med Mycol*, 56(suppl_1), S73–S82. https://doi.org/10.1093/mmy/myx119

Prasad, R., & Rawal, M. K. (2014). Efflux pump proteins in antifungal resistance. *Front Pharmacol*, 5, 202. https://doi.org/10.3389/fphar.2014.00202

du Pre, S., Birch, M., Law, D., Beckmann, N., Sibley, G. E. M., Bromley, M. J., Read, N. D., & Oliver, J. D. (2020). The dynamic influence of Olorofim (F901318) on the cell morphology and organization of living cells of *Aspergillus fumigatus*. *J Fungi*, *6*(2). https://doi.org/10.3390/jof6020047

Quindos, G., Miranda-Cadena, K., San-Millan, R., Borroto-Esoda, K., Canton, E., Linares-Sicilia, M. J., Hamprecht, A., Montesinos, I., Tortorano, A. M., Prigitano, A., Vidal-Garcia, M., Marcos-Arias, C., Guridi, A., Sanchez-Reus, F., Machuca-Barcena, J., Rodriguez-Iglesias, M. A., Martin-Mazuelos, E., Castro-Mendez, C., Lopez-Soria, L., … Eraso, E. (2022). In vitro antifungal activity of Ibrexafungerp (SCY-078) against contemporary blood isolates from medically relevant species of Candida: A European study. *Front Cell Infect Microbiol*, *12*, 906563. https://doi.org/10.3389/fcimb.2022.906563

Rabaan, A. A., Eljaaly, K., Alfouzan, W. A., Mutair, A. A., Alhumaid, S., Alfaraj, A. H., Aldawood, Y., Alsaleh, A. A., Albayat, H., Azmi, R. A., AlKaabi, N., Alzahrani, S. J., AlBahrani, S., Sulaiman, T., Alshukairi, A. N., Abuzaid, A. A., Garout, M., Ahmad, R.,

& Muhammad, J. (2023). Psychogenetic, genetic and epigenetic mechanisms in *Candida auris*: Role in drug resistance. *J Infect Public Health*, 16(2), 257–263. https://doi.org/10.1016/j.jiph.2022.12.012

Rai, M. N., & Rai, R. (2024). H(3)K(4) methylation and demethylation in fungal pathogens: The epigenetic toolbox for survival and adaptation in the host. *Pathogens*, *13*(12). https://doi.org/10.3390/pathogens13121080

Rajendran, R., Mowat, E., McCulloch, E., Lappin, D. F., Jones, B., Lang, S., Majithiya, J. B., Warn, P., Williams, C., & Ramage, G. (2011). Azole resistance of *Aspergillus fumigatus* biofilms is partly associated with efflux pump activity. *Antimicrob Agents Chemother*, *55*(5), 2092–2097. https://doi.org/10.1128/aac.01189-10

Ramage, G., Kean, R., Rautemaa-Richardson, R., Williams, C., & Lopez-Ribot, J. L. (2025). Fungal biofilms in human health and disease. *Nat Rev Microbiol*, 23(6), 355–370. https://doi.org/10.1038/s41579-025-01147-0

Reichhardt, C., Ferreira, J. A., Joubert, L. M., Clemons, K. V., Stevens, D. A., & Cegelski, L. (2015). Analysis of the *Aspergillus fumigatus* biofilm extracellular matrix by solid-state nuclear magnetic resonance spectroscopy. *Eukaryot Cell*, 14(11), 1064–1072. https://doi.org/10.1128/ec.00050-15

Revankar, S. G., & Sutton, D. A. (2010). Melanized fungi in human disease. *Clin Microbiol Rev*, 23(4), 884–928. https://doi.org/10.1128/cmr.00019-10

Robbins, N., Caplan, T., & Cowen, L. E. (2017). Molecular evolution of antifungal drug resistance. *Annu Rev Microbiol*, 71, 753–775. https://doi.org/10.1146/annurev-micro-030117-020345

Robbins, N., Wright, G. D., & Cowen, L. E. (2016). Antifungal drugs: The current armamentarium and development of new agents. *Microbiol Spectr*, *4*(5). https://doi.org/10.1128/microbiolspec.FUNK-0002-2016

Rogers, P. D., & Barker, K. S. (2002). Evaluation of differential gene expression in fluconazole-susceptible and -resistant isolates of *Candida albicans* by cDNA microarray analysis. *Antimicrob Agents Chemother*, 46(11), 3412–3417. https://doi.org/10.1128/aac.46.11.3412-3417.2002

Rosam, K., Monk, B. C., & Lackner, M. (2020). Sterol 14alpha-demethylase ligand-binding pocket-mediated acquired and intrinsic azole resistance in fungal pathogens. *J Fungi*, *7*(1). https://doi.org/10.3390/jof7010001

Sant, D. G., Tupe, S. G., Ramana, C. V., & Deshpande, M. V. (2016). Fungal cell membrane-promising drug target for antifungal therapy. *J Appl Microbiol*, 121(6), 1498–1510. https://doi.org/10.1111/jam.13301

Santana, D. J., Anku, J. A. E., Zhao, G., Zarnowski, R., Johnson, C. J., Hautau, H., Visser, N. D., Ibrahim, A. S., Andes, D., Nett, J. E., Singh, S., & O'Meara, T. R. (2023). A *Candida auris*-specific adhesin, Scf1, governs surface association, colonization, and virulence. *Science*, 381(6665), 1461–1467. https://doi.org/10.1126/science.adf8972

Satish, S., & Perlin, D. S. (2019). Echinocandin resistance in *Aspergillus fumigatus* has broad implications for membrane lipid perturbations that influence drug-target interactions. *Microbiol Insights*, 12, 1178636119897034. https://doi.org/10.1177/1178636119897034

Schneider, S., & Morschhauser, J. (2015). Induction of *Candida albicans* drug resistance genes by hybrid zinc cluster transcription factors. *Antimicrob Agents Chemother*, 59(1), 558–569. https://doi.org/10.1128/AAC.04448-14

Schubert, S., Barker, K. S., Znaidi, S., Schneider, S., Dierolf, F., Dunkel, N., Aid, M., Boucher, G., Rogers, P. D., Raymond, M., & Morschhauser, J. (2011). Regulation of efflux pump expression and drug resistance by the transcription factors Mrr1, Upc2, and Cap1 in *Candida albicans*. *Antimicrob Agents Chemother*, 55(5), 2212–2223. https://doi.org/10.1128/AAC.01343-10

Segéňová, I., Víglaš, J., Pagáč, T., & Olejníková, P. (2025). Adaptation under pressure: Resistance and stress response interplay in clinical *Aspergillus fumigatus* isolates. *J Fungi*, 11(6), 428. https://www.mdpi.com/2309-608X/11/6/428

Singh, S., Fatima, Z., Ahmad, K., & Hameed, S. (2018). Fungicidal action of geraniol against *Candida albicans* is potentiated by abrogated CaCdr1p drug efflux and fluconazole synergism. *PLoS ONE*, *13*(8), e0203079. https://doi.org/10.1371/journal.pone.0203079

Singh, S. D., Robbins, N., Zaas, A. K., Schell, W. A., Perfect, J. R., & Cowen, L. E. (2009). Hsp90 governs echinocandin resistance in the pathogenic yeast *Candida albicans* via calcineurin. *PLoS Pathog*, *5*(7), e1000532. https://doi.org/10.1371/journal.ppat.1000532

Sofjan, A. K., Mitchell, A., Shah, D. N., Nguyen, T., Sim, M., Trojcak, A., Beyda, N. D., & Garey, K. W. (2018). Rezafungin (CD101), a next-generation echinocandin: A systematic literature review and assessment of possible place in therapy. *J Glob Antimicrob Resist*, 14, 58–64. https://doi.org/10.1016/j.jgar.2018.02.013

Song, L., Wang, S., Zou, H., Yi, X., Jia, S., Li, R., & Song, J. (2025). Regulation of ergosterol biosynthesis in pathogenic fungi: Opportunities for therapeutic development. *Microorganisms*, *13*(4). https://doi.org/10.3390/microorganisms13040862

Stempinski, P. R., Gerbig, G. R., Greengo, S. D., & Casadevall, A. (2023). Last but not yeast-The many forms of *Cryptococcus neoformans*. *PLoS Pathog*, *19*(1), e1011048. https://doi.org/10.1371/journal.ppat.1011048

Stewart, G. G. (2017). The structure and function of the yeast cell wall, plasma membrane and periplasm. In G. G. Stewart (Ed.), *Brewing and Distilling Yeasts* (pp. 55–75). Springer International Publishing. https://doi.org/10.1007/978-3-319-69126-8_5

Su, Y., Li, Y., Yi, Q., Xu, Y., Sun, T., & Li, Y. (2025). Insight into the mechanisms and clinical relevance of antifungal heteroresistance. *J Fungi*, *11*(2). https://doi.org/10.3390/jof11020143

Sudbery, P., Gow, N., & Berman, J. (2004). The distinct morphogenic states of *Candida albicans*. *Trends Microbiol*, 12(7), 317–324. https://doi.org/10.1016/j.tim.2004.05.008

Summers, D. K., Perry, D. S., Rao, B., & Madhani, H. D. (2020). Coordinate genomic association of transcription factors controlled by an imported quorum sensing peptide in *Cryptococcus neoformans*. *PLOS Genet*, 16(9), e1008744. https://doi.org/10.1371/journal.pgen.1008744

Sundstrom, P. (2002). Adhesion in Candida spp. *Cell Microbiol*, 4(8), 461–469. https://doi.org/10.1046/j.1462-5822.2002.00206.x

Szymankiewicz, M., Kamecki, K., Jarzynka, S., Koryszewska-Bagińska, A., Olędzka, G., & Nowikiewicz, T. (2021). Case report: Echinocandin-resistance *Candida glabrata* FKS mutants from patient following radical cystoprostatectomy due to muscle-invasive bladder cancer. *Front Oncol*, *11*, 794235. https://doi.org/10.3389/fonc.2021.794235

Tanwar, S., Kalra, S., & Bari, V. K. (2024). Insights into the role of sterol metabolism in antifungal drug resistance: A mini-review. *Front Microbiol*, 15, 1409085. https://doi.org/10.3389/fmicb.2024.1409085

Tian, X., He, G.-J., Hu, P., Chen, L., Tao, C., Cui, Y.-L., Shen, L., Ke, W., Xu, H., Zhao, Y., Xu, Q., Bai, F., Wu, B., Yang, E., Lin, X., & Wang, L. (2018). *Cryptococcus neoformans* sexual reproduction is controlled by a quorum sensing peptide. *Nat Microbiol*, 3(6), 698–707. https://doi.org/10.1038/s41564-018-0160-4

Timmermans, B., De Las Peñas, A., Castaño, I., & Van Dijck, P. (2018). Adhesins in *Candida glabrata*. *J Fungi*, *4*(2). https://doi.org/10.3390/jof4020060

Trzoss, M., Covel, J. A., Kapoor, M., Moloney, M. K., Soltow, Q. A., Webb, P. J., & Shaw, K. J. (2019). Synthesis of analogs of the Gwt1 inhibitor manogepix (APX001A) and in vitro evaluation against Cryptococcus spp. *Bioorg Med Chem Lett*, *29*(23), 126713. https://doi.org/10.1016/j.bmcl.2019.126713

Upadhyay, S. K., Mahajan, L., Ramjee, S., Singh, Y., Basir, S. F., & Madan, T. (2009). Identification and characterization of a laminin-binding protein of *Aspergillus fumigatus*: Extracellular thaumatin domain protein (AfCalAp). *J Med Microbiol*, 58(Pt 6), 714–722. https://doi.org/10.1099/jmm.0.005991-0

Uppuluri, P., Chaturvedi, A. K., Srinivasan, A., Banerjee, M., Ramasubramaniam, A. K., Köhler, J. R., Kadosh, D., & Lopez-Ribot, J. L. (2010). Dispersion as an important step in the *Candida albicans* biofilm developmental cycle. *PLoS Pathog*, *6*(3), e1000828. https://doi.org/10.1371/journal.ppat.1000828

Vanbiervliet, Y., Van Nieuwenhuyse, T., Aerts, R., Lagrou, K., Spriet, I., & Maertens, J. (2024). Review of the novel antifungal drug olorofim (F901318). *BMC Infect Dis*, *24*(1), 1256. https://doi.org/10.1186/s12879-024-10143-3

Vasicek, E. M., Berkow, E. L., Flowers, S. A., Barker, K. S., & Rogers, P. D. (2014). UPC2 is universally essential for azole antifungal resistance in *Candida albicans*. *Eukaryot Cell*, 13(7), 933–946. https://doi.org/10.1128/EC.00221-13

Vermes, A., Guchelaar, H. J., & Dankert, J. (2000). Flucytosine: A review of its pharmacology, clinical indications, pharmacokinetics, toxicity and drug interactions. *J Antimicrob Chemother*, 46(2), 171–179. https://doi.org/10.1093/jac/46.2.171

Vermeulen, P., Gruez, A., Babin, A. L., Frippiat, J. P., Machouart, M., & Debourgogne, A. (2022). CYP51 Mutations in the *Fusarium solani* species complex: First clue to understand the low susceptibility to azoles of the genus Fusarium. *J Fungi*, *8*(5). https://doi.org/10.3390/jof8050533

Verweij, P. E., Chowdhary, A., Melchers, W. J., & Meis, J. F. (2016). Azole resistance in *Aspergillus fumigatus*: Can we retain the clinical use of mold-active antifungal azoles? *Clin Infect Dis*, 62(3), 362–368. https://doi.org/10.1093/cid/civ885

Vu, B. G., Stamnes, M. A., Li, Y., Rogers, P. D., & Moye-Rowley, W. S. (2021). The *Candida glabrata* Upc2A transcription factor is a global regulator of antifungal drug resistance pathways. *PLoS Genet*, 17(9), e1009582. https://doi.org/10.1371/journal.pgen.1009582

Walker, L. A., Gow, N. A., & Munro, C. A. (2010). Fungal echinocandin resistance. *Fungal Genet Biol*, 47(2), 117–126. https://doi.org/10.1016/j.fgb.2009.09.003

Walker, L. A., Munro, C. A., de Bruijn, I., Lenardon, M. D., McKinnon, A., & Gow, N. A. (2008). Stimulation of chitin synthesis rescues *Candida albicans* from echinocandins. *PLoS Pathog*, *4*(4), e1000040. https://doi.org/10.1371/journal.ppat.1000040

Wang, T. W., Sofras, D., Montelongo-Jauregui, D., Paiva, T. O., Carolus, H., Dufrêne, Y. F., Alfaifi, A. A., McCracken, C., Bruno, V. M., Van Dijck, P., & Jabra-Rizk, M. A. (2024). Functional redundancy in *Candida auris* cell surface adhesins crucial for cell-cell interaction and aggregation. *Nat Commun*, 15(1), 9212. https://doi.org/10.1038/s41467-024-53588-5

Whaley, S. G., Berkow, E. L., Rybak, J. M., Nishimoto, A. T., Barker, K. S., & Rogers, P. D. (2016). Azole antifungal resistance in *Candida albicans* and emerging non-albicans Candida species. *Front Microbiol*, 7, 2173. https://doi.org/10.3389/fmicb.2016.02173

White, T. C., Marr, K. A., & Bowden, R. A. (1998). Clinical, cellular, and molecular factors that contribute to antifungal drug resistance. *Clin Microbiol Rev*, 11(2), 382–402. https://doi.org/10.1128/CMR.11.2.382

Wiederhold, N. P. (2017). Antifungal resistance: Current trends and future strategies to combat. *Infect Drug Resist*, 10, 249–259. https://doi.org/10.2147/IDR.S124918

Wiederhold, N. P. (2018). The antifungal arsenal: Alternative drugs and future targets. *Int J Antimicrob Agents*, 51(3), 333–339. https://doi.org/10.1016/j.ijantimicag.2017.09.002

Wuyts, J., Van Dijck, P., & Holtappels, M. (2018). Fungal persister cells: The basis for recalcitrant infections? *PLoS Pathog*, *14*(10), e1007301. https://doi.org/10.1371/journal.ppat.1007301

Yau, K. P. S., Weerasinghe, H., Olivier, F. A. B., Lo, T. L., Powell, D. R., Koch, B., Beilharz, T. H., & Traven, A. (2023). The proteasome regulator Rpn4 controls antifungal drug tolerance by coupling protein homeostasis with metabolic responses to drug stress. *PLoS Pathog*, *19*(4), e1011338. https://doi.org/10.1371/journal.ppat.1011338

Yue, H., Bing, J., Zheng, Q., Zhang, Y., Hu, T., Du, H., Wang, H., & Huang, G. (2018). Filamentation in *Candida auris*, an emerging fungal pathogen of humans: Passage through the mammalian body induces a heritable phenotypic switch. *Emerg Microbes Infect*, 7(1), 188. https://doi.org/10.1038/s41426-018-0187-x

Zaragoza, O., Garcia-Rodas, R., Nosanchuk, J. D., Cuenca-Estrella, M., Rodriguez-Tudela, J. L., & Casadevall, A. (2010). Fungal cell gigantism during mammalian infection. *PLoS Pathog*, *6*(6), e1000945. https://doi.org/10.1371/journal.ppat.1000945

Zarnowski, R., Noll, A., Chevrette, M. G., Sanchez, H., Jones, R., Anhalt, H., Fossen, J., Jaromin, A., Currie, C., Nett, J. E., Mitchell, A., & Andes, D. R. (2021). Coordination of fungal biofilm development by extracellular vesicle cargo. *Nat Commun*, 12(1), 6235. https://doi.org/10.1038/s41467-021-26525-z

Zhang, Y., Zeng, L., Huang, X., Wang, Y., Chen, G., Moses, M., Zou, Y., Xiong, S., Xue, W., Dong, Y., Tian, Y., Guan, M., Hu, L., Yin, Z., Zhou, D., Huang, X., & Chen, C. (2025). Targeting epigenetic regulators to overcome drug resistance in the emerging human fungal pathogen *Candida auris*. *Nat Commun*, 16(1), 4668. https://doi.org/10.1038/s41467-025-59898-6

Zhong, W., Jeffries, M. W., & Georgopapadakou, N. H. (2000). Inhibition of inositol phosphorylceramide synthase by aureobasidin A in Candida and Aspergillus species. *Antimicrob Agents Chemother*, 44(3), 651–653. https://doi.org/10.1128/AAC.44.3.651-653.2000

7 Deciphering the Role of Biofilm in Fungal Pathogenesis

Tushar Bandhu,
Patel Manish Chandrashekhar Kumar,
Himanshi Bihaniya, Chaitanya Ullhas Gatkal,
Garima Gangwar, and Anil Thakur

7.1 INTRODUCTION

Biofilms are one of the most emerging and critical factors in modern medical mycology that plays a central role in promoting drug resistance, treatment failure, and patient mortality. The sessile form of fungi associated with such biofilms is the member of complex and structured community embedded in protective, self-synthesized matrix (building block of biofilm). Fungal infections usually disregarded threat to global health; however, they have been associated with broad spectrum of clinical conditions. The severity and outcome of these infections are majorly dependent on multiple factors, which include the strength of host's immune system, virulence strategies linked to the fungi, environmental stress, and predisposing risk factors like implanted medical devices or underlying illnesses (Ramage et al., 2025).

Infections caused associated with fungi are broadly classified into several groups:

Superficial, subcutaneous, and invasive mycoses: Cutaneous and superficial mycoses involve infections of the mucous membrane, hair, skin, and nails, which are majorly caused by dermatophytes. Though not life-threatening, these infections can act as reservoirs and cause severe illness in immunocompromised patients. Conditions such as sporotrichosis, chromoblastomycosis, and mycetoma fall into this group.

Invasive mycoses are the most critical group clinically, which affect internal organs and disseminate throughout the body, often leading to life-threatening disease.

7.1.1 Global Health Burden and Prevalence

Once considered rare and primarily opportunistic, fungal infections have now emerged as major clinical challenges in modern medicine, profoundly impacting patient survival and healthcare costs. Fungal infection affects nearly 300 million individual around the globe, which results in more than 1.5 million deaths each year (Joshi et al., 2024). These mortality rates are comparable to those of tuberculosis and even surpass malaria. Beyond this, fungal infections inflict a huge amount of

DOI: 10.1201/9781003633952-9

financial damage on healthcare systems. Invasive candidiasis, for example, can cost over $40,000 per case in the United States. Combining the intensive care requirements and the cost of antifungals with the number of patients, the global cost of fungal diseases easily reaches billions of dollars yearly (Benedict et al., 2019).

Several fungal species have become especially concerning due to their capability to produce biofilms and develop resilience against antifungal drugs:

Candida spp.: Members of this genus are among the top four causes of hospital-acquired bloodstream infections, with incidence rates ranging between 2 and 14 per 1,000 ICU admissions depending on the population and region (Wisplinghoff et al., 2004). Importantly, non-*albicans* species, such as *Candida tropicalis, Candida glabrata*, and *Candida parapsilosis,* and the emerging multidrug-resistant *Candida auris* are being increasingly reported. These yeasts are highly adept at forming biofilms on mucosal tissues and indwelling devices (Romero & Geddes-McAlister, 2025). Hospital-acquired candidemia carries a mortality rate of 25%–35%, prolongs hospitalization by 10–30 days and adds over $40,000 in healthcare costs per episode (Parambath et al., 2024).

Aspergillus spp.: *Aspergillus fumigatus* is the causative agent of invasive aspergillosis, especially in patients with neutropenia or chronic lung disease. Its ability to form biofilm in the respiratory tract and on medical devices significantly reduces antifungal efficacy, complicating treatment (Kowalski et al., 2020). Globally, invasive aspergillosis affects 200,000–300,000 people annually, with mortality rates ranging from 30% to 95% depending on host status and infection site (Patterson et al., 2016).

Cryptococcus spp.: *Cryptococcus neoformans* remains a major cause of life-threatening meningoencephalitis in HIV-infected individuals. The disease affects an estimated 220,000 individuals each year, resulting in 181,000 fatalities (Rajasingham et al., 2017). The related species *Cryptococcus gattii* has expanded the epidemiology of cryptococcosis by infecting otherwise healthy individuals. Biofilm formation of Cryptococcus within the central nervous system is believed to contribute to relapse and treatment failure.

Emerging fungal pathogens such as *Scedosporium* and *Lomentospora* are now increasingly recognized as important causes of life-threatening disease, particularly in patients with chronic lung conditions or following near-drowning episodes. These fungi are highly adept at forming biofilms and often display intrinsic or acquired resistance to multiple antifungal drugs, complicating treatment strategies (Mello et al., 2022).

Aspergillus fumigatus, *Candida albicans*, *Candida auris*, and *Cryptococcus neoformans* have all been listed as organisms of "critical priority" by the WHO in the initial fungal priority pathogen list published in 2022 (World Health Organization, 2022). It is estimated that climate change will expand the distribution of endemic mycoses and facilitate the establishment of pathogenic fungi in previously unsuitable environments (Fisher et al., 2012), thus increasing the spread of fungal diseases.

7.1.2 The Increasing Danger of Antifungal Resistance

Fungi have cellular machinery mechanisms resembling those of eukaryotic organisms, hence their mechanism of resistance is more complex than that of comparatively well-understood bacteria (Fisher et al., 2024). Resistance can develop through several pathways, including mutations in drug targets such as the ergosterol biosynthesis enzyme (affecting azole efficacy) or β-glucan synthase (reducing echinocandin susceptibility), higher expression of efflux pumps from the Major facilitator superfamily (MFS) and ATP-binding cassette (ABC), metabolic reprogramming, and enhanced drug tolerance in biofilm-associated states.

Biofilms, in particular, are a major driver of antifungal treatment failure. Cells within biofilms differ markedly from their free-living counterparts, both structurally and functionally. The ECM acts as a barrier, which limits antifungal penetration and sometimes directly binding or inactivating drug molecules (Kowalski et al., 2020; Roy & Gow, 2023; Shelar et al., 2023).

7.1.3 The Present Restrictions on Antifungal Treatments

The limited availability of antifungal drugs further amplifies this problem. Currently, only three main systemic drug classes, viz., azoles, polyenes, and echinocandins, are widely used to counter invasive fungal pathogens. This narrow pharmacological arsenal is further constrained by issues of toxicity and poor efficacy against biofilm-associated infections (Müller, 2014; Pereira et al., 2021). They fail to penetrate biofilms, resulting in inefficient killing and recurrent infections. When compared to the huge antibacterial industry, drug development in fungi remains limited and lagging due to a small market and long development time (Niño-Vega et al., 2024).

Clinically, azoles are commonly used as they exhibit broad-spectrum action and oral bioavailability; however, increasing cases of azole-resistant *A. fumigatus* resistance and cytochrome P450 inhibition hamper their favorability (Pappas et al., 2015).

Echinocandins are given intravenously and show good antifungal activity against *Candida* species; however, cases of resistance in *C. glabrata* and poor activity against *Cryptococcus* limit their use (Kriegl et al., 2025). The lack of oral preparations further complicates long-term therapy, as patients often require hospitalization or complex outpatient parenteral administration.

The lack of oral preparations further complicates long-term therapy, as patients often require hospitalization or complex outpatient parenteral administration.

Polyenes, such as amphotericin B, maintain broad-spectrum efficacy with relatively low resistance rates. However, their use is constrained by notable toxicity, most significantly nephrotoxicity with conventional amphotericin B deoxycholate (Kraft et al., 2023). Lipid-based amphotericin B formulations have improved safety profiles, but their high cost limits accessibility, especially in resource-constrained healthcare systems.

7.1.3.1 Emergence of Drug-Resistant Fungus Strains

A particularly concerning development is the rise of multidrug-resistant (MDR) fungal pathogens, with *C. auris* standing as the most alarming example (Jamil et al., 2024). Since its identification in 2009, *C. auris* has been involved in multiple healthcare-associated outbreaks globally with alarming rates of mortality and frequent resistance across antifungal classes. Many isolates are resistant to fluconazole, and some show pan-resistance to azoles, echinocandins, and polyenes. Its ability to persist on medical devices and hospital surfaces, partly due to robust biofilm formation, further complicates infection control (Romero & Geddes-McAlister, 2025). Another growing concern is azole-resistant *A. fumigatus*. Resistance rates, driven by both medical azole usage and exposure to triazole fungicides in agriculture, have surpassed 20% in some European countries, significantly compromising the management of invasive aspergillosis (Kowalski et al., 2020). Similarly, echinocandin-resistant *C. glabrata* has emerged as a critical clinical challenge. This species already demonstrates reduced azole susceptibility, and rising echinocandin resistance narrows available therapeutic options. Its propensity for biofilm development on medical devices compounds treatment difficulties (Shelar et al., 2023).

7.2 BIOFILMS AND THEIR IMPORTANCE IN MICROBIAL PATHOGENESIS AS A CRUCIAL VIRULENCE FACTOR

Biofilm formation is recognized as a key virulence factor among fungal pathogens, enabling them to persist under hostile conditions and evade antifungal treatments (Roy & Gow, 2023). Structurally, fungal biofilms are majorly composed of polysaccharides, proteins, and extracellular DNA complex formed by the multicellular community that are enclosed within a protective ECM. This matrix not only provides physical shielding but also sequesters antifungal drugs and supports intercellular communication through quorum-sensing (QS) pathways (Ramage et al., 2012). Biofilm-linked fungal infections are notoriously difficult to treat. The extracellular matrix, coupled with the diverse metabolic states of cells within the biofilm, significantly reduces antifungal susceptibility. Consequently, minimum inhibitory concentrations (MICs) often far exceed levels achievable with standard therapeutic dosing. Studies show that biofilm growth can increase antifungal tolerance by several orders of magnitude compared to planktonic cells. For instance, MICs for *C. albicans* in biofilm states can be up to 1,000 times higher than those for free-living cells, providing a clear explanation for persistent resistance and high relapse rates in biofilm-related infections (Shelar et al., 2023).

Several mechanisms underlie this phenomenon:

Matrix-mediated protection: The extracellular polymeric substance (EPS) surrounding biofilm cells restricts antifungal drug penetration. This barrier generates concentration gradients within the biofilm, promoting survival and potential selection of resistant subpopulations.

Metabolic heterogeneity: The three-dimensional biofilm architecture creates microenvironments with variable oxygen and nutrient levels. These gradients drive the emergence of metabolically diverse subpopulations, many of which exhibit altered antifungal susceptibility profiles (Kowalski et al., 2020).

Persister cells: Biofilms harbor a subset of dormant or slowly dividing cells, termed persister cells. These cells are inherently tolerant to antifungal drugs and play a central role in treatment failures and infection recurrence.

Given the limitations of the current antifungal arsenal, the possibility of biofilms to evade both host immunity and drug activity necessitates a deeper understanding of their biology.

This chapter introduces the clinical relevance of fungal biofilms, examining their contribution to the rising global burden of fungal infections and highlighting their status as key virulence determinants in today's healthcare settings.

The following sections of this chapter will therefore:

- Examine the structural and molecular foundations of fungal biofilms.
- Explore the regulatory mechanisms and signaling networks that govern their development.
- Highlight in vivo models that advance our understanding of host–pathogen interactions.
- Discuss current and emerging therapeutic approaches, as well as future strategies aimed at mitigating the global health burden posed by fungal biofilms.

7.3 DEVELOPMENT AND STRUCTURE OF FUNGAL BIOFILM FORMATION

7.3.1 Biofilms: Definition and Features

Biofilms are structured communities of microorganisms that irreversibly adhere to biotic or abiotic surfaces and become encased in a self-produced extracellular matrix (ECM) (Giammarino et al., 2025). In fungi, biofilms consist of fungal cells embedded in EPS, which provide protection against host immune defenses and confer resistance to antifungal agents. Unlike planktonic cells, which grow freely in suspension, fungal cells within biofilms synthesize their own matrix and adopt specialized phenotypes (Desai et al., 2014). Biofilms are the predominant mode of microbial growth during infections and are especially associated with medical devices such as catheters, prosthetic implants, and dentures. Their ability to persist contributes to chronic and recurrent infections.

7.3.2 Phases of Biofilm Development

Fungal biofilm formation proceeds through four distinct but interconnected stages: (1) adhesion, (2) proliferation, (3) maturation, and (4) dispersion (Giammarino et al., 2025).

7.3.2.1 Adhesion

The process begins with the attachment of fungal cells to a host tissue or abiotic surface. Adhesion is mediated by cell-surface molecules, including sugars, proteins, and lipids, which enhance aggregation and surface binding (Wang et al., 2024). In *C. albicans*, adhesive proteins belonging to the glycosylphosphatidylinositol (GPI)-anchored family are central to this stage. The *Als* (agglutinin-like sequence) family of adhesins is the best-characterized group, with orthologs also identified in *Candida tropicalis*, *C. parapsilosis*, and *C. auris* (Lagree & Mitchell, 2017).

7.3.2.2. Proliferation and Microcolony Formation

Following attachment, fungal cells undergo rapid growth and multiplication, leading to the establishment of microcolonies. This phase is characterized by increased transcriptional and metabolic activity, alongside the activation of QS mechanisms that coordinate community development. In *C. albicans* and filamentous molds, this stage is further marked by extensive filamentation, producing complex and interconnected three-dimensional biofilm structures (Ramage et al., 2025).

7.3.2.3 EPS Production and Maturation

As the biofilm matures, cells become embedded in a dense EPS, forming a highly organized three-dimensional structure with microcolonies, filaments, and water channels. These channels facilitate nutrient distribution, waste removal, and cell–cell communication. In *C. albicans* and *A. fumigatus*, hyphal and mycelial growth are prominent features of this stage. The ECM composition varies across fungi but generally includes polysaccharides, proteins, and nucleic acids. In *C. albicans*, mannans and glucans are predominant components. This matrix is the defining hallmark of mature biofilms, providing mechanical stability and shielding cells from antifungal drugs and immune clearance (Ramage et al., 2025).

7.3.2.3.1 Dispersion

The final stage of the biofilm life cycle is dispersion, which enables the colonization of new surfaces and initiates a fresh cycle of biofilm formation. Dispersion occurs in two main forms:

- *Active dispersal:* It is a genetically regulated process in which cells reduce matrix production, express enzymes that degrade the biofilm, and regain motility, thereby facilitating the release of planktonic cells.
- *Passive dispersal:* It is driven by environmental stressors such as nutrient depletion, oxygen limitation, or mechanical disruption, which weaken the biofilm and allow cells to detach.

Once dispersed, fungal cells can adhere to new niches and establish secondary biofilms. This dynamic cycle ensures persistence under fluctuating environmental conditions (Wang et al., 2024) (Figure 7.1).

7.4 COMPOSITION OF THE EXTRACELLULAR POLYMERIC SUBSTANCE

Fungal biofilms are characterized by their EPS, as it is an important key component of the biofilm. These components provide structural support and stability, while also helping in maintaining the architecture of the biofilms. Unlike the planktonic cells, the EPS surrounds sessile cells, which create highly systematic and protective microenvironment. Polysaccharides, proteins, lipids, and extracellular DNA (eDNA) are the major constituents of EPS that facilitate the development and stability to the biofilms (Wang et al., 2024).

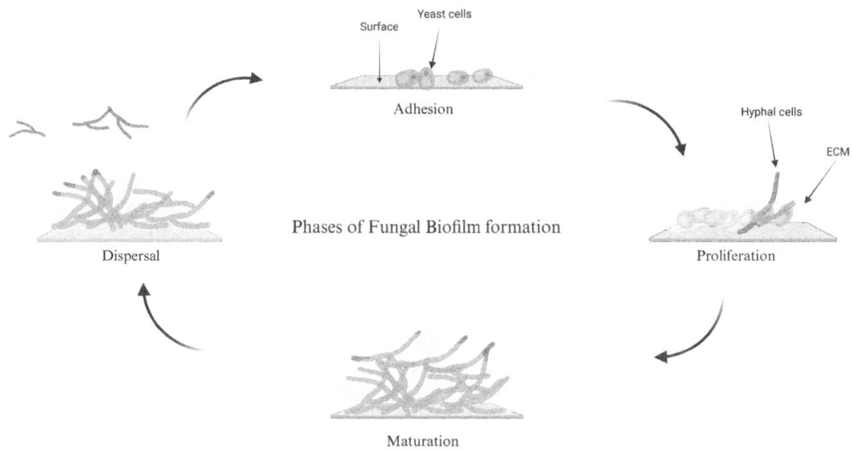

FIGURE 7.1 Different phases of fungal biofilm development: (1) Adhesion (involves attachment of cells onto the surface); (2) proliferation (here the adherent cells rapidly proliferate and show morphological alternation); (3) maturation (involves massive buildup of ECM around the cells); (4) dispersion (here a part of biofilm gets detached to start new cycle). (Created in Biorender.)

7.4.1 Polysaccharides

The major structural components of EPS are polysaccharides, which play a significant role in promoting intercellular and cell-matrix adhesion. Chitins, β-1,3-glucans, and β-1,6-glucans are the primary constituents of polysaccharides. Among them, β-1,3-glucans serves as a key fibrillar component that provides stiffness and fosters cell-cell cohesiveness. In addition, it lowers the effectivity of the drugs by entrapping the antifungal drug from the infected area. On the other hand, chitins support the biofilm structure and its integrity under stress. Mannans and glucans work together to create an intricate branching polymer in *C. albicans*.

7.4.2 The Proteins

The EPS proteins have a variety of structural and functional functions. Numerous proteins that stabilize connections between cells and surfaces are adhesins and cell-wall-associated proteins. Hydrolases, proteases, and lipases are examples of enzymatic proteins that alter the environment and aid in the uptake of nutrients. Crucially, the hydrophobicity of biofilms is influenced by their protein content; biofilms with a higher protein content typically have stronger hydrophobic contacts, which increase their resistance to host defenses and shear forces.

7.4.3 The Lipids

Despite being comparatively less prevalent, lipids have significant modulatory effects. They frequently combine with polysaccharides to create structures that resemble glycolipids or lipopolysaccharides, which function as organic surfactants. Through the regulation of surface tension and the encouragement of the development of cohesive layers, these lipid–polysaccharide complexes sustain the biofilm. Additionally, lipid components are involved in stress adaptation and signaling, especially when nutrients are scarce.

7.4.4 DNA Outside of Cells (eDNA)

The importance of eDNA as an EPS component in fungal biofilms is becoming more widely acknowledged. It is a structural "glue" that holds proteins and polysaccharides together and is produced by the autolysis of a subpopulation of cells. Additionally, eDNA promotes horizontal gene transfer, which may aid in the dissemination of resistance determinants. The integrity of biofilms in *Candida* species depends on eDNA, and it has been demonstrated that enzymatic eDNA breakdown (e.g., using DNase) alters biofilm structure and increases susceptibility to antifungals.

7.4.5 Metabolites and Tiny Molecules

Biofilm dynamics are regulated by small diffusible molecules that serve as chemical messengers. Morphogenesis, filamentation, and population density are all regulated by QS chemicals, including farnesol in *C. albicans*. Other secondary metabolites influence nutrition distribution, waste disposal, and redox balance by interacting with macromolecules to establish the internal chemical gradients of the biofilm.

These molecules work together to provide the chemical diversity that distinguishes sessile cells within the biofilm from planktonic cultures in terms of their physiological makeup.

7.5 MECHANISM OF BIOFILM-MEDIATED PATHOGENICITY IN FUNGI

Fungi are opportunistic pathogens that naturally infect immunocompromised individuals, and they can survive within the host via different strategies such as immune evasion, metabolic flexibility within nutrient-limited host niches, and tolerance to oxidative or thermal stress, which further enhance their pathogenic potential. Importantly, many of these virulence traits, especially biofilm formation, are tightly regulated by intricate signalling networks such as the cAMP–protein kinase A cascade and MAP kinase signaling, which are known to control adhesion, morphogenesis, and matrix production during biofilm development (Cottier & Hall, 2020; Katsipoulaki et al., 2024).

7.5.1 Key Features of Biofilm-Mediated Pathogenicity

Biofilms, a highly ordered way of microbial life, are formed when cells embed themselves in an extracellular polymeric matrix (EPS) that they create on their own. This matrix binds the cells to biotic and abiotic surfaces and encourages interactions between cells (Pierce et al., 2017). Single-species populations, clusters of closely related cells, or intricate multispecies consortia can all make up these hierarchical communities. Biofilm-associated fungi exhibit different physical, metabolic, and regulatory traits than free-living, planktonic cells, which enable them to adopt a more energy-efficient, sessile lifestyle (Ramage et al., 2025). In addition to mechanical stability, the 3D arrangement of the biofilm provides a controlled microenvironment. This microenvironment produces waste disposal channels, oxygen niches, and nutritional gradients that improve cell survival and promote intercellular communication (Liu et al., 2022). Additionally, the EPS matrix acts as a barrier that protects implanted fungal cells from environmental stressors such as oxidative stress, pH changes, desiccation, and antifungal medications (Donlan & Costerton, 2002; Fanning & Mitchell, 2012).

Fungi experience significant physiological and molecular alterations within this protected habitat. In order to survive host-imposed stresses, metabolic pathways are modified to deal with food scarcity, stress response pathways are triggered, and gene expression patterns are reprogrammed to promote biofilm-associated phenotypes (Lohse et al., 2017; Nett et al., 2007).

Furthermore, persister subpopulations—metabolically inactive but extremely resistant to antifungal medications—can be produced by biofilm cells, making treatment more challenging and encouraging infection recurrence (LaFleur et al., 2006) (Figure 7.2).

There are the key characteristics that highlight the pathogenic potential of biofilms.

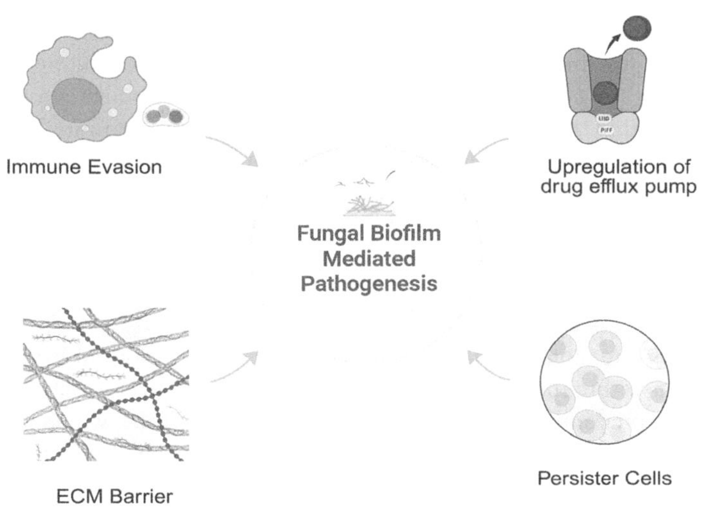

FIGURE 7.2 Key mechanisms contributing to fungal biofilm-mediated pathogenicity, including immune evasion, ECM barrier, upregulation of drug efflux pumps, and formation of persister cells.

7.5.2 Enhanced Drug Resistance

There are three major classes of antifungal agents that are used clinically for systemic infection, which are azoles, echinocandins, and polyenes (Kaur & Nobile, 2023). In planktonic populations, resistance mechanisms commonly include overexpression of efflux pumps and mutations in drug target genes such as ERG11, which encodes the azole target lanosterol demethylase, and FKS1, which encodes the echinocandin target, i.e., β-1,3 glucan synthase (Nett et al., 2010). Additionally, loss-of-function genetic mutations that are responsible for the ergosterol biosynthesis pathway, such as ERG3, also confer resistance to azole (Branco et al., 2017). The sessile form of fungi shows different modes of drug resistance along with the abovementioned mechanisms; one of the common ways is to produce ECM, which acts as a physical barrier or glue that limits the drug availability by capturing it within the biofilm and hampers the drug penetration; this upregulates the efflux pumps and increases the number of persister cells. All these factors are the major contributors for resistance development (Chang et al., 2018; Johnson et al., 2016; Nobile & Johnson, 2015).

7.5.3 Role of Extracellular Matrix

ECM acts as a platform that facilitates adhesion and grouping of fungal cells. It has high water retention capability and nutrient absorption. The fungal cells secrete some enzymes that can break ECM to release carbon and nitrogen in a nutrient-deprived condition. In a field of clinical mycology matrix plays an important role to protect cells from different stresses. Matrix material synthesis is influenced by environmental conditions and is a key component of *C. albicans* biofilms, which consist of proteins (55%), carbohydrates (25%), lipids (15%), and eDNA (5%) (Hawser et al., 1998). *C. neoformans* biofilm consists of glucuronoxylomannan (GXM) and sugars. Similarly, *Aspergillus* has Galactomannan, α-1,3 glucans, monosaccharides, polyols, melanins, and proteins such as hydrophobins in its biofilm (Rajendran et al., 2013). Matrix proteins are composed of both secretome components and cytoplasmic proteins. eDNA within the matrix plays an imperative role in maintaining biofilm architecture and influencing the drug susceptibility (Rajendran et al., 2013). It has been seen in different studies that matrix β-1,3-glucan can sequester antifungal agents, which contributes to drug resistance (Mitchell et al., 2013).

7.5.4 Role of Persister Cells and Efflux Pumps

Persister cells are non-dividing, phenotypic variants of normal wild-type cells subpopulation within fungal biofilms that can withstand high concentrations of antifungal agents (predominantly Amphotericin B and miconazole), which play a central role in the persistence and drug resistance of biofilm-associated infections (Brauner et al., 2016). These cells usually show a biphasic killing pattern when exposed to fungicidal drugs, where most of the population is eliminated, but a small fraction manages to survive (Lewis, 2010). Their survival strategies differ depending on the antifungal. For instance, when treated with miconazole, persister cells produce high levels of superoxide dismutases, which neutralize ROS and prevent cell death. In the case of amphotericin B, the persister cell produces Hsp21, which increases glycogen storage to reduce oxidative damage and Hsp90 that acts as a central modulator, influencing the cellular decision to initiate apoptosis or adopt persistence (Diezmann et al., 2015; Robbins et al., 2011). Along with persister cells, efflux pumps also contribute to the antifungal resistance. These are membrane transport proteins that maintain the cellular homeostasis by regulating the metabolites level inside the cell. They are typically involved in removing the foreign substances (drugs, toxins, etc.) out of the cell. All these contribute to the lower efficacy toward antifungal drug (Chang et al., 2018). CDR1 and CDR2 of ABC transporters families and MDR1 of the MFS transporter superfamily are observed to be upregulated in *C. albicans*, which enhances the tolerance against azoles and other antifungal drugs (White, 1997).

7.5.5 Immune Evasion and Long-Term Colonization

Fungi exhibit different approaches to escape host immune system, which include the morphogenic transition between yeast to hyphae, altering pathogen-associated molecular patterns (PAMPs), secretion of various degrading enzymes, and formation of biofilms. Host immune cells recognize PAMPs present on pathogens, which initiates downstream signaling cascades that normally lead to an immunological response such as ROS generation against the invading pathogens. Fungal cell adapts to such ROS-induced stress by modulating transcription and translational activity (Rana & Thakur, 2025). Biofilm cells also undergo transcriptional and translational reprogramming, which boosts stress resistance and promotes survival in hostile conditions, including during antifungal exposure.

Different theoretical models have been proposed by different groups to explain how fungal biofilms evade host immune defenses. The first theoretical model is known as *immunological silence*. Biofilm acts as a protective barrier that prevents phagocytic cells like macrophages and dendritic cells from effectively reaching the fungal community (Atriwal et al., 2021; Nett, 2016). Biofilms trigger only weak reactive oxygen species (ROS) responses, allowing them to effectively escape neutrophil-mediated killing (Coste et al., 2007). Hyphal cells are invasive in nature, which helps them to escape immune responses by physically penetrating host cells (Atriwal et al., 2021). The second theoretical model is known as *immune deviation*. This model explains how fungi diverge the host immune response with the help of various factors, which ultimately results in the generation of weak host immune response (Netea et al., 2008). The third theoretical model is known as immune resistance. This model hypothesizes that biofilm acts as a protective barrier, which may have been resulted due to the accumulation of structural or molecular adaptations. These adaptations help in resisting the drugs penetration and also the immune cells breaching within the biofilm (Netea et al., 2008). Along with immune evasion, fungal biofilms also regulate the host cytokine responses. For example, *C. albicans* disrupts the equilibrium between the pro- and anti-inflammatory response within host. Hyphal cells reduce the pro-inflammatory cytokines by blocking TLR4-mediated Th1 response activation. On the other hand, it represses the host immune response by activating the Th2 response through TLR2, which raises anti-inflammatory cytokines like IL-10 (Luo et al., 2013).

7.6 SIGNALING PATHWAYS INVOLVED IN FUNGAL BIOFILM PATHOGENESIS

Biofilm formation is a well-regulated and intricate process, controlled by multiple signaling pathways. These pathways act as a communicating system that helps fungi to respond to the external stimuli as well as adapt to it.

7.6.1 QUORUM-SENSING MOLECULES

Regulated expression of virulence genes is one of the key aspects associated with the pathogenesis of fungi (Casadevall & Pirofski, 1999). One of the effective strategies used by microorganism to regulate its pathogenic and virulent trait is QS. In this process, cells act as a single unit and communicate in a coordinated manner with the help of the chemical inducers called as QS molecules. These molecules help in the regulation of virulent traits in fungi and many other microorganisms. The concentration of QSM is directly proportional to the cell number, which triggers the coordinated responses. This results in the regulation at genetic levels. Also, based on the population density, the cells also coordinate the expression of virulence factors, which enable their survival within the host.

This synchronized behavior allows microorganisms to function as a single system, which leads to increase in their survival and virulence strategies (Bassler, 2002; Williams, 2007). It was discovered in bacteria for regulating processes like motility, biofilm formation, secretion of virulence factors, sporulation, etc. Farnesol, a naturally occurring sesquiterpene alcohol ($C_{15}H_{26}O$; molecular weight ~222.37 g/mol), was the first QS molecule discovered in eukaryotes. It is synthesized as an intermediate in the ergosterol biosynthetic pathway (Navarathna et al., 2007; Nickerson & Atkin, 2017) and plays a crucial role in regulating biofilm development and fungal adaptation within the host. Its primary function lies in controlling the yeast-to-hyphae transition, which is a crucial step in biofilm maturation and pathogenicity. High concentrations of farnesol act as a negative regulator by inhibiting hyphal formation, thereby restricting excessive biofilm expansion and maintaining microbial population balance (Navarathna et al., 2007). This inhibition prevents uncontrolled filamentation, which could otherwise trigger stronger immune responses. Other than morphogenesis, farnesol also affects interactions between different kingdoms and responses to stress. Research shows that it can modify oxidative stress pathways and trigger processes similar to apoptosis in fungal cells. This process helps to maintain population balance (Shchepin et al., 2003). Furthermore, farnesol disrupts bacterial QS systems and influences polymicrobial biofilms, therefore emphasizing its role in communication between kingdoms. In host–pathogen relationships, farnesol may help with long-term colonization. By suppressing hyphal growth and changing immune recognition patterns, it lowers host inflammatory responses and aids fungal survival. However, its ability to inhibit biofilm formation can make fungal cells more susceptible to antifungal treatments. This suggests a complex relationship between managing virulence and survival (Chen et al., 2018; Navarathna et al., 2007; Chen et al., 2018; Gaálová-Radochová et al., 2023; Rodrigues & Černáková, 2020) (Figure 7.3).

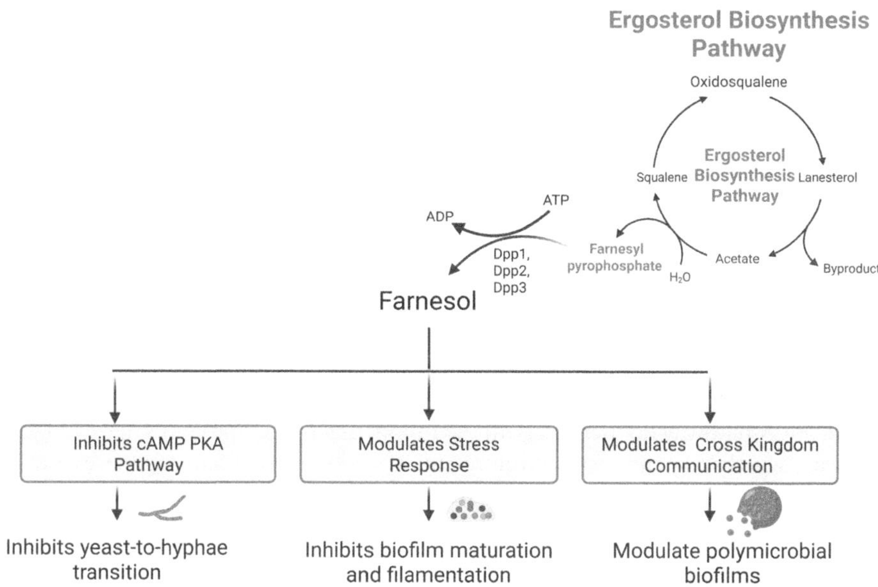

FIGURE 7.3 Role of farnesol (a key intermediate of ergosterol biosynthesis pathway) in fungal regulation and pathogenicity.

7.6.2 cAMP-PKA Pathway

The cAMP-PKA signaling pathway is one of the key regulating pathways in biofilm formation of *Saccharomyces cerevisiae* via control of the adhesin gene FLO11. When sucrose or maltose is available, sucrose is hydrolyzed by Suc2, a sucrose converting enzyme secreted by yeast, which hydrolyzes sucrose into glucose and fructose (Lengeler et al., 2000). These sugars, along with maltose, are taken up by hexose transporters Hxt and phosphorylated by hexokinases Hxk to glucose-6-phosphate, which increases the activity of adenylate cyclase (Cyrl). In parallel, glucose and sucrose are detected by G-protein coupled receptors Gprl–Gpa2, providing an additional activating signal enhancement to Cyrl (Di Como & Arndt, 1996). Activated Cyrl produces cAMP, which binds to Bcyl (a regulatory subunit of PKA), releasing the catalytic subunits Tpk1, Tpk2, and Tpk3 (Forehand et al., 2022; Loeb et al., 1999). These have opposing effects on FLO11, that is Tpk2 promotes expression by inactivating the repressor Sfl1 and activating Flo8, while Tpk1 and Tpk3 inhibit FLO11 by repressing Flo8. The balance between these subunits determines FLO11 expression (Herranz et al., 2005; Li et al., 1997). When FLO11 is active, it produces Flo11p, a cell-surface adhesin that mediates adhesion, cell–cell aggregation, invasive growth, and biofilm formation. Negative regulators like Ira1, Ira2, and Sfl1–Hda1 complex inhibit Cyrl or directly repress FLO11 (Xu et al., 2004).

In *C. albicans*, the cAMP-PKA cascade plays a central role in the management of biofilm formation and filamentous growth. Gprl, Gpr2, and Nce103 are principal upstream regulators of this cascade that activates adenylate cyclase Cyrl (Cabello et al., 2019; García-Sánchez et al., 2004; Kakade et al., 2016). This leads to the activation of cAMP, which further activates the catalytic components of PKA, i.e., Tpk1 and Tpk2, which further phosphorylate Efg1. Efg1 remains in active form in phosphorylated state, triggering the expression of the ume6 hyphal elongation gene, which in turn regulates downstream targets such as Hgc1 (Kaplan et al., 2003; Vylkova & Lorenz, 2014). Hgc1 regulates cell secretion in hyphae, whereas Eed1 is required for ume6 expression and thus for hyphal elongation (Shapiro et al., 2009). ume6, Eed1, and Hgc1 together constitute a core regulatory module for hyphal development. Moreover, ume6 and Eed1 repress the negative regulators Tup1 and Nrg1, thus indirectly inducing hyphal growth (Konwar et al., 2025). HWP1 and ALS3, two hyphal-specific genes vital for adhesion and biofilm formation, are controlled by Efg1 expression. Its importance is indicated by the defective biofilm formation observed in mutants that do not have Efg1. The cAMP-PKA pathway also crosstalks with other signaling pathways to facilitate the biosynthesis of biofilms. For instance, the cAMP-PKA pathway cooperates with the mitogen-activated protein kinase (MAPK) pathway, which responds to environmental stresses, in regulating gene expression during biofilm development (Khari et al., 2023).

7.6.3 MAP Kinase Pathway

MAPK signaling is among the most evolutionarily conserved eukaryotic pathways, which has a pivotal role in sensing environmental stresses, as well as in adaptive cellular response. In *Saccharomyces cerevisiae*, the MAPK pathway activates downstream transcription factors such as Dig1, Ste12, and Tec1, which regulate FLO11 expression. It is an important gene that promotes adhesion, which is a critical step in biofilm formation. The signaling pathway is activated by sensor protein Msb2, which has been believed to sense extracellular stimuli like alcohols (Chen & Lan, 2015; Nobile et al., 2008). Upon activation, Msb2 activates the protein complex Sho1-Cdc42-Ste20, which transduces the signal to central kinases Ste11 and Ste7. These kinases in turn phosphorylate and activate Kss1, a downstream MAP kinase that controls Dig1 (repressor of Ste12 and Tec1). Upon relief of this repression, Ste12 and Tec1 induce FLO11, which results in increased cell adhesion and biofilm development (Forehand et al., 2022). Likewise, in pathogenic fungi, such as *C. albicans*, Rho GTPases perceive cell wall disturbances via FKS1/2 and trigger the Smi1 and Slt2-mediated MAP kinase pathway. This further activates the MADS-box transcription factor RLM1, whose regulation oversees the synthesis of β-1,3-glucans and is involved in biofilm formation (Martin-Yken et al., 2003). MAPK pathway is activated by Msb2 and the MAP kinase Cek1. Msb2 is further cleaved by Secreted aspartyl proteases (SAPs), leaving Msb2 heavily glycosylated extracellular domain intact, which is required for the adhesion and transferring of the signal to phosphorylate Cek1. Activated Cek1 further stimulates the transcription factor Ace2, which regulates the hyphae-associated gene expression (Chen et al., 2020; Puri et al., 2012). Additionally, Ras1 regulates MAPK signaling by activating Cek1, and in parallel, it also activates Bcr1, Als1, Tec1, and Efg1 via PKA signaling pathway through Cyr1-Cap1 protein complex. Comparative analysis reveals that Smi1, Slt2, Rlm1, Msb2, and Cek1 in *C. albicans* share 55.39%, 69.70%, 64.44%, 47.92%, and 79.67% sequence identity, with their orthologs in *C. auris*, with Slt2 being the most conserved, showing 69.70% similarity with *S. cerevisiae* (Khari et al., 2023)

7.7 IN VIVO MODEL TO STUDY FUNGAL PATHOGENESIS

Fungi can infect a wide range of organisms, which include both vertebrates and invertebrates. Pathogenic fungi are mainly classified as invasive and non-invasive; they may result in both acute and chronic infections in human. Non-invasive fungi mainly cause superficial infections on skin, hair, and nail, and it requires long-term therapy for treatment. The invasive fungi are more pathogenic as they have the ability to invade the host tissue and can cause systemic infection, where they can spread and colonize to different vital organs such as the kidney, liver, lungs, and heart. When fungi adhere to and colonize any surface, they start to synthesize biofilm. It acts as a barrier or protective wall that prevents drugs from reaching fungi and also protects it from host immune responses (Dijck et al., 2018). So, it is important to study how the fungi evade the host immune system, colonize, and infect. Both invasive and non-invasive fungi have the ability to form biofilm. In many literatures, they have described the use of different model systems to investigate biofilm development and colonization, the molecular elements that contribute fungal pathogenesis such as genes involved, their regulation, and the different signaling pathways linked with colonization and biofilm development, which need to be studied in order to fully comprehend it (Kucharíková et al., 2015; Dijck et al., 2018; Wang et al., 2024 : Wang et al., 2022).

Figure 7.4 shows few such model systems.

7.7.1 Subcutaneous Catheter Mouse Model

Many of the of the Fungal infections are linked to Biomedical implants, catheter, and grafts. They adhere themselves to these materials, divide, and form biofilm that results in systemic infection in the implanted area (Martinez & Fries, 2010). In vitro

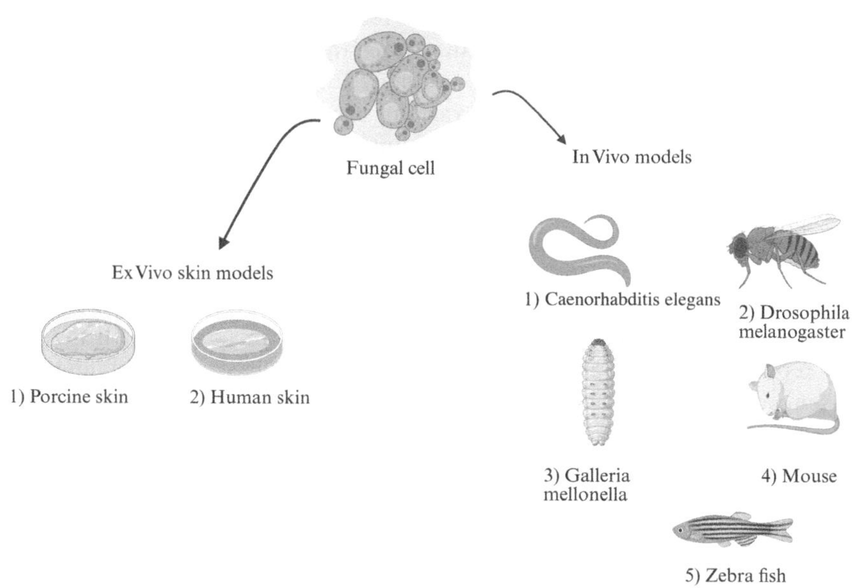

FIGURE 7.4 Various model systems that are used in investigating fungal biofilm development: (a) Ex vivo model and (b) in vivo model.

models were developed earlier to understand the biofilm formation, which mainly involves the use of synthetic materials involved in making of implants. However, this model is unable to replicate the in vivo conditions. To address such problems, i.e., fungal infection caused due to implants, Subcutaneous catheter mouse model has been developed that mimics the fungal infection caused due to catheter devices (Kucharíková et al., 2015). In order to study the catheter-associated fungal biofilm development in mice, a catheter seeded with fungal cell is implanted in the dorsal side of the mice. Additionally, this model gives an opportunity to implant multiple catheters in order to investigate the development of multiple biofilms in a single mouse, thus reducing the overuse of laboratory animals and adhering to the 3Rs in animal research (Costa-Orlandi et al., 2017). This model facilitates a great advantage of being compatible with bioluminescence imaging, which is a non-invasive method used for detecting the cell load and biofilm formation (Avci et al., 2018). In this model, mice are anesthetized beforehand, and the intended site for catheter placement was disinfected. Catheter pre-seeded with fungal cells was inserted subcutaneously into the dorsal region of the mice. After that, the wound was closed and disinfected. A reversal of anesthesia was given to mice, and they were transferred to a fresh and pre-warmed cage. After a certain defined time, i.e., second and sixth day, the mice were anesthetized for imaging. During this time, they were placed in an imaging chamber that acquires images and quantifies photon flux. Later to determine the fungal cell load on catheter, the catheters were surgically removed and the CFU was measured to estimate the fungal cell burden on them (Kucharíková et al., 2015; Vande Velde et al., 2014).

7.7.2 Porcine Skin Ex Vivo Model

Skin acts as an initial or innate barrier that restricts the entry of many pathogens inside the body. Many of the fungal infections in human are associated with the skin, which acts as a primary target for many fungi to get attached, where they form biofilm that shields them from various stresses and facilitates in their dispersion. As porcine skin resembles with human skin in different aspects such as width, vascular system, and repair system, this model can be used as a substitute in infection biology (Eix et al., 2022). In this model, Porcine skin was excised and subsequently washed with DPBS and 70% ethanol, followed by shaving to remove hair. The excised skin was then sectioned into small samples using a biopsy punch and subsequently transferred into DMEM medium containing FBS and Pen-Strap mix. Samples were incubated for 18 hours in a CO_2 incubator, with the media being replaced every 6 hours. The samples were then washed, dried, and placed on a semisolid medium. Later, fungal cells were inoculated onto the surface of the tissue and kept in a CO_2 incubator. At the end, SEM and fungal cell burden present in biofilm can be assessed (Horton et al., 2020). One can study polymicrobial biofilm formation and also determine the thickness by Gram and Hucker-Twort staining. Such a model can be used to comprehend the structure and content of the biofilm. Along with helping in studying the association of biofilm within polymicrobial community, it also aids in determining the effect of antifungal and antibiofilm (Pati et al., 2021).

7.7.3 Human Skin Ex Vivo Model

Numerous human models have been employed in order to investigate diverse research areas, including infections, diseases, and disorders. Ex vivo skin models have been used to study superficial skin infection caused by many microorganisms, especially fungi and bacteria. This model has been applied in many studies like colonization, epidemiology, infection biology, toxicological assessment, and process of wound healing (Corzo-León et al., 2019). This model initially involves the collection of a skin biopsy sample and remove all adipose tissues or fats followed by disinfection of skin sample before sectioning it into a small biopsy punch sample (Xu et al., 2012). This sample is then placed on top of semisolid media; further, fungal cells were applied on the top of this skin sample and incubated under CO_2 conditions. This approach enables the assessment of fungal cell burden and thickness of biofilm. Also, it helps in the evaluation of invasive characteristics of the infection (Eix et al., 2022).

7.7.4 *Galleria mellonella* Infection Model

It's a wax moth, a common pest of bee hives, which belongs to lepidoptera. It has a life span of 8–13 weeks, which is divided into four main phases, i.e., eggs, larva, pupa, and adult stage (insect). In recent years, this model has been adapted for investigating many infectious diseases and has been reported in many research publications. The use of such insect model has been encouraged by many research agencies and institutes, which correspond to 3R guidelines (Marena et al., 2025). *Galleria mellonella* has short life span and can produce large progeny that can be used for research. In many peer-reviewed publications, the use of this model to study fungal infection, evolutionary aspects, and antifungal drug screening has been mentioned. Its immune system resembles to mammalian immune system, as its immune cells can differentiate among various microbial invaders and can induce specific responses against them. It is easily available, i.e., can be purchased or can be cultured in a laboratory. Many phenotypic abnormalities are often seen during the period of infection; this acts as a marker or indicator to any infection or stress. Oral gavage, intra-hemocoel injection, and inoculation at last proleg are the sites of administration that are commonly employed in this model. The site of infection is selected in accordance to the study to be performed (Kavanagh & Sheehan, 2018; Freitak et al., 2014; Curtis et al., 2022). Sheehan and its colleague have investigated *Madurella mycetomatis* infection using this model by inoculating certain number of fungal cells into the left pro-leg of the larvae. They have observed dark melanin pigmentation, Grain or granuloma formation, and an increase in enzymes and proteins that aid in combating infection (Sheehan et al., 2020). *Beauveria bassiana* is widely used as an insecticide; this model has been used to study the effect of this insecticide. When this insect was injected with *B. bassiana* after certain days of post-infection, the insect develops dark pigmentation, and alternation in immune cells present in hemolymph (Vertyporokh et al., 2020; Pereira & Rossi, 2020) increases the production of phospholipase, which is directly linked to the virulence of fungus (Curtis et al., 2022).

7.7.5 Murine Infection Model

Many fungal infections in human are limited to a specific site such as oral, mucosal, subcutaneous, and invasive. Many model systems have been developed that help in investigating such site-restricted infections. However, the murine model has a greater advantage among them, as this model can be used to investigate all these site-restricted infections. It provides the same conditions that are present in many mammals, i.e., immune response, vasculature system, repair, and physiological conditions. The route of administration of the fungal pathogen varies depending upon the type of infection to be studied. Mainly three to four ways of fungal infection are currently been employed, i.e., subcutaneously catheter implantation, oral gavage, tongue swabbing, intraperitoneal, and intravenous (Vila et al., 2020). As previously described, the subcutaneous catheter model has been used to study implant-associated infection. In one study, to mimic oropharyngeal candidiasis, oral gavage is given to induce immunocompromised mice; later, CFU load of the candida was determined; also, stained tongue tissue biopsies were observed to analyze fungal colonization on the tongue (Solis & Filler, 2012). Intraperitoneal infection is frequently utilized to study systemic infection, as this site provides a better vascular system and provides an opportunity to fungus to move to various vital organs. For exploring the vaginal and oral colonization, fungal cells were rubbed in the mouth, teeth, and vagina. Later, after certain period of time, tissue biopsies were taken from mice in this part and were examined for the colonization and abundance of fungal cells (Rahman et al., 2007) (Figure 7.5).

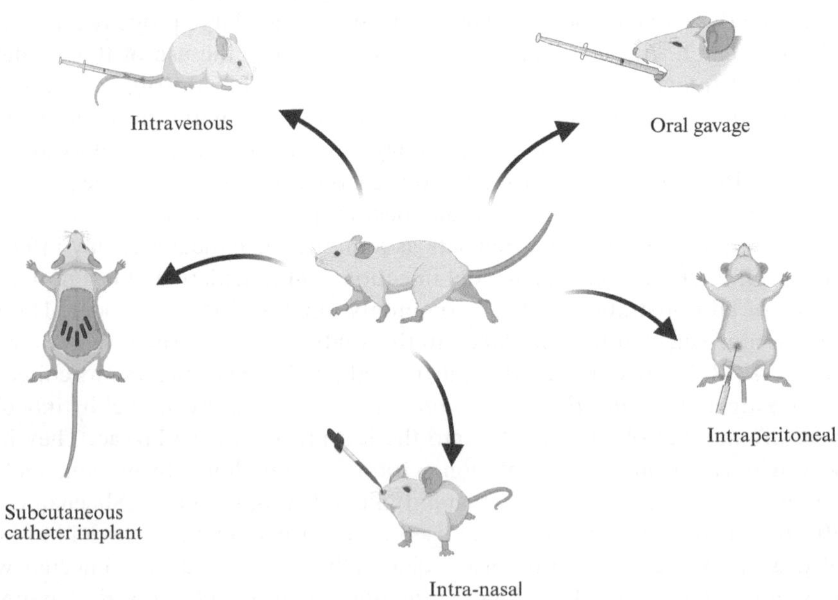

FIGURE 7.5 Different routes of administration of fungal infection in the mice model system: (a) Intravenous, (b) oral gavage, (c) intraperitoneal, (d) intra-nasal, and (e) subcutaneous catheter implants.

7.8 GENETIC REGULATION OF FUNGAL BIOFILM

During a fungal infection, fungal cells colonize to different tissues or body parts where they divide and form biofilm that protects fungal cell from host immune system, increases the pathogenesis/virulence, and makes them resistant against different antifungal drugs also it. Discovering more about biofilm, its composition, the genes that involves in the regulation of biofilm formation, could assist in understanding more about fungal pathogenesis, virulence, and resistance development. Such study helps in screening and development of antifungal strategy formulation. There are many genes that are differentially expressed in many fungi during biofilm formation process; some genes among them are highly conserved in many fungi. FLO11, EED1, HWP1, HWP2, and ECE1 genes are conserved among *Saccharomyces cerevisiae* (Khari et al., 2023). Biofilm formation is differentiated into four distinct stages, i.e., (1) Adherence, (2) Initiation/Proliferation, and (3) Dispersion. Different stages of biofilm development correspond to distinct gene expression. The adhesion in fungi is majorly promoted by early and late genes. Some of the early genes are ALS1, ALS2, ALS4, and PGA6, while the later expressed genes are PGA32, IFF4, PGA55, and IFF6 (Fox et al., 2015). Following adhesion, they start to secrete ECM, and there are several genes and factors that involve in synthesizing these ECM components, i.e., exopolysaccharide, mannan, glucan, and lipids. Rlm1 and Zap1 are the two most important transcriptional regulators in ECM production; among them, Zap 1 is a negative regulator, i.e., it inhibits Gca1 and Gca2, which code for glucoamylase enzyme that converts long-chain polysaccharide into short-chain polysaccharide, whereas Rlm1 promotes ECM synthesis by upregulating genes involved in mannan and glucan synthesis (Gsc1, AlG11, MNN1, KRE1, etc.). Efflux pump is one of the structural elements of the cell membrane that aid in boosting the virulence and resistance in fungi. As the biofilm formation progresses, its number rises (Khari et al., 2023; Horton & Nett, 2020; Lohse et al., 2017). The last stage is dispersion of the biofilm, and several studies have proven that Ume6, Nrg1, and Pes1/Nop7 are the important factors that regulate dispersion of the biofilm. Among three factors, Nrg1 and Pse1/Nop7 positively influence biofilm dispersion, whereas Ume6 negatively influences the biofilm development effect (Uppuluri et al., 2010). The activation and inactivation of these genes have mainly been mediated by certain signaling pathways such as GPCR, mTOR, QS, and pH-temperature stress. This pathway's ultimate outcome is the overexpression of genes linked to food deprivation, stress flexibility, and biofilm formation. Flo11 has primary regulatory effect in biofilm development in *S. cerevisiae*. In *Candida* species, Als1, Eap1, Brg1, Hgc1, Wor1, Adh5, Gca1, and Gca2 regulate different stages of biofilm development (Wang et al., 2024).

7.9 THERAPEUTIC APPROACHES AND FUTURE STRATEGIES

Biofilms often form on medical implants like pacemakers, prosthetic joints, dentures, and catheters. They can also develop on living tissues, such as mucosal linings. From the three major classes of antifungal drugs, azoles, such as fluconazole, are the majorly used for both mild and serious infections; these drugs inhibit the enzyme in ergosterol biosynthetic pathway. From the class of Polyenes, amphotericin B is one of the earliest and widely used for severe cases. Echinocandins, including caspofungin, are the newest group of antifungal drugs, but biofilms can resist antifungal drugs through several mechanisms, such as making more efflux pumps after drug exposure, mutations in target enzymes like ERG11, and changes in ergosterol and the protective ECM (Kaur & Nobile, 2023).

These current antifungal drugs mainly act by disrupting ergosterol synthesis (azoles), binding to ergosterol itself (polyenes), or inhibiting cell wall synthesis (echinocandins). Researchers are now investigating several new targets for the development of next-generation antifungal agents. One of the targets under investigation is fungal sphingolipids, which are essential components of cell membrane that play an important role in virulence; different studies suggested that disrupting sphingolipid biosynthesis can make Candida avirulent. Mutations affecting GPI anchor synthesis cause abnormal morphology, weakened cell walls, and reduced virulence in fungi such as *C. albicans* and *A. fumigatus*. Hsp90, a molecular chaperone essential for protein stability and stress responses, is being explored as a drug target (Robbins et al., 2016). Microbial attachment depends on the properties of the material used in the medical devices like catheters. Microbes interact with hydrophobic surfaces of devices by different adhesion proteins like fibronectin and fibrins. To avoid the formation of biofilm, hydrophilic coatings like hyaluronic acids and antimicrobial catheters are used (Francolini & Donelli, 2010).

7.10 COMBINATION THERAPIES

Combination therapies are needed because of the resistance to the conventional antifungals. It enhances the antifungal efficacy by synergistic interactions and allows dose reduction, minimizing the toxicity. Several studies have been performed that show the combination of antifungals with different drugs and natural products effective against fungal pathogens. Fluconazole, for example, when coupled with the calcium channel blocker verapamil, was found to downregulate ALS3 adhesin expression, lowering the biofilm adhesion,

whereas its combination with the antiviral ribavirin inhibited biofilm formation and hyphal growth. When combined with fluconazole, corticosteroids such as budesonide impaired phospholipase activity, interfered with efflux pumps, and caused death in Candida cells. Similarly, when amphotericin B was combined with the natural chemical eugenol, it enhances oxidative stress and calcium signaling in fungal cells. Biofilm inhibition has improved. Statins, such as lovastatin, acted synergistically with itraconazole by inhibiting ergosterol production, hence compromising membrane integrity (Tits et al., 2020).

These techniques, which combine antifungal medicines with repurposed pharmaceuticals, natural products, or innovative delivery vehicles, have the potential to improve treatment results, minimize medication dosages, and restrict toxicity. However, translation into clinical practice necessitates rigorous investigation of synergistic mechanisms, safety, and efficacy via well-designed studies.

Microbial attachment depends on the properties of the material used in the medical devices like catheters. Microbes interact with hydrophobic surfaces of devices by different adhesion proteins like fibronectin and fibrins. To avoid the formation of biofilm, hydrophilic coatings like hyaluronic acids and antimicrobial catheters are used (Francolini & Donelli, 2010).

7.11 IMMUNE-BASED STRATEGIES TARGETING BIOFILM FORMATION

Conventional antifungal medicines are frequently insufficient due to their toxicity, limited range, and high tolerance for biofilm-associated infections. These issues underscore the importance of immune-based interventions, such as vaccinations, which can provide long-term and widespread protection. However, vaccine development against fungi is uniquely difficult. Fungi are eukaryotes, sharing many molecular features with humans, which limits safe antigen selection; their remarkable antigenic diversity across species and morphotypes (yeast, hyphae, spores, and biofilms) complicates the design of universal vaccines; and the patients who need them most are often immunocompromised, raising safety concerns (Oliveira et al., 2021; Cassone & Casadevall, 2012). Despite these challenges, significant advances have been achieved. The NDV-3A vaccine uses the *C. albicans* adhesin Als3p, which has been shown to elicit protective Th1/Th17 immunity in early trials. Other subunit candidates, such as the protease Sap2 or mannan-, β-glucan-, and chitin-based glycoconjugates, aim to disrupt adhesion and biofilm formation. Novel strategies are emerging, including β-glucan-induced trained immunity to reprogram innate responses, glucan particle platforms for antigen delivery, powerful adjuvants like CpG or glucan-chitin particles to bias protective T-cell responses, and adoptive T-cell therapies to restore antifungal immunity in high-risk patients (Cassone & Casadevall, 2012; Oliveira et al., 2021).

Antibodies can perform numerous protective roles, according to research on *C. neoformans*, a biofilm-forming fungus. Antibodies against the polysaccharide capsule promote not only phagocytosis but also complement activation, cytokine response, polysaccharide antigen clearance, and antigen presentation. They can also control the expression of immune-related molecules. More recently, particular antibodies were shown to prevent the release of GXM from the fungal capsule, most likely by cross-linking capsule components. Because *C. neoformans* biofilms on medical devices rely on polysaccharide structures, this study implies that antibodies targeting GXM may interfere with biofilm growth (Martinez & Casadevall, 2005).

7.12 CONCLUSION

Fungal biofilms constitute a major challenge in contemporary medical mycology, particularly in immunocompromised populations, where colonization of indwelling medical devices such as catheters frequently results in bloodstream infections and high mortality. In contrast to planktonic fungal cells, biofilm-associated cells are embedded within a structurally complex ECM that restricts antifungal penetration and fosters drug tolerance. Matrix components, most notably β-1,3-glucan, function as molecular reservoirs that sequester antifungal agents, while the upregulation of matrix-related genes, such as FKS1, confers insensitivity to azoles. The regulation of biofilm development is orchestrated by sophisticated signaling networks, including QS systems and intracellular cascades such as the cAMP-PKA, MAPK, and mTOR pathways. These circuits govern key processes such as adhesion, biofilm maturation, and dispersal, while simultaneously promoting phenotypic heterogeneity through the induction of persister cells, increased efflux pump activity, and enhanced adhesin expression. Collectively, these mechanisms aid in remarkable recalcitrance of fungal biofilms toward conventional antifungal therapies. In addition to antifungal resistance, biofilms exert profound effects on host immune responses. For example, *C. albicans* biofilms have been shown to skew cytokine production, downregulating TNF-α and IL-6. This immunomodulatory profile suppresses protective inflammatory responses and facilitates fungal persistence within the host.

The rapid emergence of antifungal resistance among the fungal strains demands the exploration of alternative and innovative antifungal strategies. Combination therapies offer promising avenues, such as pairing azoles with echinocandins to simultaneously disrupt membrane integrity and weaken the cell wall, or using polyenes with echinocandins in cases of rapidly progressive infections. Similarly, coupling antifungal drugs with matrix-disrupting enzymes like β-1,3-glucanase or DNase (where clinically safe) can enhance drug penetration through the protective biofilm barrier. Targeting key virulence determinants, such as adhesins and QS pathways, also provides opportunities for the development of novel anti-biofilm agents. On the preventive front, coating medical devices with anti-adhesive polymers like PEG or zwitterionic compounds can limit initial fungal attachment, while incorporating contact-killing agents such as silver, quaternary ammonium, or chitosan into hospital surfaces, and coatings may further reduce nosocomial fungal colonization. Immune-based Vaccines (NDV-3A) and Trained immunity (β-glucan) are also being developed.

Beyond the clinic, the extensive use of antifungals in agriculture represents a major driver of resistance. Developing alternative plant-protection strategies is therefore critical to reducing cross-resistance between clinical and environmental fungi. Advanced drug delivery systems, including liposomes and nanomaterials, offer additional means to bypass the biofilm barrier and deliver antifungals directly to fungal cells with improved efficacy.

Despite these advances, substantial challenges persist. The biological heterogeneity of biofilms across species, strains, and environmental conditions continues to hinder the identification of universal therapeutic targets. As antimicrobial resistance escalates and new fungal pathogens emerge, deeper insights into the molecular regulation of biofilms will be essential to shift the field from reactive treatment toward proactive, preventive strategies.

Ultimately, overcoming fungal biofilms will require an integrated approach—one that combines novel therapeutics, preventive technologies, stewardship in agriculture and medicine, and a systems-level understanding of biofilm biology to transform fungal infections from life-threatening complications into manageable conditions.

REFERENCES

Atriwal, T., Azeem, K., Husain, F. M., Hussain, A., Khan, M. N., Alajmi, M. F., & Abid, M. (2021). Mechanistic understanding of *Candida albicans* biofilm formation and approaches for its inhibition. *Frontiers in Microbiology*, 12. https://doi.org/10.3389/fmicb.2021.638609

Avci, P., Karimi, M., Sadasivam, M., Antunes-Melo, W. C., Carrasco, E., & Hamblin, M. R. (2018). In-vivo monitoring of infectious diseases in living animals using bioluminescence imaging. *Virulence*, 9(1), 28–63. https://doi.org/10.1080/21505594.2017.1371897

Bassler, B. L. (2002). Small talk: Cell-to-cell communication in bacteria. *Cell*, 109(4), 421–424. https://doi.org/10.1016/S0092-8674(02)00749-3

Benedict, K., Jackson, B. R., Chiller, T., & Beer, K. D. (2019). Estimation of direct healthcare costs of fungal diseases in the United States. *Clinical Infectious Diseases*, 68(11), 1791–1797. https://doi.org/10.1093/CID/CIY776,

Branco, J., Fonseca, E., Gomes, N. C., Martins-Cruz, C., Silva, A. P., Silva-Dias, A., Pina-Vaz, C., Rodrigues, A. G., Miranda, I. M., Erraught, C., Brennan, L., Silva, R. M., Silva, A. P., Silva-Dias, A., Pina-Vaz, C., & Miranda, I. M. (2017). Impact of ERG3 mutations and expression of ergosterol genes controlled by UPC2 and NDT80 in *Candida parapsilosis* azole resistance. *Clinical Microbiology and Infection*, 23(8), 575.e1–575.e8. https://doi.org/10.1016/J.CMI.2017.02.002

Brauner, A., Fridman, O., Gefen, O., & Balaban, N. Q. (2016). Distinguishing between resistance, tolerance and persistence to antibiotic treatment. *Nature Reviews Microbiology*, 14(5), 320–330. https://doi.org/10.1038/nrmicro.2016.34

Cabello, L., Gómez-Herreros, E., Fernández-Pereira, J., Maicas, S., Martínez-Esparza, M. C., De Groot, P. W. J., & Valentín, E. (2019). Deletion of GLX3 in *Candida albicans* affects temperature tolerance, biofilm formation and virulence. *FEMS Yeast Research*, *19*(2). https://doi.org/10.1093/femsyr/foy124

Casadevall, A., & Pirofski, L. (1999). Host-pathogen interactions: Redefining the basic concepts of virulence and pathogenicity. *Infection and Immunity*, 67(8), 3703–3713. https://doi.org/10.1128/iai.67.8.3703-3713.1999

Cassone, A., & Casadevall, A. (2012). Recent progress in vaccines against fungal diseases. *Current Opinion in Microbiology*, 15(4), 427–433. https://doi.org/10.1016/J.MIB.2012.04.004

Chang, W., Liu, J., Zhang, M., Shi, H., Zheng, S., Jin, X., Gao, Y., Wang, S., Ji, A., & Lou, H. (2018). Efflux pump-mediated resistance to antifungal compounds can be prevented by conjugation with triphenylphosphonium cation. *Nature Communications*, 9(1). https://doi.org/10.1038/s41467-018-07633-9

Chen, H. F., & Lan, C. Y. (2015). Role of SFP1 in the regulation of *Candida albicans* biofilm formation. *PLoS ONE*, *10*(6). https://doi.org/10.1371/journal.pone.0129903

Chen, H., Zhou, X., Ren, B., & Cheng, L. (2020). The regulation of hyphae growth in *Candida albicans*. *Virulence*, 11(1), 337–348. https://doi.org/10.1080/21505594.2020.1748930

Chen, S., Xia, J., Li, C., Zuo, L., & Wei, X. (2018). The possible molecular mechanisms of farnesol on the antifungal resistance of *C. albicans* biofilms: The regulation of CYR1 and PDE2. *BMC Microbiology*, *18*(1). https://doi.org/10.1186/s12866-018-1344-z

Corzo-León, D. E., Munro, C. A., & MacCallum, D. M. (2019). An ex vivo human skin model to study superficial fungal infections. *Frontiers in Microbiology*, 10(JUN), 453235. https://doi.org/10.3389/fmicb.2019.01172

Costa-Orlandi, C. B., Sardi, J. C. O., Pitangui, N. S., de Oliveira, H. C., Scorzoni, L., Galeane, M. C., Medina-Alarcón, K. P., Melo, W. C. M. A., Marcelino, M. Y., Braz, J. D., Fusco-Almeida, A. M., & Mendes-Giannini, M. J. S. (2017). Fungal biofilms and polymicrobial diseases. *Journal of Fungi*, 3(2), 22. https://doi.org/10.3390/JOF3020022

Coste, A., Selmecki, A., Forche, A., Diogo, D., Bougnoux, M. E., D'Enfert, C., Berman, J., & Sanglard, D. (2007). Genotypic evolution of azole resistance mechanisms in sequential *Candida albicans* isolates. *Eukaryotic Cell*, 6(10), 1889–1904. https://doi.org/10.1128/EC.00151-07

Cottier, F., & Hall, R. A. (2020). Face/off: The interchangeable side of *Candida albicans*. *Frontiers in Cellular and Infection Microbiology*, 9. https://doi.org/10.3389/fcimb.2019.00471

Curtis, A., Binder, U., & Kavanagh, K. (2022). Galleria mellonella Larvae as a Model for Investigating Fungal—Host Interactions. *Frontiers in Fungal Biology*, *3*. https://doi.org/10.3389/ffunb.2022.893494

Desai, J. V., Mitchell, A. P., & Andes, D. R. (2014). Fungal biofilms, drug resistance, and recurrent infection. *Cold Spring Harbor Perspectives in Medicine*, *4*(10), a019729. https://doi.org/10.1101/CSHPERSPECT.A019729

Di Como, C. J., & Arndt, K. T. (1996). Nutrients, via the Tor proteins, stimulate the association of Tap42 with type 2A phosphatases. *Genes & Development*, *10*(15), 1904–1916. https://doi.org/10.1101/gad.10.15.1904

Diezmann, S., Leach, M. D., & Cowen, L. E. (2015). Functional divergence of Hsp90 genetic interactions in biofilm and planktonic cellular states. *PLoS ONE*, *10*(9). https://doi.org/10.1371/journal.pone.0137947

Dijck, P. V., Sjollema, J., Camue, B. P. A., Lagrou, K., Berman, J., d'Enfert, C., Andes, D. R., Arendrup, M. C., Brakhage, A. A., Calderone, R., Cantón, E., Coenye, T., Cos, P., Cowen, L. E., Edgerton, M., Espinel-Ingroff, A., Filler, S. G., Ghannoum, M., Gow, N., … Thevissen, K. (2018). Methodologies for in vitro and in vivo evaluation of efficacy of antifungal and antibiofilm agents and surface coatings against fungal biofilms. *Microbial Cell*, 5(7), 300–326. https://doi.org/10.15698/mic2018.07.638

Donlan, R. M., & Costerton, J. W. (2002). Biofilms: Survival mechanisms of clinically relevant microorganisms. *Clinical Microbiology Reviews*, 15(2), 167–193. https://doi.org/10.1128/CMR.15.2.167-193.2002

Eix, E. F., Johnson, C. J., Wartman, K. M., Kernien, J. F., Meudt, J. J., Shanmuganayagam, D., Gibson, A. L. F., & Nett, J. E. (2022). Ex vivo human and porcine skin effectively model *Candida auris* colonization, differentiating robust and poor fungal colonizers. *The Journal of Infectious Diseases*, *2022*, 1791. https://doi.org/10.1093/infdis/jiac094

Fanning, S., & Mitchell, A. P. (2012). Fungal biofilms. *PLOS Pathogens*, *8*(4), e1002585. https://doi.org/10.1371/JOURNAL.PPAT.1002585

Fisher, M. C., Burnett, F., Chandler, C., Gow, N. A. R., Gurr, S., Hart, A., Holmes, A., May, R. C., Quinn, J., Soliman, T., Talbot, N. J., West, H. M., West, J. S., White, P. L., Bromley, M., & Armstrong-James, D. (2024). A one health roadmap towards understanding and mitigating emerging fungal antimicrobial resistance: fAMR. *Npj Antimicrobials and Resistance*, 2(1), 1–10. https://doi.org/10.1038/S44259-024-00055-2

Fisher, M. C., Henk, D. A., Briggs, C. J., Brownstein, J. S., Madoff, L. C., McCraw, S. L., & Gurr, S. J. (2012). Emerging fungal threats to animal, plant and ecosystem health. *Nature*, 484(7393), 186–194. https://doi.org/10.1038/NATURE10947

Forehand, A. L., Myagmarsuren, D., Chen, Z., & Murphy, H. A. (2022). Variation in pH gradients and FLO11 expression in mat biofilms from environmental isolates of the yeast *Saccharomyces cerevisiae*. *MicrobiologyOpen*, *11*(2). https://doi.org/10.1002/mbo3.1277

Fox, E. P., Bui, C. K., Nett, J. E., Hartooni, N., Mui, M. C., Andes, D. R., Nobile, C. J., & Johnson, A. D. (2015). An expanded regulatory network temporally controls *Candida albicans* biofilm formation. *Molecular Microbiology*, *96*(6), 1226–1239. https://doi.org/10.1111/MMI.13002

Francolini, I., & Donelli, G. (2010). Prevention and control of biofilm-based medical-device-related infections. *FEMS Immunology and Medical Microbiology*, 59(3), 227–238). https://doi.org/10.1111/j.1574-695X.2010.00665.x

Freitak, D., Schmidtberg, H., Dickel, F., Lochnit, G., Vogel, H., & Vilcinskas, A. (2014). The maternal transfer of bacteria can mediate trans-generational immune priming in insects. *Virulence*, 5(4), 547–554. https://doi.org/10.4161/VIRU.28367.

Gaálová-Radochová, B., Kendra, S., Jordao, L., Kursawe, L., Kikhney, J., Moter, A., & Bujdáková, H. (2023). Effect of Quorum Sensing Molecule Farnesol on Mixed Biofilms of Candida albicans and Staphylococcus aureus. Antibiotics, 12(3), 441. https://doi.org/10.3390/antibiotics12030441.

García-Sánchez, S., Aubert, S., Iraqui, I., Janbon, G., Ghigo, J. M., & D'Enfert, C. (2004). *Candida albicans* biofilms: A developmental state associated with specific and stable gene expression patterns. *Eukaryotic Cell*, 3(2), 536–545. https://doi.org/10.1128/EC.3.2.536-545.2004

Giammarino, A., Verdolini, L., Simonetti, G., & Angiolella, L. (2025). Fungal biofilm: An overview of the latest nano-strategies. *Antibiotics*, 14(7), 718. https://doi.org/10.3390/ANTIBIOTICS14070718

Hawser, S. P., Baillie, G. S., & Douglas, L. J. (1998). Production of extracellular matrix by *Candida albicans* biofilms. *Journal of Medical Microbiology*, 47(3), 253–256. https://doi.org/10.1099/00222615-47-3-253

Herranz, S., Rodríguez, J. M., Bussink, H.-J., Sánchez-Ferrero, J. C., Arst, H. N., Jr., Peñalva, M. A., & Vincent, O. (2005). Arrestin-related proteins mediate pH signaling in fungi. *Proceedings of the National Academy of Sciences*, *102*(34), 12141–12146. https://doi.org/10.1073/pnas.0504776102

Horton, M. V., Johnson, C. J., Kernien, J. F., Patel, T. D., Lam, B. C., Cheong, J. Z. A., Meudt, J. J., Shanmuganayagam, D., Kalan, L. R., & Nett, J. E. (2020). *Candida auris* forms high-burden biofilms in skin niche conditions and on porcine skin. *MSphere*, *5*(1). https://doi.org/10.1128/MSPHERE.00910-19

Horton, M. V., & Nett, J. E. (2020). *Candida auris* infection and biofilm formation: Going beyond the surface. *Current Clinical Microbiology Reports*, 7(3), 51–56. https://doi.org/10.1007/s40588-020-00143-7

Jamil, S., Adepoju, V. A., & Chowdhury, A. A. (2024). Fungal apocalypse: The emergence of drug-resistant *Candida auris* in the United States. *Microbes and Infectious Diseases*. https://doi.org/10.21608/MID.2024.302586.2059

Johnson, C. J., Cabezas-Olcoz, J., Kernien, J. F., Wang, S. X., Beebe, D. J., Huttenlocher, A., Ansari, H., & Nett, J. E. (2016). The extracellular matrix of *Candida albicans* biofilms impairs formation of neutrophil extracellular traps. *PLoS Pathogens*, *12*(9). https://doi.org/10.1371/journal.ppat.1005884

Joshi, P., Navale, A., Shelke, A., & Patel, M. (2024). Exploring the arsenal of novel antifungal drug targets for combating fungal infections. *Current Pharmaceutical Biotechnology*, 26(13), 2097–2110. https://doi.org/10.2174/0113892010304880240828075411

Kakade, P., Sadhale, P., Sanyal, K., & Nagaraja, V. (2016). ZCF32, a fungus specific Zn(II)2 Cys6 transcription factor, is a repressor of the biofilm development in the human pathogen *Candida albicans*. *Scientific Reports*, *6*. https://doi.org/10.1038/srep31124

Kaplan, J. B., Ragunath, C., Ramasubbu, N., & Fine, D. H. (2003). Detachment of *Actinobacillus actinomycetemcomitans* biofilm cells by an endogenous β-hexosaminidase activity. *Journal of Bacteriology*, 185(16), 4693–4698. https://doi.org/10.1128/JB.185.16.4693-4698.2003

Katsipoulaki, M., Stappers, M. H. T., Malavia-Jones, D., Brunke, S., Hube, B., & Gow, N. A. R. (2024). *Candida albicans* and *Candida glabrata* : Global priority pathogens. *Microbiology and Molecular Biology Reviews*, 88(2). https://doi.org/10.1128/mmbr.00021-23

Kaur, J., & Nobile, C. J. (2023). Antifungal drug-resistance mechanisms in Candida biofilms. *Current Opinion in Microbiology*, 71. https://doi.org/10.1016/j.mib.2022.102237

Kavanagh, K., & Sheehan, G. (2018). The use of *Galleria mellonella* larvae to identify novel antimicrobial agents against fungal species of medical interest. *Journal of Fungi*, 4(3), 113. https://doi.org/10.3390/jof4030113

Khari, A., Biswas, B., Gangwar, G., Thakur, A., & Puria, R. (2023). *Candida auris* biofilm: A review on model to mechanism conservation. *Expert Review of Anti-Infective Therapy*, 21(3), 295–308. https://doi.org/10.1080/14787210.2023.2179036

Konwar, A., Mathur, K., Pandey, S., Bhorali, K., Thakur, A., & Puria, R. (2025). Insights into the evolution of Candidalysin and recent developments. *Archives of Microbiology*, 207(9), 1–16. https://doi.org/10.1007/S00203-025-04414-Z

Kowalski, C. H., Morelli, K. A., Schultz, D., Nadell, C. D., & Cramer, R. A. (2020). Fungal biofilm architecture produces hypoxic microenvironments that drive antifungal resistance. *Proceedings of the National Academy of Sciences of the United States of America*, 117(36), 22473–22483. https://doi.org/10.1073/PNAS.2003700117

Kraft, L., Ribeiro, V. S. T., Gonçalves, G. A., Suss, P. H., & Tuon, F. F. (2023). Comparison of amphotericin B lipid complex, deoxycholate amphotericin B, fluconazole, and anidulafungin activity against *Candida albicans* biofilm isolated from breakthrough candidemia. *Enfermedades Infecciosas y Microbiologia Clinica (English Ed.)*, 41(10), 596–603. https://doi.org/10.1016/j.eimce.2022.07.009

Kriegl, L., Egger, M., Boyer, J., Hoenigl, M., & Krause, R. (2025). New treatment options for critically important WHO fungal priority pathogens. *Clinical Microbiology and Infection*, 31(6), 922–930. https://doi.org/10.1016/J.CMI.2024.03.006

Kucharíková, S., Vande Velde, G., Himmelreich, U., & Van Dijck, P. (2015). *Candida albicans* biofilm development on medically-relevant foreign bodies in a mouse subcutaneous model followed by bioluminescence imaging. *Journal of Visualized Experiments*, 95. https://doi.org/10.3791/52239

LaFleur, M. D., Kumamoto, C. A., & Lewis, K. (2006). *Candida albicans* biofilms produce antifungal-tolerant persister cells. *Antimicrobial Agents and Chemotherapy*, 50(11), 3839–3846. https://doi.org/10.1128/AAC.00684-06

Lagree, K., & Mitchell, A. P. (2017). Fungal biofilms: Inside out. *Microbiology Spectrum*, 5(2). https://doi.org/10.1128/MICROBIOLSPEC.FUNK-0024-2016

Lengeler, K. B., Davidson, R. C., D'souza, C., Harashima, T., Shen, W.-C., Wang, P., Pan, X., Waugh, M., & Heitman, J. (2000). Signal transduction cascades regulating fungal development and virulence. *Microbiology and Molecular Biology Reviews*, 64(4), 746–785. https://doi.org/10.1128/mmbr.64.4.746-785.2000

Lewis, K. (2010). Persister cells. *Annual Review of Microbiology*, 64, 357–372. https://doi.org/10.1146/annurev.micro.112408.134306

Li, W., Mitchell, A. P., & Mitchell, A. (1997). *Proteolytic activation of Rimlp, a positive regulator of yeast sporulation and invasive growth.* https://academic.oup.com/genetics/article/145/1/63/6018152

Liu, S., Mauff, F. L., Sheppard, D. C., & Zhang, S. (2022). Filamentous fungal biofilms: Conserved and unique aspects of extracellular matrix composition, mechanisms of drug resistance and regulatory networks in *Aspergillus fumigatus*. *npj Biofilms and Microbiomes*, 8(1). https://doi.org/10.1038/s41522-022-00347-3

Loeb, J. D. J., Kerentseva, T. A., Pan, T., Sepulveda-Becerra, M., & Liu, H. (1999). *Saccharomyces cerevisiae G1 cyclins are differentially involved in invasive and pseudohyphal growth independent of the filamentation mitogen-activated protein kinase pathway*. https://academic.oup.com/genetics/article/153/4/1535/6035037

Lohse, M. B., Gulati, M., Johnson, A. D., & Nobile, C. J. (2017). Development and regulation of single- and multi-species *Candida albicans* biofilms. *Nature Reviews Microbiology*. https://doi.org/10.1038/nrmicro.2017.107

Luo, S., Skerka, C., Kurzai, O., & Zipfel, P. F. (2013). Complement and innate immune evasion strategies of the human pathogenic fungus *Candida albicans*. *Molecular Immunology*, 56(3), 161–169. https://doi.org/10.1016/J.MOLIMM.2013.05.218

Marena, G. D., Thomaz, L., Nosanchuk, J. D., & Taborda, C. P. (2025). *Galleria mellonella* as an invertebrate model for studying fungal infections. *Journal of Fungi*, 11(2), 157. https://doi.org/10.3390/jof11020157

Martin-Yken, H., Dagkessamanskaia, A., Basmaji, F., Lagorce, A., & Francois, J. (2003). The interaction of Slt2 MAP kinase with Knr4 is necessary for signalling through the cell wall integrity pathway in *Saccharomyces cerevisiae*. *Molecular Microbiology*, 49(1), 23–35. https://doi.org/10.1046/j.1365-2958.2003.03541.x

Martinez, L. R., & Casadevall, A. (2005). Specific antibody can prevent fungal biofilm formation and this effect correlates with protective efficacy. *Infection and Immunity*, 73(10), 6350–6362. https://doi.org/10.1128/IAI.73.10.6350-6362.2005

Martinez, L. R., & Fries, B. C. (2010). Fungal biofilms: Relevance in the setting of human disease. *Current Fungal Infection Reports*, 4(4), 266–275. https://doi.org/10.1007/s12281-010-0035-5

Mello, T. P., Oliveira, S. S. C., Branquinha, M. H., & Santos, A. L. S. (2022). Decoding the antifungal resistance mechanisms in biofilms of emerging, ubiquitous and multidrug-resistant species belonging to the Scedosporium/Lomentospora genera. *Medical Mycology*, 60(6), 36. https://doi.org/10.1093/MMY/MYAC036

Mitchell, K. F., Taff, H. T., Cuevas, M. A., Reinicke, E. L., Sanchez, H., & Andes, D. R. (2013). Role of matrix β-1,3 glucan in antifungal resistance of non-albicans Candida biofilms. *Antimicrobial Agents and Chemotherapy*, 57(4), 1918–1920. https://doi.org/10.1128/AAC.02378-12

Müller, F. M. C. (2014). Biofilm formation and its impact on antifungal therapy. *Current Fungal Infection Reports*, 8(3), 235–241. https://doi.org/10.1007/s12281-014-0194-x

Navarathna, D. H. M. L. P., Hornby, J. M., Krishnan, N., Parkhurst, A., Duhamel, G. E., & Nickerson, K. W. (2007). Effect of farnesol on a mouse model of systemic candidiasis, determined by use of a DPP3 knockout mutant of *Candida albicans*. *Infection and Immunity*, 75(4), 1609–1618. https://doi.org/10.1128/IAI.01182-06

Netea, M. G., Brown, G. D., Kullberg, B. J., & Gow, N. A. R. (2008). An integrated model of the recognition of *Candida albicans* by the innate immune system. *Nature Reviews Microbiology*, 6(1), 67–78. https://doi.org/10.1038/nrmicro1815

Nett, J. E. (2016). The host's reply to Candida biofilm. *Pathogens*, 5(1). https://doi.org/10.3390/pathogens5010033

Nett, J., Lincoln, L., Marchillo, K., Massey, R., Holoyda, K., Hoff, B., VanHandel, M., & Andes, D. (2007). Putative role of β-1,3 glucans in *Candida albicans* biofilm resistance. *Antimicrobial Agents and Chemotherapy*, 51(2), 510–520. https://doi.org/10.1128/AAC.01056-06

Nett, J. E., Sanchez, H., Cain, M. T., & Andes, D. R. (2010). Genetic basis of Candida biofilm resistance due to drug-sequestering matrix glucan. *Journal of Infectious Diseases*, 202(1), 171–175. https://doi.org/10.1086/651200

Nickerson, K. W., & Atkin, A. L. (2017). Deciphering fungal dimorphism: Farnesol's unanswered questions. *Molecular Microbiology*, 103(4), 567–575. https://doi.org/10.1111/mmi.13601

Niño-Vega, G. A., Padró-Villegas, L., & López-Romero, E. (2024). New ground in antifungal discovery and therapy for invasive fungal infections: innovations, challenges, and future directions. *Journal of Fungi*, 10(12). https://doi.org/10.3390/JOF10120871

Nobile, C. J., & Johnson, A. D. (2015). *Candida albicans* biofilms and human disease. *Annual Review of Microbiology*, 69(1), 71–92. https://doi.org/10.1146/annurev-micro-091014-104330

Nobile, C. J., Schneider, H. A., Nett, J. E., Sheppard, D. C., Filler, S. G., Andes, D. R., & Mitchell, A. P. (2008). Complementary adhesin function in *C. albicans* biofilm formation. *Current Biology*, *18*(14), 1017–1024. https://doi.org/10.1016/j.cub.2008.06.034

Oliveira, L. V. N., Wang, R., Specht, C. A., & Levitz, S. M. (2021). Vaccines for human fungal diseases: Close but still a long way to go. *Npj Vaccines*, 6(1), 1–8. https://doi.org/10.1038/s41541-021-00294-8

Pappas, P. G., Kauffman, C. A., Andes, D. R., Clancy, C. J., Marr, K. A., Ostrosky-Zeichner, L., Reboli, A. C., Schuster, M. G., Vazquez, J. A., Walsh, T. J., Zaoutis, T. E., & Sobel, J. D. (2015). Clinical practice guideline for the management of candidiasis: 2016 Update by the Infectious Diseases Society of America. *Clinical Infectious Diseases*, 62(4), e1–e50. https://doi.org/10.1093/CID/CIV933

Parambath, S., Dao, A., Kim, H. Y., Zawahir, S., Izquierdo, A. A., Tacconelli, E., Govender, N., Oladele, R., Colombo, A., Sorrell, T., Ramon-Pardo, P., Fusire, T., Gigante, V., Sati, H., Morrissey, C. O., Alffenaar, J. W., & Beardsley, J. (2024). *Candida albicans*—A systematic review to inform the World Health Organization Fungal Priority Pathogens List. *Medical Mycology*, *62*(6). https://doi.org/10.1093/MMY/MYAE045

Pati, B. A., Kurata, W. E., Horseman, T. S., & Pierce, L. M. (2021). Antibiofilm activity of chitosan/epsilon-poly-L-lysine hydrogels in a porcine ex vivo skin wound polymicrobial biofilm model. *PierceWound Repair and Regeneration*, 29(2), 316–326. https://doi.org/10.1111/WRR.12890

Patterson, T. F., Thompson, G. R., Denning, D. W., Fishman, J. A., Hadley, S., Herbrecht, R., Kontoyiannis, D. P., Marr, K. A., Morrison, V. A., Nguyen, M. H., Segal, B. H., Steinbach, W. J., Stevens, D. A., Walsh, T. J., Wingard, J. R., Young, J. A. H., & Bennett, J. E. (2016). Practice guidelines for the diagnosis and management of aspergillosis: 2016 update by the infectious diseases society of America. *Clinical Infectious Diseases*, 63(4), e1–e60. https://doi.org/10.1093/CID/CIW326

Pereira, R., dos Santos Fontenelle, R. O., de Brito, E. H. S., & de Morais, S. M. (2021). Biofilm of *Candida albicans*: Formation, regulation and resistance. *Journal of Applied Microbiology*, 131(1), 11–22. https://doi.org/10.1111/JAM.14949.

Pereira, M. F., Rossi, C. C., da Silva, G. C., Rosa, J. N., & Bazzolli, D. M. S. (2020). Galleria mellonella as an infection model: an in-depth look at why it works and practical considerations for successful application. *Pathogens and Disease*, *78*(8). https://doi.org/10.1093/femspd/ftaa056

Pierce, C. G., Vila, T., Romo, J. A., Montelongo-Jauregui, D., Wall, G., Ramasubramanian, A., & Lopez-Ribot, J. L. (2017). The *Candida albicans* biofilm matrix: Composition, structure and function. *Journal of Fungi*, 3(1). https://doi.org/10.3390/jof3010014

Puri, S., Kumar, R., Chadha, S., Tati, S., Conti, H. R., Hube, B., Cullen, P. J., & Edgerton, M. (2012). Secreted aspartic protease cleavage of *Candida albicans* Msb2 activates Cek1 MAPK signaling affecting biofilm formation and oropharyngeal candidiasis. *PloS One*, *7*(11). https://doi.org/10.1371/journal.pone.0046020

Rahman, D., Mistry, M., Thavaraj, S., Challacombe, S. J., & Naglik, J. R. (2007). Murine model of concurrent oral and vaginal *Candida albicans* colonization to study epithelial host–pathogen interactions. *Microbes and Infection*, 9(5), 615–622. https://doi.org/10.1016/J.MICINF.2007.01.012

Rajasingham, R., Smith, R. M., Park, B. J., Jarvis, J. N., Govender, N. P., Chiller, T. M., Denning, D. W., Loyse, A., & Boulware, D. R. (2017). Global burden of disease of HIV-associated cryptococcal meningitis: An updated analysis. *The Lancet Infectious Diseases*, 17(8), 873–881. https://doi.org/10.1016/S1473-3099(17)30243-8

Rajendran, R., Williams, C., Lappin, D. F., Millington, O., Martins, M., & Ramage, G. (2013). Extracellular DNA release acts as an antifungal resistance mechanism in mature *Aspergillus fumigatus* biofilms. *Eukaryotic Cell*, 12(3), 420–429. https://doi.org/10.1128/EC.00287-12

Ramage, G., Kean, R., Rautemaa-Richardson, R., Williams, C., & Lopez-Ribot, J. L. (2025). Fungal biofilms in human health and disease. *Nature Reviews Microbiology*, 23(6), 355–370. https://doi.org/10.1038/s41579-025-01147-0

Ramage, G., Rajendran, R., Sherry, L., & Williams, C. (2012). Fungal biofilm resistance. *International Journal of Microbiology*, *2012*. https://doi.org/10.1155/2012/528521

Rana, A., & Thakur, A. (2025). Translation regulation promotes stress adaptation in the human fungal pathogen *Candida glabrata*. *Genetics*, *231*(1). https://doi.org/10.1093/GENETICS/IYAF134

Robbins, N., Uppuluri, P., Nett, J., Rajendran, R., Ramage, G., Lopez-Ribot, J. L., Andes, D., & Cowen, L. E. (2011). Hsp90 governs dispersion and drug resistance of fungal biofilms. *PLoS Pathogens*, *7*(9). https://doi.org/10.1371/journal.ppat.1002257

Robbins, N., Wright, G. D., & Cowen, L. E. (2016). Antifungal drugs: The current armamentarium and development of new agents. *Microbiology Spectrum*, 4(5). https://doi.org/10.1128/MICROBIOLSPEC.

Rodrigues, C. F., & Černáková, L. (2020). Farnesol and tyrosol: Secondary metabolites with a crucial quorum-sensing role in candida biofilm development. *Genes*, *11*(4), 444. https://doi.org/10.3390/genes11040444

Romero, O., & Geddes-McAlister, J. (2025). Breaking down biofilms across critical priority fungal pathogens: Proteomics and computational innovation for mechanistic insights and new target discovery. *MBio*. https://doi.org/10.1128/mbio.02303-24

Roy, S., & Gow, N. A. R. (2023). The role of the Candida biofilm matrix in drug and immune protection. *The Cell Surface*, 10, 100111. https://doi.org/10.1016/J.TCSW.2023.100111

Shapiro, R. S., Uppuluri, P., Zaas, A. K., Collins, C., Senn, H., Perfect, J. R., Heitman, J., & Cowen, L. E. (2009). Hsp90 Orchestrates temperature-dependent *Candida albicans* morphogenesis via Ras1-PKA signaling. *Current Biology*, 19(8), 621–629. https://doi.org/10.1016/j.cub.2009.03.017

Shchepin, R., Hornby, J. M., Burger, E., Niessen, T., Dussault, P., & Nickerson, K. W. (2003). Quorum sensing in *Candida albicans*: Probing Farnesol's mode of action with 40 natural and synthetic Farnesol analogs. *Chemistry & Biology*, 10, 743–750. https://doi.org/10.1016/s1074-5521(03)00158-3

Sheehan, G., Konings, M., Lim, W., Fahal, A., Kavanagh, K., & van de Sande, W. W. J. (2020). Proteomic analysis of the processes leading to Madurella mycetomatis grain formation in *Galleria mellonella* larvae. *PLOS Neglected Tropical Diseases*, *14*(4), e0008190. https://doi.org/10.1371/JOURNAL.PNTD.0008190

Shelar, A., Didwal, P. N., & Patil, R. (2023). Recent advances in antifungal nanomaterials for combating biofilm infection caused by *Candida albicans*. In *Applications of Nanotechnology in Microbiology* (pp. 271–290). Springer Nature Switzerland. https://doi.org/10.1007/978-3-031-49933-3_10

Solis, N. V., & Filler, S. G. (2012). Mouse model of oropharyngeal candidiasis. *Nature Protocols*, 7(4), 637–642. https://doi.org/10.1038/nprot.2012.011

Tits, J., Cammue, B. P. A., & Thevissen, K. (2020). Combination therapy to treat fungal biofilm-based infections. *International Journal of Molecular Sciences*, 21(22), 8873. https://doi.org/10.3390/IJMS21228873

Uppuluri, P., Chaturvedi, A. K., Srinivasan, A., Banerjee, M., Ramasubramaniam, A. K., Köhler, J. R., Kadosh, D., & Lopez-Ribot, J. L. (2010). Dispersion as an important step in the *Candida albicans* biofilm developmental cycle. *PLoS Pathogens*, 6(3), 1000828. https://doi.org/10.1371/JOURNAL.PPAT.1000828

Vande Velde, G., Kucharíková, S., Schrevens, S., Himmelreich, U., & Van Dijck, P. (2014). Towards non-invasive monitoring of pathogen–host interactions during *Candida albicans* biofilm formation using in vivo bioluminescence. *Cellular Microbiology*, 16(1), 115–130. https://doi.org/10.1111/CMI.12184

Vertyporokh, L., Hułas-Stasiak, M., & Wojda, I. (2020). Host–pathogen interaction after infection of *Galleria mellonella* with the filamentous fungus *Beauveria bassiana*. *Insect Science*, 27(5), 1079–1089. https://doi.org/10.1111/1744-7917.12706

Vila, T., Montelongo-Jauregui, D., Ahmed, H., Puthran, T., Sultan, A. S., & Jabra-Rizk, M. A. (2020). Comparative evaluations of the pathogenesis of *Candida auris* phenotypes and *Candida albicans* using clinically relevant murine models of infections. *Msphere*, *5*(4). https://doi.org/10.1128/MSPHERE.00760-20

Vylkova, S., & Lorenz, M. C. (2014). Modulation of phagosomal pH by *Candida albicans* promotes hyphal morphogenesis and requires Stp2p, a regulator of amino acid transport. *PLoS Pathogens*, *10*(3). https://doi.org/10.1371/journal.ppat.1003995

Wang, C., Mcclelland, E. E., Geddes-Mcalister, J., Kavanagh, K., Curtis, A., & Binder, U. (2022). Galleria mellonella larvae as a model for investigating fungal—host interactions. *Frontiers in Fungal Biology*, *3*, 893494. https://doi.org/10.3389/ffunb.2022.893494

Wang, D., Zeng, N., Li, C., Li, Z., Zhang, N., & Li, B. (2024). Fungal biofilm formation and its regulatory mechanism. *Heliyon*, 10(12), e32766. https://doi.org/10.1016/j.heliyon.2024.e32766

White, T. C. (1997). Increased mRNA levels of ERG16, CDR, and MDR1 correlate with increases in azole resistance in Candida albicans isolates from a patient infected with human immunodeficiency virus. *Antimicrobial Agents and Chemotherapy*, *41*(7), 1482–1487. https://doi.org/10.1128/aac.41.7.1482

Williams, P. (2007). Quorum sensing, communication and cross-kingdom signalling in the bacterial world. *Microbiology*, 153(12), 3923–3938. https://doi.org/10.1099/mic.0.2007/012856-0

Wisplinghoff, H., Bischoff, T., Tallent, S. M., Seifert, H., Wenzel, R. P., & Edmond, M. B. (2004). Nosocomial bloodstream infections in US hospitals: Analysis of 24,179 cases from a prospective nationwide surveillance study. *Clinical Infectious Diseases*, 39(3), 309–317. https://doi.org/10.1086/421946

World Health Organization. (2022). *WHO fungal priority pathogens list to guide research, development and public health action. Licence:* CC BY-NC-SA 3.0 IGO, 1, 1–48. https://www.who.int/publications/i/item/9789240060241

Xu, W., Jong Hong, S., Jia, S., Zhao, Y., Galiano, R. D., & Mustoe, T. A. (2012). Application of a partial-thickness human ex vivo skin culture model in cutaneous wound healing study. *Laboratory Investigation*, 92(4), 584–599. https://doi.org/10.1038/labinvest.2011.184

Xu, W., Smith, F. J., Subaran, R., & Mitchell, A. P. (2004). Multivesicular body-ESCRT components function in pH response regulation in *Saccharomyces cerevisiae* and *Candida albicans*. *Molecular Biology of the Cell*, 15, 5528–5537. https://doi.org/10.1091/mbc.e04-08-0666

8 Metagenomic Perspectives on the Human Gut Ecosystem and Disease Associations

Bablu Kumar, Ashwinipriyadarshini Megur, Nishtha Kaushik, and Shweta Singh

8.1 INTRODUCTION

Advancement in microbiology, especially through high-throughput metagenomic sequencing and computational analysis, has enabled a deeper understanding of the study of the human gut virome. The gut viral community includes viruses that can directly infect human cells but mostly dominated by bacteriophages, which infect bacteria (i.e., phages). Bacteriophages promote the transfer of genetic material among their host cells (e.g., bacteria) by horizontal gene transfer (HGT), contributing to bacterial evolution and adaptation within the dynamic gut environment (Barr, 2017; Cao et al., 2022; Dalmasso et al., 2014). Phage-mediated HGT can spread antibiotic resistance toxins genes, duplicate virulence factors, enhance host cell fitness, and regulate host population size, indicating a complex role of phages (Dalmasso et al., 2014; Shang et al., 2022).

Shotgun metagenomic sequencing enables a comprehensive study of all genetic materials present in a sample, leveraging advantage of access not only to the genomes of prokaryotes but also to fungi and viruses, including bacteriophages (Camarillo-Guerrero et al., 2021; Rahimian & Panahi, 2024; Kumar et al., 2024). In the case of bacteriophages, metagenomic datasets are employed to discern both lytic (virulent) and temperate (prophage) phages, providing vital understanding of the various phages in bacterial life cycles. Temperate phages integrate into bacterial genomes as prophages, enabling their dormancy within the host and consequently affecting the fitness of the bacteria. Lytic (virulent) refers to bacteriophages that infect and rapidly destroy host bacteria by replicating within them and causing cell lysis to release new phage particles (Townsend et al., 2021). The roles of genes encoded by phages can be anticipated through metagenomic sequence analysis, advancing understanding of their biological functions and potential effects on both the host and the microbial physiology.

This chapter discusses the role of shotgun metagenomics in the advancement of microbiology research, with particular emphasis on computational methods for

DOI: 10.1201/9781003633952-10

analysing intestinal Fungi and bacteriophages. It reviews the methods and tools used for phage identification, host prediction, and functional inference, as well as the roles of fungi and phages in the human gut ecosystem. In general, this chapter reviews advanced resources and theoretical insights on fungi and bacteriophages, highlighting the significance of metagenomic techniques in mycobiome and virome research related to human health and disease.

8.2 ROLE OF METAGENOMIC IN ADVANCING MICROBIOLOGY RESEARCH

It is well established that traditional microbiology has been fundamentally limited by the fact that fewer than 1% of bacterial species and less than 5% of fungal species are currently known to cultivation under standard laboratory conditions (Staley et al., 1997). Starting in the mid-2000s and continuing until today, metagenomics overcomes the limitation of requiring pure cultures by enabling the direct extraction of genetic material from environmental samples (Yang et al., 2025). Metagenomics is a sequence-based approach to study all genetic elements from all organisms (bacteria, viruses, and fungi) in a particular environment. This enables the identification of genomic fragments of dominant and rare species from samples that are non-cultivating microorganisms (Kumar et al., 2024), generating a large number of short reads that are then subsequently assembled into contigs.

The shotgun sequencing approach begins with the extraction of nucleic acid (DNA or RNA) from an environmental sample in laboratory settings. For whole-genome shotgun sequencing, the extracted DNA are then fragmented into small pieces of DNA, which are then amplified by PCR. This process generates millions of short reads per run (typically 150–350 base pairs), which are then assembled into contigs (Quince et al., 2017). For viral genome identification, metagenomic contigs are analysed using advance computational tools that utilized sequence-based features or maker-based signatures, providing access to the high resolution of viral genomic diversity (Nayfach et al., 2021). Furthermore, host prediction tools are used on the assembled metagenomes to elucidate host–phage interactions. The main purpose of metagenomic sequencing is to sequence all genetic elements present in an environmental sample, with the aim of characterizing the taxonomic composition of the microbial communities that describe which microbes are in that sample and how many are there. Metagenomics also aims to infer their functional capabilities, giving us an idea of what microorganisms are doing in that sample.

8.2.1 The Concept of Mycobiome

The variety of microorganisms, including bacteria, viruses, Fungi, and archaea, inhabits the human gastrointestinal (GI) tract. Bacteria are the major focus of most studies because of their numbers and well understood roles in host physiology. The "rare biosphere," in which microorganisms are not as prevalent but gaining better acknowledgement, is being studied (Chin et al., 2020). Fungi are present on and within human bodies as components of the human microbiome, which is not intriguing considering their widespread existence in the environment. A relatively unexplored aspect of the human microbiome is the fungal microbiome termed as mycobiome (Nash et al., 2017). While fungi are considered to contribute for less than 1% of the human gut microbiome, this statistical representation in all likelihood understates their biological role (Lai et al., 2023).

The fungal communities that reside in luminal compartments and mucosal layers of the gut are intricate and diverse, as brought to knowledge by such advancements. Fungi also contribute to pathophysiology of various gut-associated and systemic metabolic diseases, an evidence that they play a significant role in maintaining intestinal homeostasis (Chin et al., 2020). Several diseases, such as inflammatory bowel disease (IBD), irritable bowel syndrome (IBS), celiac disease, autoimmune and metabolic disorders, neurological diseases, and cancer, have been associated with dysbiosis (Gaspar et al., 2025).

Almost 70% of healthy individuals harbour fungal communities in their gut, illustrating the stable presence of fungi in human gut microbiology. Ascomycota is the most widespread phylum of fungi, contributing between 48% and 99% of all identified species. Relative abundance of Basidiomycota lies between 0.5% and 14% (Belvoncikova et al., 2022; Sam et al., 2017), while Mucoromycota are present in smaller quantities. These phyla combined more than 400 different species of fungi. The complexity of the gut mycobiome and its potential roles in host immunity, metabolism, and overall gut homeostasis are emphasized by this taxonomic diversity (Belvoncikova et al., 2022). Firstly, immediately after birth, fungal colonization of the newborn gut is simultaneous with bacterial colonization. Fungal genera in the Saccharomycetales and Malasseziales orders are naturally dominant early in infancy (Zhang et al., 2022a). Delivery mode, gestational age, type of lactation, nutritional status of the mother, exposure to ambient microbes, and host genotypic susceptibility are certain of the perinatal and postnatal determinants that have a bearing on the initial composition and diversity of gut mycobiome (Wu et al., 2021). Dietary consumption is a key regulator of the structure and diversity of the gut mycobiome during the entire human life cycle. The mycobiome modulates metabolism as some of the fungi show positive as well as negative correlations with the lipid and carbohydrate metabolism of the host (Sun et al., 2021b).

8.2.2 Metagenomics Approach to Study Mycobiome

Mycobiome studies were initially limited by several technological constraints (Wu et al., 2021). The absence of universally accepted genetic markers and large reference databases hampered molecular methods, while culture methods were inadequate to reveal the entire diversity of fungi (Fiers et al., 2019; Soverini et al., 2019). However, the introduction of next generation sequencing, metagenomics, and amplicon-based technologies has transformed the detection and characterization of fungi (Xie & Manichanh, 2022). In addition, precise quantification and visualization of Fungi in situ is revolutionized through culture independent methods such as fluorescent in situ hybridization and quantitative PCR (Iyer & Ojcius, 2024). Direct genetic examination of microbial communities from environmental samples without culturing is referred to as metagenomics. High-throughput sequencing technology integrated with advanced bioinformatics tools are the pillars of this field. The two major techniques employed are shot gun sequencing and marker gene analysis (El Jaddaoui et al., 2025).

Sample collection is the primary stage of characterizing the mycobiome. Faecal matter is the most commonly used in the studies of human gut mycobiome since it is easy to acquire and non-invasive (Thielemann et al., 2022). The vast majority of studies utilize an amplicon or marker sequencing approach to study faecal mycobiome. The amplicon method is widely accepted from using the fungal rRNA gene locus. Within this region are the genes for ribosomal small subunit and large subunit, which are separated by the internal transcribed spacers, ITS1 and ITS2 (De Filippis et al., 2017; Hadrich et al., 2025; Hoggard et al., 2018). ITS amplicon sequencing involves selectively amplifying and sequencing the region and aligning that short genomic area of fungal DNA to reference sequences of a known identification (Balderramo et al., 2023). Primers of marker-based methods are specifically designed to amplify conserved regions of target genes, thereby enabling microbes to be taxa classified. These methods enable one to study the structure and organization of communities within different habitats in broad terms (El Jaddaoui et al., 2025; Kedia & Ahuja, 2023). The sequences are initially grouped into Operational Taxonomic Units (OTUs) by similarity and subsequently searched against curated reference databases for taxonomic annotation. Multivariate statistical comparisons are subsequently carried out to analyse OTU abundance patterns (Sam et al., 2017; Wang et al., 2024b). Multiple PCR-generated amplicons across a variety of samples have been sequenced using sequencing platforms including Nanopore, Ion Torrent, Roche 454, and Illumina HiSeq/MiSeq. These sequencing technologies allow high-throughput formats to process the amplified regions (El Jaddaoui et al., 2025).

8.2.3 Bioinformatics and Data Analysis Pipelines

DNA sequencing is followed by bioinformatics analysis where a variety of computational algorithms are used to rebuild the microbial community of the sample. There are various approaches depending on the particular research goals and the end result desired for the study (Fox et al., 2024). For the comparison of metagenomes, the sequences are compared against sequences in current databases (Table 8.1) or against a known activity. Pipelines (Table 8.2) have been built for end-to-end metagenome analysis, which are executed automatically, with an ordered set of software programs executed sequentially in order to generate results that are interpretable (Navgire et al., 2022). In metagenomic assembly, DNA sequences are reconstructed into reads, and it is possible due to an approach of de Bruijn graph. In this approach, reads are split into overlapping k-mers that will act as edges in the assembly graph (Quince et al., 2017).

To precisely identify fungi and advance diagnostic application, there must be high-quality, well-curated reference databases. Clinical mycobiome science is complemented by integrating the databases to Next-Generation Sequencing technology. Determining the biological function of fungi requires effective bioinformatics tools since NGS platforms generate a vast amount of data (Gaspar et al., 2025). There are numerous tools applied in metagenomic and amplicon data analysis, and those are needed to determine the quality and interpretation of the data (Anslan et al., 2018).

TABLE 8.1
List of Databases for Mycobiome Analysis

S. No.	Database	Description	References
1.	UNITE	This web-based molecular identification platform of eukaryotes offers around 10 million reference sequences for taxonomic analysis and classification	Abarenkov et al. (2024)
2.	ITSoneDB	GenBank annotations and/or an HMM-based method were employed to identify the start and end positions of ITS1 in each of the 405,433 taxonomically classified sequence entries that comprise ITSoneDB	Fosso et al. (2012)
3.	Findley	Fungal richness at human skin locations can be comprehensively described due to the high-throughput sequencing data, which is primarily based on the IYS1 site of fungal rRNA genes	Tang et al. (2015)
4.	RefSeq targeted loci	ITS region sequences from INSD are reviewed and re-annotated. The sequences can be edited, have improved taxonomy, and provide more metadata for more accurate fungal identification	(*NCBI RefSEQ Targeted LOCI Project*)
5.	ITS2 database IV	It facilitates sequence structure-based alignments and the building of phylogenetic trees using over 2,80,000 ITS2 sequences and their corresponding structures contained in the ITS2 database	Koetschan et al. (2012)
6.	Targeted host associated fungi (THF) database	1,817 ITS1 sequences reflecting the sequence diversity of fungal species previously discovered in the mouse and human gut are archived	Forbes et al. (2019)
7.	The International Society for Human and Animal Mycology (ISHAM)	It is committed to enhancing the research, identification, and management of fungal infections affecting both humans and animals. It is responsible for developing evidence-based clinical guidelines in that context	Agarwal et al. (2025)
8.	ITS database SILVA	It is primarily based on phylogenetic analysis of 16SrRNA for prokaryotes and 18SrRNA for eukaryotes offering taxonomy for bacteria, archaea, and eukarya according to Bergey's Taxonomic Outlines, List of Prokaryotic Names with Standing in Nomenclature (LPSN)	Park and Won (2018)
9.	Ribosomal Database Project	It provides aligned, annotated rRNA gene sequence data and tools for analysing their own rRNA sequences	El Jaddaoui et al. (2025)

Through metagenomic data, function-driven research identifies genes and enzymes with specific functions. Various useful approaches that allow detailed annotations and comparisons can provide significant biological insights across large sequencing datasets (Navgire et al., 2022).

TABLE 8.2
Tools Used in Bioinformatics Analysis of Mycobiome

Software	Description	References
QIIME	For sequencing data processing, it is an open-source pipeline based on PyCogent that supports database deposition, diversity estimation, analysis of microbial communities, core microbiome discovery, and interactive graphical representation	Halwachs et al. (2017)
CLoVR-ITS	It processes fungal ITS pyro sequences from metagenomic DNA. Based on metadata provided by the user, it performs quality filtering, chimera removal, OUT clustering, taxonomic identification, and statistical analysis between sample groups	Forbes et al. (2019)
CONSTAX	CONSTAX (CONSensus TAXonomy classifier) merges RDP, SINTAX, and UTAX algorithms for enhanced fungal OUT taxonomic classification. For ecological inference, it provides accurate, reproducible annotation at the species or genus level	Liber et al. (2021)
HumaMycobiomeScan	It is a fast and accurate bioinformatics pipeline for mycobiome profiling from raw NGS data. Validated on metagenomics and synthetic datasets, it categorizes reads into species more than 10,000 times faster than comparable methods based on hierarchical fungal databases	Soverini et al. (2019)
LotuS	With highly advanced proprietary algorithms for denoising, clustering, and SDM-based demultiplexing and filtering, the cutting-edge LotuS pipeline confidently manages small to large 16S datasets, merges several sequencing runs, and boasts superior computing efficiency	Özkurt et al. (2022)
DADA2	DADA2 applies quality filtering in order to reduce sequence bias and enables precise discrimination of ITS1 sequences with a single nucleotide difference	Rolling et al. (2022)
FunOMIC	3.4 million non-reductant proteins and 1.6 million marker genes are the part of FunOMIC. It identifies fungal species that have been linked to disease, body locations, and geography when tested against over 2,600 human metagenomes	Xie and Manichanh (2022)

8.2.4 Limitations and Challenges in Mycobiome Analysis

Although metagenomics has revolutionized human gut mycobiome research, it has inherent limitations that prevent a full understanding of microbial ecosystems. Barriers comprise methodological biases, interpretative challenges, repeatability, and issues with functional annotation that impact the capacity of the technique to accurately characterize diversity and activity of microbes (Yang et al., 2025). A major concern in mycobiome research is the reliance on BLAST-based fungal identification systems, which can ultimately misinterpret diversity. Closest BLAST hits commonly suffer from low resolution at the species level and host many human gut mycobiome that still remain undescribed. By expressing the need for improved bioinformatics tools, muscular-standardized workflows, and increased repositories of reference databases, it could improve the mycobiome analysis (El Jaddaoui et al., 2025). Reproducibility still represents a formidable issue in metagenomic studies due to differences during sample collection, DNA extraction, sequencing platforms, and bioinformatic pipelines, which contribute to variations that hinder the ability to conduct cross-study comparisons. There is a critical need for reproducible workflows and standardized protocols (Nearing et al., 2021). Another significant limitation when studying mycobiomes is functional redundancy. Functional redundancy is when different species of microbes conduct the same metabolic function, allowing for ecological resilience during disturbances. While metagenomics does an excellent job of depicting the genetic potential of microbial communities, the approach still limits our ability to understand active gene expression or metabolic activity (Yang et al., 2025).

Accurate species level identification using ITS regions is challenging, yet no other potential substitute markers provide the same level of lineage resolution. The taxonomic thresholds for filamentous fungi report 94.3% (genus), 88.5% (family), 81.2% (order), and 80.9% (class). Third-generation long read sequencing has the potential to resolve intron, repetitive and non- coding regions as well as improve amplicon-targeted ITS sequencing (Nilsson et al., 2019; Tiew et al., 2020). Despite being used extensively in large-scale microbiome studies, the efficacy of ITS primers is still unknown. According to studies, these primers undergo primer bias and amplification bias, which can often lead to incorrect and inconsistent taxonomic profiling results and misrepresented synthetic fungal communities (Ali et al., 2019). With a wide range of opportunities for amplicon sequence analysis of barcoding datasets, analytic tools for ITS-derived amplicon sequences are still in sparse, although demand and interest are growing in this new area of research (Tiew et al., 2020).

The filtering steps and parameters chosen will heavily influence the number of OTUs recovered; applying the most stringent filtering may underestimate the diversity of that species richness in the dataset being examined (Halwachs et al., 2017). The limited fungal reference databases, which leave many reads unclassified or OTUs inaccurately classified, have been an enduring problem in the study of mycobiome metagenomics. By generating more high-quality metagenomic and whole fungal genome assemblies, this could be mitigated (Tiew et al., 2020).

8.2.5 Human Health: Gut Mycobiome and Diseases

8.2.5.1 Inflammatory Bowel Disease (IBD)

IBD is a collection of recurrent, chronic inflammatory GI disorders due to dysregulated T-cell mediated immune responses to gut microbial populations (Hsu et al., 2023). It is of mainly two subtypes: ulcerative colitis (UC) and Crohn's disease (CD) (Chin et al., 2020). Various studies have demonstrated that the gut mycobiome of IBD patients looks different from that of healthy individuals. Catalán-Serra et al. (2024) marked the depletion of Zygomycota in IBD patients; independent of the activity of disease, it is an association that has not previously been reported. These IBD samples also showed higher abundance of *Candida* and lower abundance of *Aspergillus, Debaryomyces,* and *Cladosporium* (Catalán-Serra et al., 2024). One of the studies reported that gut mycobiome samples belonging to patients showed significant taxonomic changes and specific alterations including increased *Saccharomyces cerevisiae* counts and increased alpha diversity compared to healthy controls (Yu et al., 2023). Interkingdom interactions possibly have an effect on IBD pathophysiology through concurrent study of both the bacterial and fungal microbiota, and therefore, targeting and manipulating the gut microbiota may provide a new opportunity to treat IBD (Chin et al., 2020). In addition, fungi can also modulate intestinal inflammation both directly, through metabolites affecting pathways such as T cell signalling and NF-κB activation, and indirectly via interactions with other microorganisms, contributing to the pathogenesis of IBD.

8.2.5.2 Obesity

Obesity compounds major associated comorbidities such as diabetes and heart disease, as well as musculoskeletal ailments (Zhan et al., 2024). Fungi and obesity were associated in a patient with a BMI of 48.9 suggesting that fungi dysbiosis may be contributing factor to or reflection of metabolic alterations related to extreme obesity (Chin et al., 2020). The main fungal phyla in studies were Ascomycota and Basidiomycota, with notable genera being *Saccharomyces, Candida*, and *Malassezia.* Overweight individuals consumed higher carbohydrates, which likely encouraged *Candida,* as it metabolizes polysaccharides of fermentable sugars, disrupting the gut flora balance (García-Gamboa et al., 2024). In a study, the guts of obese people showed lower levels of *Nakaseomyces, Rhodotorula*, and *Schizosaccharomyces* but higher levels of *Dematiopleospora.* These differences might be attributed to differences in lifestyle, eating habits, characteristics of populations, and geographical conditions (Zhan et al., 2024). Mucor species belongs to the gut microbiota of slender, healthy individuals, as they are negatively correlated with obesity. Vavreckova et al. (2022) studied that Mucor was substantially enriched in pregnant women with good health.

8.3 DEVELOPMENT AND DYNAMICS NATURE OF THE HUMAN GUT PHAGEOME

Phages were first described by William Twort in 1915, and their ability to infect prokaryotic hosts, particularly bacteria, was further explained by Félix d'Herelle in 1917 (Clokie et al., 2011). Bacteriophages bind to specific receptors on bacterial cell surfaces using receptor-binding proteins (RBPs), and this attachment process initiates their infection cycle. Phage receptors, known as a cell surface-exposed molecule (protein), are located on the surface of bacterial cells, thereby facilitating bacteriophage attachment with the host. These interactions play a crucial role in determining the host range of bacteriophages during the infection process (Dowah & Clokie, 2018).

The human gut "phageome" refers to the entire community of phages in the gut, which contain either DNA or RNA and may exist in double-stranded or single-stranded forms. The gut phageome begins to develop in childhood and continues to change gradually throughout life (Shamash & Maurice, 2022). However, phages influence the gut ecosystem by modulating taxonomic diversity and function. Their effects stem from replication mechanisms, the "lytic" and "lysogenic" cycles, defining the phage "life cycle". Phage life cycles significantly regulate bacterial populations, shaping the microbiome and maintaining gut ecosystem balance and functionality (Barr, 2017; Dalmasso et al., 2014; Lang et al., 2025).

The lytic infection cycle is a hallmark of lytic bacteriophages, enabling rapid infection and replication within bacterial hosts. The infection process starts immediate after entering; these phages capture the host's cellular machinery to replicate their genetic material and produce new viral progeny. This process of lytic infection cycle reaches to a final stage of lysis of the bacterial cell and enables to release a newly formed virions into the extracellular environment, where they can cause subsequent rounds of infection (Campbell, 2003; Zhang et al., 2024). The replication cycle of the lytic phage involves sequential steps such as attachment, penetration, replication, packaging, and exit (burst out) from the host cell. Moving beyond the core function of viral propagation, the lytic cycle plays a crucial role in the ecological role by maintaining the rate of bacterial population dynamics and preventing the overgrowth of dominant species, thereby contributing to the maintenance of microbial diversity within the ecosystem (Wheatley et al., 2025).

Unlike the lytic phage, temperate phages can possess the ability to enter the lysogenic cycle from the lytic cycle during the infection mode during which a temperate phage integrates its genetic material into the chromosome of the host bacterium as a prophage. Prophages mostly exit in a dormant or quiescent state and replicate their genomes along with the host genome, conferring genomic stability under favourable conditions (Zhang et al., 2022b). However, environmental stressors, such as DNA damage, nutrient limitation, or antibiotic exposure, can trigger prophage induction, often by activating the bacterial SOS response. Prophages are initiated at the beginning of induction, transitioning into the lytic cycle, resulting in host cell lysis (self-amplifying biological weapons) and the release of newly formed virions. Temperate phages are extremely common in the genomes of bacteria that reside in the human gut, which not only contributes to the dissemination of phages, but also plays a critical role in HGT, facilitating the spread of genetic elements, including virulence determinants and antibiotic resistance genes, between microbial populations (Wheatley et al., 2025).

Metagenomic analysis based on more than 11,800 human gut samples revealed that prophages make up approximately 45% of the human gut bacterial genome. Temperate phages demonstrate the widespread nature of phage–host interactions within the intestinal microbiota. Temperate phage activation stimulates by various environmental stresses, such as exposure to antibiotic treatment, inflammation, or oxidative stress. Temperate phages play an important role in mobilization of antibiotic resistance genes that improve pathogen survival (Shamash & Maurice, 2022). The spread of Shiga toxin genes and β-lactamase genes in *Escherichia coli* is one of the common examples of mobilization of antibiotic resistance gene via HGT. Despite the fact that HGT promotes the bacterial fitness under extremely unfavourable conditions, it disrupts the balanced microbial ecosystem by promoting dysbiosis (known as an unbalanced gut ecosystem) (Rodrguez-Rubio et al., 2021; Keen & Dantas, 2018).

The dual-nature lytic-lysogenic life cycle of bacteriophages forces selective pressures on bacterial hosts, shaping key microbial traits such as metabolic activity, virulence, and antibiotic resistance. By lysing susceptible bacteria or integrating as prophages, phages contribute to microbial evolution and influence gut microbiota diversity (Barr, 2017; Wheatley et al., 2025). Although their regulatory roles in modulating community composition such as the suppression of pathogenic taxa or the dissemination of adaptive genes are increasingly acknowledged, the exact molecular mechanisms underlying phage-driven microbiota remodelling remain insufficiently characterized. Most current metagenomic investigations prioritize bacterial community profiling, often overlooking distinct contributions and mechanistic roles of bacteriophages, thereby highlighting a critical gap and an opportunity for deeper virome-centric research within the human microbiome(Kurilovich & Geva-Zatorsky., 2025).

Phages are known to be bacterial parasites that contribute to the essential role in modulation of human immune systems. During the lytic cycle, the destruction of bacterial cells by phages leads to the release of various bacterial components, including cell wall fragments, lipopolysaccharides, and nucleic acids. This molecular part, collectively referred to as pathogen-associated molecular patterns (PAMPs), is directly recognized by host pattern recognition receptors (PRRs), particularly Toll-like receptors (TLRs), which serve as key mediators of innate immunity. The interaction between PAMPs and PRRs initiates downstream signalling cascades that influence inflammatory responses, immune cell activation, and broader immunological homeostasis within the GI tract. For instance, phage DNA has been shown to activate TLR-9 on dendritic cells, a mechanism implicated in the production of interferon-γ (IFN-γ) and the exacerbation of intestinal inflammation in murine models. Inflammatory (pro-inflammatory) responses are the most common event in IBD, where disruption of phage communities leads to alterations in bacterial populations, resulting in increased immune activation (Federici et al., 2021; Hibstu et al., 2022; Lin et al., 2020).

T4 phages demonstrate the ability to directly interact with immune cells such as neutrophils, thus reducing the production of harmful reactive oxygen species (ROS). In addition, phages stabilize mucosal immunity by adhering to gut mucins through immunoglobulin-like domains on their capsid proteins, limiting bacterial colonizations and subsequent inflammation (de Souza et al., 2023). Phages boost intestinal barrier integrity by selectively lysing pathobionts and preventing their translocation across the epithelial lining, thus reducing systemic inflammation and maintaining gut homeostasis.

8.3.1 Metagenomic Approaches and Advantages of Shotgun Sequencing in Studying Human Gut Bacteriophages

The discovery of bacteriophages in the laboratory, using culture-dependent methods, is time intensive, expensive, and technically complex. Culture-dependent methods rely on specific environmental conditions, such as media required for cultivation, precise temperature and pH ranges, and defined ion concentrations. Therefore, culture-independent methods, especially shotgun metagenomics, overcome these limitations and provide a high-taxon-resolution and cost-effective method to analyse microorganisms by direct sequencing of nucleotide sequences (DNA/RNA) (Hayes et al., 2017; Shkoporov et al., 2019).

In this modern era, where sequencing techniques are advancing, shotgun metagenomic sequencing is a high-throughput excellent sequencing technology that enables comprehensive characterization of the viral components of gut microbiota. Unlike amplicon-based methods, in which we target 16S bacterial rRNA markers for bacteria and 18sRNA gene markers for fungi, shotgun metagenomics typically provides in-depth information on the entire DNA contents present in a sample, giving a holistic view of the microbial ecosystem (Kumar et al., 2024; Zhang et al., 2024). In these methods, DNA sequences are randomly fragmented into many small fragments, followed by sequencing and metagenomic assembly. The two most common methods are implemented to explore the diversity of viruses, bacteria, and other microorganisms: (1) The reference-based method aligns sequencing reads against known species present in a pre-existing reference database. (2) Reference-independent method or de novo assembly, in which unknown viruses or rare viral species are analysed by reconstructing their entire genome. Avoiding solely dependence on marker genes, shotgun metagenomics provides an unbiased and comprehensive assessment of the total DNA content present in a sample (Kumar et al., 2024; Nayfach et al., 2021).

In recent years, several studies have supported the effectiveness of machine learning-based models in predicting disease status through reference-based virome profiling using metagenomic datasets. The reference approach allows for rapid and cost-effective assessment of known phage diversity (taxonomic diversity) and abundance from metagenomic data, including integrated prophages, and also captures sequences from both extracellular and intracellular viruses. In recent years, several comprehensive virome databases have been developed on a large scale using metagenomic datasets, providing valuable resources for researchers investigating virome diversity and ecological dynamics (Table 8.3).

Shotgun metagenomics has transformed virome profiling by enabling comprehensive characterization of intestinal viral communities through untargeted short-read sequencing, capturing both known and unknown viral populations. The main advantage of this method for bacteriophage studies is its capability to identify novel phages via de novo assembly, allowing the discovery of phages not present in existing reference databases (Roux et al., 2021; Sun et al., 2021a). Metagenomic approaches are not limited to the identification of taxonomy information or diversity of phage genomes but also provide a deeper understanding of the functional proteins encoded by these genomes and the total quantity of phages. Most importantly, metagenomic techniques instead offer the possibility to simultaneously investigate the presence and potential interactions between phages and bacterial taxa (phage–host interactions) within the same dataset (Lang et al., 2025; Sun et al., 2021a.b). Furthermore, shotgun sequencing is capable of identifying phages integrated into bacterial chromosomes (e.g., prophages). The detection of these integrated elements provides critical insight into the lysogenic cycle, a crucial aspect of phage biology that can significantly impact the fitness of bacterial cells and the stability of microbial communities (Venturini et al., 2022).

TABLE 8.3
List of Virome Databases Developed Based on Metagenomic Datasets

Database Name	Description	Reference
Metagenomic Gut Virus (MGV)	The extensive viral genome repository comprises 142,809 distinct gut phages, identified through the examination of 11,810 bulk metagenomic samples collected from human stool, and compiled from 61 preceding studies	Nayfach et al. (2021)
Gut Phage Database (GPD)	The Gut Phage Database (GPD) contains approximately 142,000 viral genomes developed from 28,060 metagenomes	Camarillo-Guerrero et al. (2021)
IMG/VR v4	The database is committed to the genomic study of uncultivated viruses, providing insights into viral diversity that has not been cultured under laboratory conditions	Camargo et al. (2023)
Gut Virome Database (GVD)	The GVD was developed using 2,697 viral particles or microbial metagenomes collected from 1,986 people in 16 various geographic locations	Gregory et al. (2020)
Hadza Phage Catalog	The assembly of thousands of novel human gut bacteria, archaea, eukaryotes, and phages recovered 91,662 genomes of bacteria, archaea, bacteriophages, and eukaryotes	Carter et al. (2023)
Infant Gut Atlas (COPSAC)	The COPSAC study categorizes 10,000 viral species into 248 family-level clades, shedding light on the early development and evolution of the infant gut microbiome	Shah et al. (2023)
Early-Life Gut Virome (ELGV)	The Early-Life Gut Virome (ELGV) contains 160,478 distinct viral sequences, including DNA and RNA viruses. These were obtained from 8,130 virus-like particles (VLPs) from metagenomic samples collected in early development	Zeng et al. (2024)
Aggregated Gut Viral Catalogue (AVrC)	A comprehensive resource comprising 1,018,941 unique viral sequences, categorized into 449,859 species-level vOTUs	Galperina et al. (2024)
Chinese Gut Virome Catalogue (cnGVC)	A gut virome database, cnGVC, was created by analysing over 11,000 metagenomes from Chinese faecal samples, containing 93,462 unique viral genomes, with over 70% previously unknown	Yan et al. (2025)
Chinese Gut Viral Reference (CGVR)	The CGVR database contains 120,568 viral genomes from 3,234 sequenced faecal samples collected nationwide, organized into 72,751 unique viral species-level clusters, with about 90% being novel sequences not found in existing databases	Wang et al. (2024a)

8.4 COMPUTATIONAL TOOLS FOR PHAGE DETECTION IN METAGENOMICS

Over the past several decades, various computational tools have emerged to identify and classify bacteriophages from metagenomic datasets. In this chapter, computational tools are categorized into three different groups based on working principles:

(1) homology-based methods (database-dependent tools), (2) feature-based tools (sequence-driven approaches), and (3) hybrid tools (integrating homology, sequence-based methods, and advanced computational algorithms). To provide a brief overview and facilitate comparison, we have summarized the features of the tools in Table 8.4.

8.4.1 Homology-Based Based Tool for the Phage Identification

Homology-based approaches are the most common and effective computational methods for identifying bacteriophages, also known as database-dependent tools. These methods use sequence alignment algorithms to compare input query sequences (nucleotides or amino acids) against existing databases of known viral genomes. Homology-based phage detection methods are predicated on the fundamental principle that viruses sharing a common ancestry or exhibiting similar functional characteristics often possess significant similarities in their genetic sequences, either at the nucleotide or amino acid level (Owen et al., 2021). The foundation of these methods is the application of sequence alignment algorithms, with the Basic Local Alignment Search Tool (BLAST) being a widely adopted example, to perform comparisons between query sequences of unknown origin (derived from sources such as metagenomic datasets or bacterial genomes) and extensive databases that house the genetic information of previously identified and characterized phages. The identification of statistically significant sequence matches, typically above a predefined threshold of similarity and coverage, operates as an indicator indicating a potential phage origin for the sequence under investigation. Homology-based tool depends upon existing databases such as the NCBI Viral Genome Database, IMG/VR (Integrated Microbial Genome and Microbiome Viral Database), and ViPhOG (Viral Phage Orthologous Group), which contain information on a large number of known viral sequences.

8.4.2 Sequence-Based Phage Identification

Sequence-based computational tools are based on the analysis of essential features of sequences rather than relying on pre-existing reference databases for phage identification. In general, sequence-based tools assume that viral genomes have distinct intrinsic features when compared to bacterial or other non-viral genomes (Guo et al., 2021; Ren et al., 2017). One of the main possibilities for these differences are because of evolutionary pressures, replication methods, or interactions with the host cell. These approaches consider a number of nucleotide features, such as GC content, gene structure, and attributes (e.g., protein size, direction of transcription strand, and gene compactness), as well as patterns within the nucleotide sequence itself (e.g., frequencies of short sequences called k-mers, prevalence of distinct short k-mers).

Examples of prominent feature-based tools include VirSorter (Guo et al., 2021), VirFinder (Ren et al., 2017), and PhiSpy (Akhter et al., 2012). VirSorter uses a scoring system based on the presence of viral hallmark genes, the absence of universal single-copy genes, and the enrichment of viral-like genes. VirFinder is developed based on the machine learning models where it uses k-mer frequencies to distinguish viral from non-viral sequences. These tools work on the empirical principle that

TABLE 8.4
Overview of Phage Identification Tools and Their Characteristics

Category	Tool Name	Methodology	Sequence Features/ Databases Used	Description	Advantages	Limitations	References
Homology-Based	PHASTER	BLAST against the phage database	Phage sequence databases, previously annotated genomes	PHASTER analyses bacterial genomes to identify prophages using BLAST to match sequences with phage databases, detecting regions that closely resemble known phages	Speed, efficiency, scalability, user-friendly	It depends on its similarity to known phages	Arndt et al. (2016)
Homology-Based	MetaPhinder	BLASTN against the whole-genome phage database integrates multiple hits	Database of whole-genome phage sequences, negative datasets	MetaPhinder detects phage-origin contigs within metagenomic data by matching them against a whole-genome bacteriophage sequence database using BLASTN, demonstrating superior performance over methods such as tBLASTx during its development.	Accounts for mosaic genomes, outperforms single-hit BLAST for longer sequences, competitive with k-mer methods for larger contigs, web service, and source code available	Performance declines for short sequences, depending on similarity to known phages	Jurtz et al. (2016)

(Continued)

TABLE 8.4 (*Continued*)
Overview of Phage Identification Tools and Their Characteristics

Category	Tool Name	Methodology	Sequence Features/ Databases Used	Description	Advantages	Limitations	References
Homology-Based	vConTACT 2.0	Gene-sharing networks, distance-based hierarchical clustering	Protein sequences from viral genomes, reference database of known viral genomes	Employs entire genome gene-sharing profiles for virus classification via distance-based hierarchical clustering	Scalable to thousands of genomes, confidence scores for predictions, near identical to ICTV taxonomy	High hardware requirements (memory), taxonomic resolution depends on database, no direct taxonomy for user sequences	Bolduc et al. (2017)
Homology-Based	PhageWeb	BLAST against phage database, G+C content, tRNA presence	Phage database (NCBI, EBI), bacterial genomes, biological databases	The identification of prophages in bacterial genomes involves searches for gene sequence similarity against a phage database, using alterations in GC content and the presence of tRNA as indicators, and enabling functional characterization of regions	Higher precision, faster processing, comprehensive methodology, functional characterization, GUI and API	Relies on database quality, challenges with novel phages, potential parameter sensitivity	de Sousa et al. (2018)

(*Continued*)

TABLE 8.4 (*Continued*)
Overview of Phage Identification Tools and Their Characteristics

Category	Tool Name	Methodology	Sequence Features/ Databases Used	Description	Advantages	Limitations	References
Homology-Based	PhageScanner	BLAST against protein databases	Protein databases (Uniprot, Entrez), metagenomic data	PhageScanner is a tool to identify phage virion proteins (PVPs) via metagenomic sequencing	High accuracy in PVP prediction	Significantly higher inference time than ML models	Albin et al. (2024)
Homology-Based	VirusSeeker	BLAST against viral databases, reference-based mapping, de novo assembly	Raw sequence data (NGS), reference viral genomes, custom microbial database, Viral Minion DB HMM	A comprehensive, automated, and modular software suite created for the purpose of analysing sequence data to recognize sequences derived from microbes	Dual functionality (novel discovery and virome composition), successfully applied in studies	Reliance on existing databases, potential computational resource requirements	Zhao et al. (2017)
Sequence-Based (Feature-Based)	PhiSpy	AT/GC skew, protein length, strand orientation, k-mer abundance, HMMs (opt)	Annotated bacterial genomes	Utilizes sequence characteristics like AT/GC skew, protein length, and transcription strand orientation to predict prophages, thus facilitating the identification of viruses do not present in databases	Automated prophage prediction, machine learning approach, utilizes multiple features, customizable, supports HMMs	Relies on annotation, potential for false positives/negatives, performance varies by species	Akhter et al. (2012)

(*Continued*)

TABLE 8.4 (*Continued*)
Overview of Phage Identification Tools and Their Characteristics

Category	Tool Name	Methodology	Sequence Features/ Databases Used	Description	Advantages	Limitations	References
Sequence-Based (Feature-Based)	virMine	Iterative exclusion of non-viral sequences	Raw reads, assembled contigs, databases of non-viral and viral sequences	Sequencing reads from short-read or long-read technologies, as well as assembled contigs, are annotated and evaluated for their potential viral origin	Automated workflow, flexible, potential for novel virus detection, broad viral range	Reliance on non-viral database, potential for false positives, computational resources needed	Johnson and Putonti (2022)
Sequence-Based (Feature-Based)	VirMiner	Random forest on genomic features	Assembled contigs, clean reads, various protein and functional databases, pre-built RF model	A web application utilizing the random forest algorithm to detect phage contigs, particularly focusing on those with high abundance, from metagenomic data	Comprehensive pipeline, sensitive for high-abundance phages, phage-host prediction	Dependency on external tools and databases, computational resources needed, model trained on human gut data	Zheng et al. (2019)

(*Continued*)

TABLE 8.4 (*Continued*)
Overview of Phage Identification Tools and Their Characteristics

Category	Tool Name	Methodology	Sequence Features/ Databases Used	Description	Advantages	Limitations	References
Hybrid (Homology + Sequence + ML)	VirSorter2	ML on genomic features (annotation, hallmark genes)	High-quality virus genomes, HMM databases, random forest models	Employs machine learning to predict viral potential by analysing genomic traits, encompassing structural, functional, and taxonomic annotations, along with viral hallmark genes	Works with diverse DNA and RNA viruses, applies machine learning, trained with high-quality data, prepares files for DRAMv	Default cutoff may lead to false positives in environmental data, performance affected by short/ incomplete sequences, computationally intensive	Guo et al. (2021)
Hybrid (Homology + Sequence + ML)	MARVEL	Random forest on gene density, strand shifts, viral hits	RefSeq microbial dataset, pVOGs database	A tool designed to predict dsDNA phage sequences within metagenomic bins, utilizing random forest algorithms to identify phage sequences through genomic features and homology	Comparable specificity, better recall than VirSorter/ VirFinder, uses simple features	Specifically for dsDNA phages in metagenomic bins	Amgarten et al. (2018)

(*Continued*)

TABLE 8.4 (*Continued*)
Overview of Phage Identification Tools and Their Characteristics

Category	Tool Name	Methodology	Sequence Features/ Databases Used	Description	Advantages	Limitations	References
Hybrid (Homology + Sequence + ML)	PhaMers	Supervised learning on k-mer frequencies	Phage and non-phage sequences for training, k-mer frequency vectors	Employs supervised learning to identify phages by utilizing k-mer frequency vectors, merging sequence features with machine learning techniques	Captures sequence signatures, potentially effective for novel phages, computationally efficient	Performance depends on k-mer size and training data, potential for overfitting	Deaton et al. (2019)
Hybrid (Homology + Sequence + ML)	VirFinder	Logistic regression on k-mer frequencies	Viral and host sequence databases for training, k-mer frequencies	The prediction method uses k-tuple word frequencies to differentiate virus from host sequences, trained on equal viral and host data. A C++ program counts k-tuple word occurrences using a hash table	Good accuracy for short and novel sequences, alignment-free, user-trainable	Performance biased by training data, lower recall for archaeal phages, ecosystem-dependent performance	Ren et al. (2017)

(*Continued*)

TABLE 8.4 (*Continued*)
Overview of Phage Identification Tools and Their Characteristics

Category	Tool Name	Methodology	Sequence Features/ Databases Used	Description	Advantages	Limitations	References
Hybrid (Heuristic search + Homology)	Phage_Finder	BLAST and HMMSEARCH for phage genes, sequence features for attachment sites	Bacterial genomes, phage sequence database, phage-specific HMMs, information on integration sites	A heuristic program identifies prophage regions in bacterial genomes by detecting phage-like gene clusters using BLAST and sequence features	High accuracy in prophage identification, predicts attachment sites	Limited to specific phage types known to integrate	
Hybrid (ML + Homology)	Prophage Hunter	ML on genetic features, BLAST for similarity	Bacterial genome assemblies, phage genome sequences, genetic features of prophages	Employs machine learning alongside homology to predict prophages, as deduced from analogous tools	Higher accuracy in predicting active prophages, comprehensive web service	NA	
Hybrid (ML + Sequence)	Phybrid	Additive boosting on gene content and k-mer patterns	Bacterial and phage genomes, gene annotations, nucleotide sequences	A tool that integrates gene content attributes (such as strand changes and gene density) with nucleotide k-mer patterns through an additive boosting model	Combines different genomic signals, accelerated in silico characterization	NA	

(*Continued*)

TABLE 8.4 (*Continued*)
Overview of Phage Identification Tools and Their Characteristics

Category	Tool Name	Methodology	Sequence Features/ Databases Used	Description	Advantages	Limitations	References
Hybrid (DL)	DeepVirFinder	CNN on genomic sequences	Viral and prokaryotic genomic sequences for training, metagenomic data	A Convolutional Neural Network (CNN) is employed to autonomously learn genomic patterns from both viral and prokaryotic sequences, facilitating precise detection of viral sequences within metagenomic datasets, with a notable enhancement in performance when handling short sequences	Significantly improved accuracy for short sequences, learns complex patterns automatically	Requires large training data, computationally intensive for training, learned patterns can be hard to interpret	Ren et al. (2020)
Hybrid (ML pipeline)	PhageScanner	CNNs on genomic signatures	Protein databases (Uniprot, Entrez) for training, genomic/ metagenomic data	Utilizes convolutional neural networks, which are trained on genomic signatures, to classify phage sequences	Reconfigurable framework, versatile input, multiple classification modes, BLAST integration, GUI for visualization	Computational cost for some feature extraction, model performance variability, BLAST classifier is slow	Albin et al. (2024)

(*Continued*)

TABLE 8.4 (*Continued*)
Overview of Phage Identification Tools and Their Characteristics

Category	Tool Name	Methodology	Sequence Features/ Databases Used	Description	Advantages	Limitations	References
Hybrid (DL)	PhaMer	Transformer model on metagenomic assemblies	Metagenomic assemblies, training data for transformer model (implicit)	Utilizes deep learning based on transformer models to identify phages within metagenomic assemblies	Effective in learning long-range dependencies, potentially high accuracy	Computationally intensive, requires large datasets for training, interpretability can be challenging	Shang et al. (2022)
Hybrid (DL)	PhaGCN	GCN integrating CNN on DNA and gene-sharing network	Assembled contigs, DNA and protein sequences, gene-sharing network data, training data	The classification of phage contigs is improved by employing a GCN, which integrates DNA sequence characteristics with protein similarity	Combines sequence and network information, semi-supervised learning	Requires viral contigs as input, issues with non-ACGT characters	Shang et al. (2021)
Hybrid (ML)	VIBRANT	Neural networks on protein signatures, v-score metric	Protein sequences, HMM databases, bacterial/ archaeal genomes (implicit)	Employs neural networks alongside a v-score metric to detect lytic viral genomes and integrated proviruses, improving the identification of various viruses	Applicable to diverse viruses (DNA, RNA, proviruses, eukaryotic)	NA	Kieft et al. (2020)

(Continued)

TABLE 8.4 (*Continued*)
Overview of Phage Identification Tools and Their Characteristics

Category	Tool Name	Methodology	Sequence Features/ Databases Used	Description	Advantages	Limitations	References
Hybrid (DL)	PPR-Meta	CNN on metagenomic sequences (nucleotide and codon paths)	Metagenomic sequence fragments, training data for CNNs	PPR-Meta identifies metagenomic sequences as phages, chromosomes, or plasmids by calculating scores for each possibility	Simultaneously identifies phages and plasmids, uses deep learning for classification	NA	Fang et al. (2019)
Hybrid (DL)	Seeker	LSTM models on DNA sequences (alignment-free)	Phage and bacterial complete genomes for training, DNA sequences for detection	Seeker is an alignment-free method that uses deep learning-based LSTM models, a type of RNN, for phase identification.	Alignment-free, unbiased, identifies long-range dependencies, good for short sequences, computationally fast	Recommended upper bound for input length (around 1 kb)	Auslander et al. (2020)

viruses and hosts share different k-mer signatures. PhiSpy is a prophage identification tool that identifies prophage sequences integrated into bacterial genomes (and probably Archaeal). PhiSpy tools are developed based on seven different characteristics of prophages such as protein length, transcription strand directionality, customized AT and GC skew, the abundance of unique phage words, phage insertion points, and the similarity of phage proteins. The main reason for using feature-based methods is that they are able to identify new phages that may not have significant sequence similarity with known viruses. The use of such tools overcomes a major limitation of database-dependent methods. Despite their advantages, feature-based methods can sometimes give false positive results, as some bacterial gene sequences share viral-like features.

8.4.3 Hybrid-Based Tools for Phage Identification

In recent years, there has been great progress in the study of microorganisms from metagenomic data using sequencing technology and artificial intelligence, as well as a significant increase in the development of artificial intelligence-based identification tools, such as hybrid-based tools. Hybrid-based tools are innovative and advanced approaches to identify phage sequences, combining both methods: database-based (reference dependent) and feature-based (reference independent) to improve the accuracy and sensitivity of tools. These tools often combine homology-based sequence comparison and feature-based intrinsic sequence properties analysis and use machine learning or deep learning algorithms to minimize false positives and improve phage prediction accuracy (Andrade-Martínez et al., 2022; Khot et al., 2020).

In other words, the basic principle behind using hybrid-based tools is that by combining multiple analytical approaches, we can overcome the limitations of single-method approaches, which provide phage identification more sensitive than reference-dependent methods and more precise than composition-based methods. For example, a hybrid tool first uses homology-based models to identify putative viral sequences and then applies feature-based models to refine these predictions based on genomic features. In contrast, some tools use feature-based methods to initially identify potentially viral regions and then use homology searches to confirm their viral origin and potentially classify them. Examples of hybrid tools include DeepVirFinder and MARVEL (Meta-analysis-based vibration element discovery) (Amgarten et al., 2018; Ren et al., 2020). DeepVirFinder predicts viral sequences based on k-mer frequencies and sequence structure using a deep learning-based integrated approach. It is a reference-free and alignment-free machine learning method. MARVEL works to classify bacterial sequences from phage sequences from metagenomic assemblages based primarily on three features: sequence similarity to known viruses, the presence of viral hallmark genes, and sequence structure.

8.5 SELECTING THE OPTIMAL PHAGE DETECTION TOOL: CONTEXTUAL CONSIDERATIONS FOR RESEARCHERS

In recent years, several tools have been developed to identify phage sequences from metagenomic samples and virome assemblies. A thorough evaluation of the advantages and limitations of each phage identification method can help researchers in choosing the most appropriate tool. Homology-based methods are effective in identifying phages closely related to known sequences in pre-existing databases, while they pose significant difficulty in detecting unknown or novel phages. Conversely, sequence-based tools can identify potentially new phage based on the genomic composition, as these tools provide higher sensitivity to phages with less representation in reference databases. Reference independent methods can be sensitive to the choice of features and result in more false positives. Hybrid-based methods combine the strengths of both approaches, often leading to higher accuracy and sensitivity by integrating different types of sequence information (Klumpp et al., 2012; Vieira et al., 2025).

The choice of an appropriate phage detection tool depends upon the specific aims and objectives of the research questions. For studies focused on identifying phages similar to known ones or for tasks like phage typing where specificity is critical, homology-based tools like PHASTER or MetaPhinder are suitable. Researchers exploring novel phages or analysing short-reads from metagenomic data may find sequence-based methods like VirFinder or DeepVirFinder, or hybrid methods like Seeker, which are more effective. When analysing metagenomic-assembled bins, especially for dsDNA phages, MARVEL's high recall makes it a good choice. In cases where computational resources are limited, simpler and faster methods may be preferable, whereas for studies requiring high accuracy and the detection of a broad range of viruses, more computationally intensive hybrid methods like VirSorter2 or PhaGCN are ideal.

The selection of phage identification tools also depends on the source of the dataset. The human guts contain a high diversity of phages, many of whose host range remain largely uncharacterized. The integration of both homology-based and sequence-based methods has potential to improve accuracy and specificity in phage detections. An underappreciated reason for sequences contamination sequencing depth, assembly quality, and eukaryotic contamination must also be carefully considered. High sequencing depth ensures more thorough read coverage, making it easier to detect rare phages and can therefore capture a more complete picture of viral diversity. High-quality assembly bins are essential because errors in viral genome assembly can lead to misclassification or underestimation of phage diversity (Shang et al., 2022).

8.6 CHALLENGES IN IDENTIFYING PHAGES FROM METAGENOMICS DATA

Despite advances in analytical tools, identifying phage sequences from metagenomic data faces several significant challenges (Table 8.5). It is important for researchers to meticulously consider and address these complexities during their analyses. In this section, we provide a detailed overview of these major difficulties that arise during analysis.

8.6.1 Distinguish Viral Genomes from the Host Genome

Distinguishing viral sequences from host-derived genetic material is one of the most difficult tasks, such as temperate phages that integrate into bacterial chromosomes as prophages. In general, temperate phages are very much closer to their respective host

TABLE 8.5
Challenges and Key Issues with Examples

Challenge	Description	Key Issues & Examples
1. Distinguishing viral from host genomes	Difficulty separating phage sequences from host (bacterial) genetic material due to sequence similarity and integration	• Temperate phages integrate as prophages into host chromosomes • Lytic phages encoding host-like metabolic genes (tRNA, ribosomal proteins) cause misclassification by sequence similarity tools
2. Dependencies on limited hallmark genes	Identification primarily uses hallmark genes (e.g., structural components), which represent only a small portion of the phage proteome	• High proportion of uncharacterized proteins makes reliable differentiation between viral and non-viral sequences difficult
3. Partial genome fragments	Metagenomic sequencing often yields incomplete phage genome fragments, hindering full analysis	• Prevents full reconstruction of infectious phages • Limits identification of essential genomic features (coding/regulatory regions) • Impedes understanding of phage life cycles, host interactions, and virulence mechanisms • The issue worsens when the host is unidentified or lacks characterization
4. Viral dark matter & RNA phage underreporting	A large fraction of viral sequences lack similarity to known viruses, and RNA phages are often missed by standard methods	• "Viral dark matter" (novel sequences) hinders homology-based detection of DNA phages • Traditional focus on DNA sequencing overlooks RNA phages • Limits comprehensive understanding of RNA virus diversity and ecological roles
5. Small genome sizes & lack of markers	Phage genomes are small and lack universal marker genes, making detection and classification difficult within complex samples	• No conserved marker genes like bacterial 16S rRNA for reliable taxonomy • Phage sequences are often a minor component compared to prokaryotic contigs, increasing risk of omission/misclassification • Eukaryotic contamination and low viral biomass further obscure phage signals

then to the enteric hosts; as a result, it shares significant degree of sequence similarity with their host genomes (bacterial genomes). For example, an investigation indicated that approximately 76% of phages with annotated hosts in the RefSeq database show alignments with their hosts' genomes (bacterial genomes), making it difficult to distinguish between phage and bacterial sequences. Additionally, lytic phages that encode host-like metabolic genes (e.g., tRNA or ribosomal proteins) exacerbate these issues, as sequence similarity tools commonly misclassify viral reads as bacterial (Paez-Espino et al., 2017).

8.6.2 Hallmark Gene Identification

Despite the lack of universal viral marker gene like bacteria (16s rRNA gene) and fungi (18s rRNA gene), phage contains many hallmark genes that are unique and specific to the virus and don't share similarity to any cellular genomes (Paez-Espino et al., 2017). Phage detection mostly depends upon hallmark genes (such as encoding capsid, portal, and terminase proteins) that are indicative of the structural components of phages, along with sequence information. It remains difficult to characterize the diversity of many new phage communities, since a large portion of their proteins lack functional annotation. In a study based on the largest order of phages *Caudovirales*, which constitutes approximately 93% of sequenced phages, there are an estimated 187,006 proteins classified as hypothetical proteins (Shang et al., 2022). In additions, the large gene set of phages is complicated by many newly identified phages having genes with many uncharacterized proteins, making it difficult to identify hallmark genes.

8.6.3 Partial Genome Fragment

Partial fragments (or short fragments) hinder complete reconstruction of infectious phages, limiting our ability to detect genomic features and clearly understand the functional roles of phages. For example, incomplete generation of phage assembly poses challenges in identifying complete coding and regulatory regions, which are essential for understanding the phage life cycle, phage–host interactions, and mechanisms of virulence. (Mallawaarachchi et al., 2023). There are several tools available to accomplish task, such as CheckV, which estimates genome completeness of the new virus based on marker genes already present in the large database of complete viral genomes.

8.6.4 Prevalence of Viral Dark Matter

A significant challenge in viral metagenomics studies is the widespread presence of "viral dark matter," which consists of novel sequences that share limited similarity to known viral genomes. This consequently hinders the identification and classification of DNA phages. Additionally, RNA-based phages are often overlooked during metagenomic sequencing due to traditional methods focusing on DNA sequencing, which limits our ability to identify RNA-based viral genomes. The diversity of RNA-based phages present as a component in the human gut and their ecological significance are still limited. Collectively, advanced sequencing strategies and expansion of reference databases are needed to improve the accurate identification and classification of both DNA and RNA phages (Dion et al., 2020).

8.6.5 Challenges with Small Genome Sizes

Phage genomes are relatively small in length (50 kbp) compared to bacterial genomes, complicating their detection and classification within sequences obtained from metagenomic datasets. Unlike bacteria, which have a 16S rRNA marker gene, phages lack a universal marker gene, making the development of a reliable pipeline for a taxonomic classification framework challenging. In general, metagenomic samples are dominated by prokaryotic reads (non-viral reads such as bacteria, fungi, and human reads) with a smaller number of viral sequences, increasing the chance of misclassification. Moreover, eukaryotic contamination and low viral biomass further obscure phage-derived signals, thereby diminishing the sensitivity and accuracy of identification tools (Dalmasso et al., 2014; Fremin et al., 2022).

8.7 PHAGE–HOST INTERACTION PREDICTION TOOLS: APPROACHES, TECHNIQUES, AND TOOLS

The specificity of bacteriophage–host interactions is principally mediated by RBPs displayed on the caudal structure of the phage. These RBPs demonstrate a high affinity for cognate molecular receptors located on the surface of the bacterial host cell (Barr, 2017; Degroux et al., 2023; Lang et al., 2025). The phage infection cycle begins with the binding between the phages RBPs (located at the end of the tail) and the host cells receptors (present on the cell surface). Receptor binding is highly specific and ensures that the phage can infect only those bacterial cells that have the appropriate receptors, thereby defining the range of hosts or bacterial strains susceptible to infection.

Phage–host infections begin with a series of sequential stages, starting with receptor recognition, penetrations of foreign DNA molecules, replications, and ending with the rupture of the host cell. During interactions, phage RBP proteins interact with essential components of the bacterial cell envelope, such as interaction with lipopolysaccharides (LPS), certain outer membrane proteins such as OmpC and FhuA, or capsular polysaccharides (Lang et al., 2025). After reversible attachment, the interaction progresses to irreversible adsorption and subsequent genome translocation. During translocation, the phage injects its nucleic acid across the host cell envelope, while the capsid structure remains extracellular. The efficiency of DNA delivery is strongly dependent on the stereospecific complementarity between the interacting domains of the RBP phage and the host receptor (Stone et al., 2019).

The interaction between phage and host is influenced by genetic diversity that determines host specificity and phage infectivity. For example, some phages utilize tail-spike proteins that demonstrate enzymatic activity or high-affinity binding towards specific bacterial surface glycans with considerable substrate specificity. While bacteria possess evolved antiphage defence systems, such as CRISPR-Cas-mediated immunity, these mechanisms primarily act post-adsorption and thus exert secondary influence on the initial determinants of host recognition compared to RBP-receptor compatibility. In essence, the molecular interaction between phage RBPs and bacterial surface receptors serves as the foundation for the mechanisms controlling phage tropism and infection dynamics (George & Hug, 2023; Wang et al., 2023).

8.7.1 Tools for Predicting Phage–Host Interactions

Bacteriophages, which are viruses that selectively infect and kill the bacteria, present a promising strategy because they have the capability to target specific pathogens and cause minimal disturbance to the host's beneficial microbial community. In this context, computational tools for predicting phage–host interactions provide potential phage–host pairs and guiding experimental validation efforts. In this section, we have described the employment of a diverse range of tools to tackle this complex biological problem. These tools can be broadly categorized based on their underlying principles and the types of data, and they utilize into four main categories: (1) Alignment-based and alignment-free methods, (2) machine learning-based methods, (3) structure-based prediction tools, and (4) hybrid-based prediction tools. Table 8.6 presents a comprehensive summary of various computational tools employed for predicting phage–host interactions.

TABLE 8.6
Hybrid Tools for Phage–Host Interaction Prediction: Principles, Advantages, and Limitations

Tool	Working Principle	Description	Advantages	Limitations
PHIStruct	Uses structure-aware embeddings (generated via SaPro—a protein language model) of phage receptor-binding proteins (RBPs) and a multilayer perceptron to predict host genus	Predicts the host genus (from ESKAPEE) based on the 3D structure of phage RBPs	Performs well even with low sequence similarity between training and test sets; balances precision and recall	Requires 3D structural data of RBPs, which might not always be available
DeepPBI-KG	Focuses on key genes and proteins to predict interactions using K-means sampling for feature selection and a deep neural network	Predicts phage–bacteria interactions based on the concept that these interactions are mediated by key genes	High performance in strain-level predictions	Not described
CoMPHI	Combines alignment scores (phage–phage, phage–host, host–host) and sequence features with a machine learning classifier	Predicts phage–host interactions across various taxonomic levels by integrating alignment-based and machine learning approaches	High performance across different taxonomic levels	Not described
PHIAF	Employs generative adversarial networks (GANs) for data augmentation and attention mechanisms with a convolutional neural network (CNN) for prediction.	Predicts phage-host interactions using a model trained on augmented data and features from both DNA and protein sequences.	Outperforms other methods in prediction accuracy.	Addresses data scarcity through GAN-based augmentation.

(*Continued*)

TABLE 8.6 (*Continued*)
Hybrid Tools for Phage–Host Interaction Prediction: Principles, Advantages, and Limitations

Tool	Working Principle	Description	Advantages	Limitations
vHULK	Utilizes a neural network framework to predict hosts based on annotated genomic features, considering alignment significance scores against a viral protein family database	Predicts host genus and species based on features extracted from phage genomes and their similarity to known viral protein families	High accuracy at both genus and species levels; user-friendly tool	Requires annotated genomic features
HostPhinder	Predicts the host of a query phage based on genomic similarity (co-occurring k-mers) to reference phages with known hosts.	Predicts bacterial host at species level of a new phage by comparing its nucleotide sequence to a database of known hosts.	Fast and accurate; performs well on an independent dataset.	This depends on the availability of similar phages with known hosts in the database.
iPHoP	A two-step framework that integrates host-based features and phage-based methods to predict the host taxonomy (at genus level) from metagenomes	Predicts the host taxonomy at the genus level for viruses identified from metagenomic data	High recall and low false discovery rate for host genus prediction	Primarily predicts at the genus level
CL4PHI	Generates whole-genome embeddings for phages and hosts using contrastive learning to facilitate zero-shot predictions for unseen hosts	Predicts phage–host relationships by learning whole-genome sequence representations through contrastive learning	Enables zero-shot predictions for unseen hosts; high accuracy across benchmark datasets	Not described
PHIST	The viral host interactions discover based on the number of k-mers shared between their genomic sequences.	Predicts prokaryotic hosts for phage (meta)genomic sequences by identifying shared k-mers.	Fast and accurate; suitable for metagenomics studies.	Not described

8.7.2 Alignment-Based and Alignment-Free Methods

Sequence homology-based (alignment-based) methods are rapid and intuitive approach to identifying known phage–host interactions. These methods commonly focus on identifying shared genomic regions, such as the identity of prophage sequences integrated into the host genome or similarity in CRISPR spacers that may indicate past or current phage infection. Methods for predicting host species through sequence alignment depend upon two main strategies: (1) comparing nucleotide similarity between a query virus and host genomes, and (2) examining similarity between a query virus and specific marker genes known to be exclusively encoded by viruses that infect certain host taxa. The BLAST is one of the most common methods that analyse sequence similarity to identify shared sequences (homologous sequences) and/or shared genes between phage and bacterial genomes. Nonetheless, their effectiveness limits when identifying novel phages or hosts that do not have match to entries in existing phage databases (Dion et al., 2021).

Alignment-free methods develop based on the analysis of sequence compositions such as k-mer frequencies, overall sequence structure, and codon usage bias to predict phage–host interactions, as phages and their hosts often share similar patterns in codon usage or small nucleotide words. These methods skip the computationally intensive step of explicit sequence alignment by focusing on inherent statistical properties of the genomic sequences. The advantage of alignment-free is useful for capturing subtle evolutionary adaptations of phage genomes to the cellular machinery of their host, such as biases in codon usage that reflect the host's translocation mechanism. Examples of alignment-free tools include VirHostMatcher, which uses oligonucleotide frequency dissimilarity to compare the similarity of k-mers frequency between a query viral genome and genomes from a host reference database (Wang et al., 2020).

8.7.3 Machine Learning-Based Methods

Machine learning-based tools leverage the power of statistical learning to identify signals indicating phage–bacterial interactions from the complex patterns present in metagenomic datasets. There are numerous machine-learning based learning algorithms such as random forest, support vector machines, and neural networks, which have been used to develop predictive tools such as PHIStruct (Gonzales et al., 2025), GSPHI (Pan et al., 2023), DeepPBI-KG (Wei et al., 2024), and vHULK (Amgarten et al., 2022). These are advanced predictive tools trained on various properties of phage sequences and often exhibit high prediction performance to accurate predations of their bacteria. However, the performance of machine learning models depends upon several important factors such as the quality of sequencing reads (nucleotide sequences) and size of the training data used.

8.7.4 Structure-Based Prediction Tools

Sequence-based methods may not always be able to fully characterize the complexity of phage–host interactions, especially in cases of low sequence similarity. Structural-based prediction tools overcome this limitation by analysing the three-dimensional structure of RBP proteins. Considering the structural compatibility between phage RBPs and host receptors, structural-based methods can provide more accurate predictions, especially for phages with new RBPs that may not have close sequence homologues in existing databases. PHIStruct is one of the most common examples that use deep learning model to incorporate structure aware embeddings of RBP, which then predicts the phage–host interaction (Gonzales et al., 2025). It is always recommended to use structure-based predictive tools to accurately predict the interaction between phage and host. This is because these tools work on the information of the structure of proteins and their structural changes, which provide information about protein–protein interactions and the interaction force between them.

8.8 HYBRID-BASED PREDICTION TOOL

Hybrid-based tools adopt integrated approaches that combine two or more methodologies to leverage the strengths of different methodologies to predict the phage–host interactions. For example, composite model for phage–host interaction (CoMPHI) is a machine learning-based model that generates multiple features from input nucleotide and protein sequences of both phage and host, enabling the use of complementary genetic signals from different levels of molecular interactions, thereby contributing to increased accuracy in phage–host prediction. By synergistically combining different prediction strategies, hybrid approaches aim to achieve a more robust and comprehensive assessment of phage–host interactions (Bodaka & Malgonde, 2024).

8.9 PHAGE-BASED BIOMARKERS FOR HUMAN GUT HEALTH

In the human gut, biomarkers refer to the microbial ecosystem, including bacteria, viruses, or metabolites, whose presence, abundance, and diversity may serve as indicators of health status and help to diagnose disease conditions. In this section, we described the potential role of phages in human GI health, based on empirical evidence demonstrating that their composition, diversity, or abundance is associated with particular conditions of intestinal health, such as IBD, IBS, obesity, or diabetes (Barr, 2017; Cao et al., 2022; Dalmasso et al., 2014; de Jonge et al., 2022; Townsend et al., 2021).

8.9.1 Distribution of Bacteriophages within the Human Gut

Bacteriophages are ubiquitously present across diverse environments such as soil, aquatic systems, and the mucosal surfaces within the human body wherever bacterial hosts are found. Phages adhere to the bacterial membrane and introduce their viral genetic material via specialized enzymes within the tail structure, which also engages with bacterial receptors through external structures such as tail fibers. The advent of high-throughput sequencing technologies has clearly improved the elucidation of complex microbial ecosystems, with particular emphasis on the human gut. The term "phageome" denotes the viral component of gut microbial communities, which is projected to account for approximately 5%–6% of normalized sequencing reads (Federici et al., 2021). However, the exact quantity of phage-derived DNA per gram of stool continues to be a subject of debate, owing to variations in extraction methodologies. CrAss-like phages are among the most notable gut phages and predominantly target Bacteroides species within the order *Bacteroidales* (Shkoporov et al., 2018; Smith et al., 2023). Most importantly, it is important to understand that high diversity, stability, and composition of both the gut phageome and gut microbiome are indicative of stable gut health. A research study based on healthy phageome identified 44 phage groups in 20%–50% of individuals, of which 23 phages were consistently present in more than half of the healthy samples (Manrique et al., 2016).

8.10 GUT VIROME DIVERSITY IN INFLAMMATORY BOWEL DISEASE (IBD)

IBD is a chronic inflammatory disease that encompasses both Crohn's disease (CD) and ulcerative colitis (UC), which have been shown to have a distinct alterations in virome diversity compositions compared to healthy individuals (Tun et al., 2024). Patients diagnosed with CD have been shown to have a lower diversity of both DNA and RNA viromes, accompanied by dysbiosis alterations involving *Torque teno virus (TTV)*, *Faecalibacterium phage, Escherichia phage*, and RNA viruses such as the tomato diet-related virus (Table 8.7). In IBD, changes in microbial species such as *Faecalibacterium prausnitzii* have been linked to a higher risk of flare recurrence, possibly through effects on host immune regulation, including T cell responses (Cornuault et al., 2018). Individuals suffering from UC present with increased incidences of DNA viruses, notably Caudovirales phages, yet exhibit reduced diversity and richness relative to healthy individuals. This alteration in phage populations, wherein virulent phages prevalent in healthy individuals are supplanted by temperate phages like *Siphoviridae* and *Myoviridae,* indicates microbial transformations associated with IBD. *Siphoviridae* phages demonstrate CRISPR interactions that correlate with a reduction in Firmicutes abundance, aligning with the dysbiotic microbial profile observed in patients with IBD. These observations underscore the potential significance of the altered virome in the mechanisms and pathogenesis of the disease (Federici et al., 2023; Zuo et al., 2022).

TABLE 8.7
Viral Families and Phages Associated with Disease Conditions

Disease	Virome Composition (Phages/Viral Families)	Roles and Implications
Inflammatory Bowel Disease (IBD)	• Reduced DNA and RNA virome diversity. *Faecalibacterium phage, Escherichia phage*, DNA eukaryotic Torque teno virus (TTV), RNA tomato diet-related virus • Higher DNA viruses (*Caudovirales*) in UC; lower diversity/richness/evenness • Increased temperate phages (*Siphoviridae, Myoviridae*), restoring virulent core in healthy	• Alters bacterial populations, worsening inflammation and leading to gut dysbiosis marked by reduced Firmicutes

(*Continued*)

TABLE 8.7 (*Continued*)
Viral Families and Phages Associated with Disease Conditions

Disease	Virome Composition (Phages/Viral Families)	Roles and Implications
Irritable Bowel Syndrome (IBS)	• Prevalent *Myoviridae, Podoviridae*, Siphoviridae in IBS and healthy • Reduced alpha diversity of viral OTUs, divergent beta diversity • Patients with IBS-D exhibit increased levels of *Microviridae, Myoviridae,* and *Podoviridae* • Patients with IBS-C exhibit an promoted presence of Microviridae and Myoviridae • Extended *Mimiviridae* and *Podoviridae* in some; reduced *Mimiviridae, Siphoviridae,* and *Podoviridae* in others • Lower *Poxviridae, Phycodnaviridae, Pandoraviridae, Adenoviridae,* and *Rudiviridae*	• Influences microbial dysbiosis, potentially contributing to IBS symptoms by altering bacterial communities • Virome profiles specific to subtypes suggest diagnostic potential for distinguishing IBS-D from IBS-C • It may also influence immune responses, with reduced diversity linked to gut dysfunction
Obesity	• Reduced richness/alpha diversity in obese adults (BMI ≥28) vs. non-obese (BMI 18.5–23), pronounced in diabetes • Increased dsDNA viral taxa richness/diversity in paediatric obesity/metabolic syndrome • Dominant Caudovirales order in obese/non-obese • Enriched temperate Caudovirales in high-sucrose diet obese mice Enriched Escherichia phage, Geobacillus phage, and Lactobacillus phage	• May contribute to metabolic dysregulation by altering gut microbiota, affecting energy harvest • The ratio of lytic to temperate phages has been suggested as a biomarker for metabolic diseases • Temperate phages may play a role in obesity risk, possibly affecting bacterial metabolic processes

8.11 GUT VIROME IN IRRITABLE BOWEL SYNDROME (IBS)

The virome plays a significant role in the pathogenesis of IBS, influencing disease mechanisms through interactions with the microbiome. Metagenomic analyses have identified *Myoviridae, Podoviridae*, and *Siphoviridae* as dominant bacteriophage taxa in both IBS and healthy individuals. However, IBS patients display reduced viral diversity and divergent beta diversity profiles. Unlike IBD, the IBS virome lacks distinct lytic-to-temperate phage ratios, which are markers of microbial shifts. Virome compositions vary across IBS subtypes: diarrhea-predominant IBS (IBS-D) features elevated *Microviridae, Myoviridae,* and *Podoviridae* abundances, while constipation-predominant IBS (IBS-C) shows higher proportions of *Microviridae* and *Myoviridae*. Phage families such as *Mimiviridae* and *Podoviridae* are enriched in some IBS cohorts, while others show reduced abundances of *Siphoviridae*. These findings emphasize the microbiome–phage interactions influencing IBS subtype differentiation and severity (Coughlan et al., 2021; Tun et al., 2024).

8.12 ALTERATIONS IN GUT VIROME DIVERSITY IN OBESITY

Obesity-associated metabolic disorders exhibit distinct alterations in gut virome diversity and microbiome composition. In human cohorts, adult obesity is associated with reduced virome richness and alpha diversity, with significant reductions observed in individuals with comorbid type 2 diabetes (Yadav et al., 2016). In contrast, pediatric obesity cohorts show increased virome richness, suggesting age-specific differences in gut viral composition. *Caudovirales* bacteriophages dominate both obese and non-obese gut viromes. Phage taxa, such as *Escherichia, Geobacillus*, and *Lactobacillus phages*, appear enriched in obese individuals, contributing to metabolic disturbances associated with obesity (Zhang & Wang, 2023).

8.13 CONCLUSION

Stool-based mycobiome studies rely on ITS amplicon sequencing targeting the 18S, 26S rRNA genes, and ITS1/ITS2 regions. Alignment with reference databases enables precise fungal identification. Alignment to curated fungal reference databases facilitates accurate strain identification, which may contribute to understanding mycobiome shifts linked to GI and systemic diseases. Gut bacteriophages are key modulators of microbiome balance and immune function in human guts. Metagenomic shotgun sequencing has enhanced our understanding of phage–host dynamics, though data analysis remains complex. Computational tools developed based on the database-dependent, feature-based, or hybrid are essential for studying phage roles and exploring their therapeutic potential. Their analysis may also offer valuable insights into disease-associated microbial disruptions.

REFERENCES

Abarenkov, K., Nilsson, R. H., Larsson, K.-H., Taylor, A. F. S., May, T. W., Frøslev, T. G., Pawlowska, J., Lindahl, B., Põldmaa, K., Truong, C., Vu, D., Hosoya, T., Niskanen, T., Piirmann, T., Ivanov, F., Zirk, A., Peterson, M., Cheeke, T. E., Ishigami, Y., … Kõljalg, U. (2024). The UNITE database for molecular identification and taxonomic communication of fungi and other eukaryotes: Sequences, taxa and classifications reconsidered. *Nucleic Acids Research*, 52(D1), D791–D797. https://doi.org/10.1093/nar/gkad1039

Agarwal, R., Sehgal, I. S., & Muthu, V. (2025). The 2024 International Society for Human and Animal Mycology (ISHAM)-ABPA working group guidelines: Transforming diagnosis and management of ABPA. *Lung India*, 42(2), 79–83. https://doi.org/10.4103/lungindia.lungindia_28_25

Akhter, S., Aziz, R. K., & Edwards, R. A. (2012). PhiSpy: A novel algorithm for finding prophages in bacterial genomes that combines similarity- and composition-based strategies. *Nucleic Acids Research*, 40(16), e126. https://doi.org/10.1093/nar/gks406

Albin, D., Ramsahoye, M., Kochavi, E., & Alistar, M. (2024). PhageScanner: A reconfigurable machine learning framework for bacteriophage genomic and metagenomic feature annotation. *Frontiers in Microbiology*, *15*. https://doi.org/10.3389/fmicb.2024.1446097

Ali, N. A. B. M., Mac Aogáin, M., Morales, R. F., Tiew, P. Y., & Chotirmall, S. H. (2019). Optimisation and benchmarking of targeted amplicon sequencing for mycobiome analysis of respiratory specimens. *International Journal of Molecular Sciences*, 20(20), 4991. https://doi.org/10.3390/ijms20204991

Amgarten, D., Braga, L. P. P., da Silva, A. M., & Setubal, J. C. (2018). MARVEL, a tool for prediction of bacteriophage sequences in metagenomic bins. *Frontiers in Genetics*, *9*. https://doi.org/10.3389/fgene.2018.00304

Amgarten, D., Iha, B. K. V., Piroupo, C. M., da Silva, A. M., & Setubal, J. C. (2022). vHULK, a new tool for bacteriophage host prediction based on annotated genomic features and neural networks. *PHAGE: Therapy, Applications, and Research*, 3(4), 204–212. https://doi.org/10.1089/phage.2021.0016

Andrade-Martínez, J. S., Camelo Valera, L. C., Chica Cárdenas, L. A., Forero-Junco, L., López-Leal, G., Moreno-Gallego, J. L., Rangel-Pineros, G., & Reyes, A. (2022). Computational tools for the analysis of uncultivated phage genomes. *Microbiology and Molecular Biology Reviews: MMBR*, 86(2), e00004-21. https://doi.org/10.1128/mmbr.00004-21

Anslan, S., Nilsson, R. H., Wurzbacher, C., Baldrian, P., Tedersoo, L., & Bahram, M. (2018). Great differences in performance and outcome of high-throughput sequencing data analysis platforms for fungal metabarcoding. *MycoKeys*, 39, 29–40. https://doi.org/10.3897/mycokeys.39.28109

Arndt, D., Grant, J. R., Marcu, A., Sajed, T., Pon, A., Liang, Y., & Wishart, D. S. (2016). PHASTER: A better, faster version of the PHAST phage search tool. *Nucleic Acids Research*, 44(W1), W16–W21. https://doi.org/10.1093/nar/gkw387

Auslander, N., Gussow, A. B., Benler, S., Wolf, Y. I., & Koonin, E. V. (2020). Seeker: Alignment-free identification of bacteriophage genomes by deep learning. *Nucleic Acids Research*, 48(21), e121. https://doi.org/10.1093/nar/gkaa856

Balderramo, D. C., Romagnoli, P. A., Granlund, A. V. B., & Catalan-Serra, I. (2023). Fecal fungal microbiota (mycobiome) study as a potential tool for precision medicine in inflammatory bowel disease. *Gut and Liver*, 17(4), 505–515. https://doi.org/10.5009/gnl220537

Barr, J. J. (2017). A bacteriophages journey through the human body. *Immunological Reviews*, 279(1), 106–122. https://doi.org/10.1111/imr.12565

Belvoncikova, P., Splichalova, P., Videnska, P., & Gardlik, R. (2022). The human mycobiome: Colonization, composition and the role in health and disease. *Journal of Fungi*, 8(10), 1046. https://doi.org/10.3390/jof8101046

Bodaka, S., & Malgonde, O. (2024). CoMPHI: A novel composite machine learning approach utilizing multiple feature representation to predict hosts of bacteriophages. *bioRxiv*. https://doi.org/10.1101/2024.07.29.604684

Bolduc, B., Jang, H. B., Doulcier, G., You, Z.-Q., Roux, S., & Sullivan, M. B. (2017). vConTACT: An iVirus tool to classify double-stranded DNA viruses that infect Archaea and Bacteria. *PeerJ*, *5*, e3243. https://doi.org/10.7717/peerj.3243

Camargo, A. P., Nayfach, S., Chen, I.-M. A., Palaniappan, K., Ratner, A., Chu, K., Ritter, S. J., Reddy, T. B. K., Mukherjee, S., Schulz, F., Call, L., Neches, R. Y., Woyke, T., Ivanova, N. N., Eloe-Fadrosh, E. A., Kyrpides, N. C., & Roux, S. (2023). IMG/VR v4: An expanded database of uncultivated virus genomes within a framework of extensive functional, taxonomic, and ecological metadata. *Nucleic Acids Research*, 51(D1), D733–D743. https://doi.org/10.1093/nar/gkac1037

Camarillo-Guerrero, L. F., Almeida, A., Rangel-Pineros, G., Finn, R. D., & Lawley, T. D. (2021). Massive expansion of human gut bacteriophage diversity. *Cell*, 184(4), 1098–1109.e9. https://doi.org/10.1016/j.cell.2021.01.029

Campbell, A. (2003). The future of bacteriophage biology. *Nature Reviews Genetics*, 4(6), 471–477. https://doi.org/10.1038/nrg1089

Cao, Z., Sugimura, N., Burgermeister, E., Ebert, M. P., Zuo, T., & Lan, P. (2022). The gut virome: A new microbiome component in health and disease. *eBioMedicine*, *81*. https://doi.org/10.1016/j.ebiom.2022.104113

Carter, M. M., Olm, M. R., Merrill, B. D., Dahan, D., Tripathi, S., Spencer, S. P., Yu, F. B., Jain, S., Neff, N., Jha, A. R., Sonnenburg, E. D., & Sonnenburg, J. L. (2023). Ultra-deep sequencing of Hadza hunter-gatherers recovers vanishing gut microbes. *Cell*, 186(14), 3111–3124.e13. https://doi.org/10.1016/j.cell.2023.05.046

Catalán-Serra, I., Thorsvik, S., Beisvag, V., Bruland, T., Underhill, D., Sandvik, A. K., & Granlund, A. V. B. (2024). Fungal microbiota composition in inflammatory bowel disease patients: Characterization in different phenotypes and correlation with clinical activity and disease course. *Inflammatory Bowel Diseases*, 30(7), 1164–1177. https://doi.org/10.1093/ibd/izad289

Chin, V. K., Yong, V. C., Chong, P. P., Amin Nordin, S., Basir, R., & Abdullah, M. (2020). Mycobiome in the gut: A multiperspective review. *Mediators of Inflammation, 2020*, 1–16. https://doi.org/10.1155/2020/9560684

Clokie, M. R., Millard, A. D., Letarov, A. V., & Heaphy, S. (2011). Phages in nature. *Bacteriophage*, 1(1), 31–45. https://doi.org/10.4161/bact.1.1.14942

Cornuault, J. K., Petit, M.-A., Mariadassou, M., Benevides, L., Moncaut, E., Langella, P., Sokol, H., & De Paepe, M. (2018). Phages infecting Faecalibacterium prausnitzii belong to novel viral genera that help to decipher intestinal viromes. *Microbiome*, 6, 65. https://doi.org/10.1186/s40168-018-0452-1

Coughlan, S., Das, A., O'Herlihy, E., Shanahan, F., O'Toole, P. W., & Jeffery, I. B. (2021). The gut virome in Irritable Bowel Syndrome differs from that of controls. *Gut Microbes*, 13(1), 1887719. https://doi.org/10.1080/19490976.2021.1887719

Dalmasso, M., Hill, C., & Ross, R. P. (2014). Exploiting gut bacteriophages for human health. *Trends in Microbiology*, 22(7), 399–405. https://doi.org/10.1016/j.tim.2014.02.010

De Filippis, F., Laiola, M., Blaiotta, G., & Ercolini, D. (2017). Different amplicon targets for sequencing-based studies of fungal diversity. *Applied and Environmental Microbiology*, 83(17). https://doi.org/10.1128/aem.00905-17

de Jonge, P. A., Wortelboer, K., Scheithauer, T. P. M., van den Born, B.-J. H., Zwinderman, A. H., Nobrega, F. L., Dutilh, B. E., Nieuwdorp, M., & Herrema, H. (2022). Gut virome profiling identifies a widespread bacteriophage family associated with metabolic syndrome. *Nature Communications*, 13(1), 3594. https://doi.org/10.1038/s41467-022-31390-5

de Sousa, A. L., Maués, D., Lobato, A., Franco, E. F., Pinheiro, K., Araújo, F., Pantoja, Y., da Costa da Silva, A. L., Morais, J., & Ramos, R. T. J. (2018). PhageWeb—Web interface for rapid identification and characterization of prophages in bacterial genomes. *Frontiers in Genetics*, 9, 644. https://doi.org/10.3389/fgene.2018.00644

de Souza, E. B., Pinto, A. R., & Fongaro, G. (2023). Bacteriophages as potential clinical immune modulators. *Microorganisms*, 11(9), 2222. https://doi.org/10.3390/microorganisms11092222

Deaton, J., Yu, F. B., & Quake, S. R. (2019). Mini-metagenomics and nucleotide composition aid the identification and host association of novel bacteriophage sequences. *Advanced Biosystems*, 3(11), 1900108. https://doi.org/10.1002/adbi.201900108

Degroux, S., Effantin, G., Linares, R., Schoehn, G., & Breyton, C. (2023). Deciphering bacteriophage T5 host recognition mechanism and infection trigger. *Journal of Virology*, 97(3), e0158422. https://doi.org/10.1128/jvi.01584-22

Dion, M. B., Oechslin, F., & Moineau, S. (2020). Phage diversity, genomics and phylogeny. *Nature Reviews Microbiology*, 18(3), 125–138. https://doi.org/10.1038/s41579-019-0311-5

Dion, M. B., Plante, P.-L., Zufferey, E., Shah, S. A., Corbeil, J., & Moineau, S. (2021). Streamlining CRISPR spacer-based bacterial host predictions to decipher the viral dark matter. *Nucleic Acids Research*, 49(6), 3127–3138. https://doi.org/10.1093/nar/gkab133

Dowah, A. S. A., & Clokie, M. R. J. (2018). Review of the nature, diversity and structure of bacteriophage receptor binding proteins that target Gram-positive bacteria. *Biophysical Reviews*, 10(2), 535–542. https://doi.org/10.1007/s12551-017-0382-3

El Jaddaoui, I., Sehli, S., Al Idrissi, N., Bakri, Y., Belyamani, L., & Ghazal, H. (2025). The gut mycobiome for precision medicine. *Journal of Fungi*, 11(4), 279. https://doi.org/10.3390/jof11040279

Fang, Z., Tan, J., Wu, S., Li, M., Xu, C., Xie, Z., & Zhu, H. (2019). PPR-Meta: A tool for identifying phages and plasmids from metagenomic fragments using deep learning. *GigaScience*, 8(6), giz066. https://doi.org/10.1093/gigascience/giz066

Federici, S., Kviatcovsky, D., Valdés-Mas, R., & Elinav, E. (2023). Microbiome-phage interactions in inflammatory bowel disease. *Clinical Microbiology and Infection*, 29(6), 682–688. https://doi.org/10.1016/j.cmi.2022.08.027

Federici, S., Nobs, S. P., & Elinav, E. (2021). Phages and their potential to modulate the microbiome and immunity. *Cellular and Molecular Immunology*, 18(4), 889–904. https://doi.org/10.1038/s41423-020-00532-4

Fiers, W. D., Gao, I. H., & Iliev, I. D. (2019). Gut mycobiota under scrutiny: Fungal symbionts or environmental transients? *Current Opinion in Microbiology*, 50, 79–86. https://doi.org/10.1016/j.mib.2019.09.010

Forbes, J. D., Bernstein, C. N., Tremlett, H., Van Domselaar, G., & Knox, N. C. (2019). A fungal world: Could the gut mycobiome be involved in neurological disease? *Frontiers in Microbiology*, *9*. https://doi.org/10.3389/fmicb.2018.03249

Fosso, B., Santamaria, M., Consiglio, A., De Caro, G., Grillo, G., Licciuli, F., Liuni, S., Marzano, M., & Pesole, G. (2012). ITSoneDB: A specialized ITS1 database for amplicon-based metagenomic characterization of environmental fungal communities. EMBnet.Journal, 18(A), 90. https://doi.org/10.14806/ej.18.a.430

Fox, J. D., Sims, A., Ross, M., Bettag, J., Wilder, A., Natrop, D., Borsotti, A., Kolli, S., Mehta, S., Verma, H., Kurashima, K., Manithody, C., Verma, A., & Jain, A. (2024). Bioinformatic methodologies in assessing gut microbiota. *Microbiology Research*, 15(4), 2554–2574. https://doi.org/10.3390/microbiolres15040170

Fremin, B. J., Bhatt, A. S., & Kyrpides, N. C. (2022). Thousands of small, novel genes predicted in global phage genomes. *Cell Reports*, 39(12), 110984. https://doi.org/10.1016/j.celrep.2022.110984

Galperina, A., Lugli, G. A., Milani, C., Vos, W. M. D., Ventura, M., Salonen, A., Hurwitz, B., & Ponsero, A. J. (2024). The aggregated gut viral catalogue (AVrC): A unified resource for exploring the viral diversity of the human gut. *bioRxiv*. https://doi.org/10.1101/2024.06.24.600367

García-Gamboa, R., Díaz-Torres, O., Senés-Guerrero, C., Gradilla-Hernández, M. S., Moya, A., Pérez-Brocal, V., Garcia-Gonzalez, A., & González-Avila, M. (2024). Associations between bacterial and fungal communities in the human gut microbiota and their implications for nutritional status and body weight. *Scientific Reports*, *14*(1). https://doi.org/10.1038/s41598-024-54782-7

Gaspar, B. S., Roşu, O. A., Enache, R.-M., Manciulea (Profir), M., Pavelescu, L. A., & Creţoiu, S. M. (2025). Gut mycobiome: Latest findings and current knowledge regarding its significance in human health and disease. *Journal of Fungi*, 11(5), 333. https://doi.org/10.3390/jof11050333

George, N. A., & Hug, L. A. (2023). CRISPR-resolved virus-host interactions in a municipal landfill include non-specific viruses, hyper-targeted viral populations, and interviral conflicts. *Scientific Reports*, 13(1), 5611. https://doi.org/10.1038/s41598-023-32078-6

Gonzales, M. E. M., Ureta, J. C., & Shrestha, A. M. S. (2025). PHIStruct: Improving phage–host interaction prediction at low sequence similarity settings using structure-aware protein embeddings. *Bioinformatics*, 41(1), btaf016. https://doi.org/10.1093/bioinformatics/btaf016

Gregory, A. C., Zablocki, O., Zayed, A. A., Howell, A., Bolduc, B., & Sullivan, M. B. (2020). The gut virome database reveals age-dependent patterns of virome diversity in the human gut. *Cell Host & Microbe*, 28(5), 724–740.e8. https://doi.org/10.1016/j.chom.2020.08.003

Guo, J., Bolduc, B., Zayed, A. A., Varsani, A., Dominguez-Huerta, G., Delmont, T. O., Pratama, A. A., Gazitúa, M. C., Vik, D., Sullivan, M. B., & Roux, S. (2021). VirSorter2: A multi-classifier, expert-guided approach to detect diverse DNA and RNA viruses. *Microbiome*, 9(1), 37. https://doi.org/10.1186/s40168-020-00990-y

Hadrich, I., Turki, M., Chaari, I., Abdelmoula, B., Gargouri, R., Khemakhem, N., Elatoui, D., Abid, F., Kammoun, S., Rekik, M., Aloulou, S., Sehli, M., Mrad, A. B., Neji, S., Feiguin, F. M., Aloulou, J., Abdelmoula, N. B., & Sellami, H. (2025). Gut mycobiome and neuropsychiatric disorders: Insights and therapeutic potential. *Frontiers in Cellular Neuroscience*, *18*. https://doi.org/10.3389/fncel.2024.1495224

Halwachs, B., Madhusudhan, N., Krause, R., Nilsson, R. H., Moissl-Eichinger, C., Högenauer, C., Thallinger, G. G., & Gorkiewicz, G. (2017). Critical issues in mycobiota analysis. *Frontiers in Microbiology*, *8*. https://doi.org/10.3389/fmicb.2017.00180

Hayes, S., Mahony, J., Nauta, A., & van Sinderen, D. (2017). Metagenomic approaches to assess bacteriophages in various environmental niches. *Viruses*, *9*(6), 127. https://doi.org/10.3390/v9060127

Hibstu, Z., Belew, H., Akelew, Y., & Mengist, H. M. (2022). Phage therapy: A different approach to fight bacterial infections. *Biologics: Targets & Therapy*, 16, 173–186. https://doi.org/10.2147/BTT.S381237

Hoggard, M., Vesty, A., Wong, G., Montgomery, J. M., Fourie, C., Douglas, R. G., Biswas, K., & Taylor, M. W. (2018). Characterizing the human mycobiota: A comparison of small subunit rRNA, ITS1, ITS2, and large subunit rRNA genomic targets. *Frontiers in Microbiology*, *9*. https://doi.org/10.3389/fmicb.2018.02208

Hsu, C., Ghannoum, M., Cominelli, F., & Martino, L. D. (2023). Mycobiome and inflammatory bowel disease: Role in disease pathogenesis, current approaches and novel nutritional-based therapies. *Inflammatory Bowel Diseases*, 29(3), 470–479. https://doi.org/10.1093/ibd/izac156

Iyer, P., & Ojcius, D. M. (2024). Unveiling the mycobiota: The fungal frontier of human health. *Biomedical Journal*, 47(3), 100751. https://doi.org/10.1016/j.bj.2024.100751

Johnson, G., & Putonti, C. (2022). virMine 2.0: Identifying viral sequences in microbial communities. *Microbiology Resource Announcements*, 11(5), e00107-22. https://doi.org/10.1128/mra.00107-22

Jurtz, V. I., Villarroel, J., Lund, O., Larsen, M. V., & Nielsen, M. (2016). MetaPhinder—Identifying bacteriophage sequences in metagenomic data sets. *PLoS ONE*, 11(9), e0163111. https://doi.org/10.1371/journal.pone.0163111

Kedia, S., & Ahuja, V. (2023). Human gut microbiome: A primer for the clinician. *JGH Open*, 7(5), 337–350. https://doi.org/10.1002/jgh3.12902

Keen, E. C., & Dantas, G. (2018). Close encounters of three kinds: Bacteriophages, commensal bacteria, and host immunity. *Trends in Microbiology*, *26*(11), 943–954. https://doi.org/10.1016/j.tim.2018.05.009

Khot, V., Strous, M., & Hawley, A. K. (2020). Computational approaches in viral ecology. *Computational and Structural Biotechnology Journal*, 18, 1605–1612. https://doi.org/10.1016/j.csbj.2020.06.019

Kieft, K., Zhou, Z., & Anantharaman, K. (2020). VIBRANT: Automated recovery, annotation and curation of microbial viruses, and evaluation of viral community function from genomic sequences. *Microbiome*, 8(1), 90. https://doi.org/10.1186/s40168-020-00867-0

Klumpp, J., Fouts, D. E., & Sozhamannan, S. (2012). Next generation sequencing technologies and the changing landscape of phage genomics. *Bacteriophage*, 2(3), 190–199. https://doi.org/10.4161/bact.22111

Koetschan, C., Hackl, T., Müller, T., Wolf, M., Förster, F., & Schultz, J. (2012). ITS2 Database IV: Interactive taxon sampling for internal transcribed spacer 2 based phylogenies. *Molecular Phylogenetics and Evolution*, 63(3), 585–588. https://doi.org/10.1016/j.ympev.2012.01.026

Kumar, B., Lorusso, E., Fosso, B., & Pesole, G. (2024). A comprehensive overview of microbiome data in the light of machine learning applications: Categorization, accessibility, and future directions. *Frontiers in Microbiology*, *15*. https://doi.org/10.3389/fmicb.2024.1343572

Kurilovich, E., & Geva-Zatorsky, N. (2025). Effects of bacteriophages on gut microbiome functionality. *Gut Microbes*, *17*(1), 2481178. https://doi.org/10.1080/19490976.2025.2481178

Lai, S., Yan, Y., Pu, Y., Lin, S., Qiu, J.-G., Jiang, B.-H., Keller, M. I., Wang, M., Bork, P., Chen, W.-H., Zheng, Y., & Zhao, X.-M. (2023). Enterotypes of the human gut mycobiome. *Microbiome*, *11*(1). https://doi.org/10.1186/s40168-023-01586-y

Lang, A. S., Buchan, A., & Burrus, V. (2025). Interactions and evolutionary relationships among bacterial mobile genetic elements. *Nature Reviews Microbiology*, 1–16. https://doi.org/10.1038/s41579-025-01157-y

Liber, J. A., Bonito, G., & Benucci, G. M. N. (2021). CONSTAX2: Improved taxonomic classification of environmental DNA markers. *Bioinformatics*, 37(21), 3941–3943. https://doi.org/10.1093/bioinformatics/btab347

Lin, Y.-T., Chen, Y.-P., Fang, C.-H., Huang, P.-Y., & Liang, S.-M. (2020). Capsid proteins of foot-and-mouth disease virus interact with TLR2 and CD14 to induce cytokine production. *Immunology Letters*, 223, 10–16. https://doi.org/10.1016/j.imlet.2020.04.008

Mallawaarachchi, V., Roach, M. J., Decewicz, P., Papudeshi, B., Giles, S. K., Grigson, S. R., Bouras, G., Hesse, R. D., Inglis, L. K., Hutton, A. L. K., Dinsdale, E. A., & Edwards, R. A. (2023). Phables: From fragmented assemblies to high-quality bacteriophage genomes. *Bioinformatics*, *39*(10), btad586. https://doi.org/10.1093/bioinformatics/btad586

Manrique, P., Bolduc, B., Walk, S. T., van der Oost, J., de Vos, W. M., & Young, M. J. (2016). Healthy human gut phageome. *Proceedings of the National Academy of Sciences of the United States of America*, 113(37), 10400–10405. https://doi.org/10.1073/pnas.1601060113

Nash, A. K., Auchtung, T. A., Wong, M. C., Smith, D. P., Gesell, J. R., Ross, M. C., Stewart, C. J., Metcalf, G. A., Muzny, D. M., Gibbs, R. A., Ajami, N. J., & Petrosino, J. F. (2017). The gut mycobiome of the Human Microbiome Project healthy cohort. *Microbiome*, *5*(1). https://doi.org/10.1186/s40168-017-0373-4

Navgire, G. S., Goel, N., Sawhney, G., Sharma, M., Kaushik, P., Mohanta, Y. K., Mohanta, T. K., & Al-Harrasi, A. (2022). Analysis and Interpretation of metagenomics data: An approach. *Biological Procedures Online*, *24*(1). https://doi.org/10.1186/s12575-022-00179-7

Nayfach, S., Páez-Espino, D., Call, L., Low, S. J., Sberro, H., Ivanova, N. N., Proal, A. D., Fischbach, M. A., Bhatt, A. S., Hugenholtz, P., & Kyrpides, N. C. (2021). Metagenomic compendium of 189,680 DNA viruses from the human gut microbiome. *Nature Microbiology*, 6(7), 960–970. https://doi.org/10.1038/s41564-021-00928-6

Nearing, J. T., Comeau, A. M., & Langille, M. G. I. (2021). Identifying biases and their potential solutions in human microbiome studies. *Microbiome*, *9*(1). https://doi.org/10.1186/s40168-021-01059-0

Nilsson, R. H., Anslan, S., Bahram, M., Wurzbacher, C., Baldrian, P., & Tedersoo, L. (2019). Mycobiome diversity: High-throughput sequencing and identification of fungi. *Nature Reviews Microbiology*, 17(2), 95–109. https://doi.org/10.1038/s41579-018-0116-y

Owen, S. V., Perez-Sepulveda, B. M., & Adriaenssens, E. M. (2021). Detection of bacteriophages: Sequence-based systems. In D. R. Harper, S. T. Abedon, B. H. Burrowes, & M. L. McConville (Eds.), *Bacteriophages: Biology, Technology, Therapy* (pp. 621–644). Springer International Publishing. https://doi.org/10.1007/978-3-319-41986-2_19

Özkurt, E., Fritscher, J., Soranzo, N., Ng, D. Y. K., Davey, R. P., Bahram, M., & Hildebrand, F. (2022). LotuS2: An ultrafast and highly accurate tool for amplicon sequencing analysis. *Microbiome*, *10*(1). https://doi.org/10.1186/s40168-022-01365-1

Paez-Espino, D., Pavlopoulos, G. A., Ivanova, N. N., & Kyrpides, N. C. (2017). Nontargeted virus sequence discovery pipeline and virus clustering for metagenomic data. *Nature Protocols*, 12(8), 1673–1682. https://doi.org/10.1038/nprot.2017.063

Pan, J., You, W., Lu, X., Wang, S., You, Z., & Sun, Y. (2023). GSPHI: A novel deep learning model for predicting phage-host interactions via multiple biological information. *Computational and Structural Biotechnology Journal*, 21, 3404–3413. https://doi.org/10.1016/j.csbj.2023.06.014

Park, S.-C., & Won, S. (2018). Evaluation of 16S rRNA databases for taxonomic assignments using a mock community. *Genomics & Informatics*, 16(4), e24. https://doi.org/10.5808/gi.2018.16.4.e24

Quince, C., Walker, A. W., Simpson, J. T., Loman, N. J., & Segata, N. (2017). Shotgun metagenomics, from sampling to analysis. *Nature Biotechnology*, 35(9), 833–844. https://doi.org/10.1038/nbt.3935

Rahimian, M., & Panahi, B. (2024). Metagenome sequence data mining for viral interaction studies: Review on progress and prospects. *Virus Research*, 349, 199450. https://doi.org/10.1016/j.virusres.2024.199450

Ren, J., Ahlgren, N. A., Lu, Y. Y., Fuhrman, J. A., & Sun, F. (2017). VirFinder: A novel k-mer based tool for identifying viral sequences from assembled metagenomic data. *Microbiome*, 5(1), 69. https://doi.org/10.1186/s40168-017-0283-5

Ren, J., Song, K., Deng, C., Ahlgren, N. A., Fuhrman, J. A., Li, Y., Xie, X., Poplin, R., & Sun, F. (2020). Identifying viruses from metagenomic data using deep learning. *Quantitative Biology*, 8(1), 64–77. https://doi.org/10.1007/s40484-019-0187-4

Rodríguez-Rubio, L., Jofre, J., & Muniesa, M. (2017). Is genetic mobilization considered when using bacteriophages in antimicrobial therapy? *Antibiotics*, *6*(4), 32. https://doi.org/10.3390/antibiotics6040032

Rolling, T., Zhai, B., Frame, J., Hohl, T. M., & Taur, Y. (2022). Customization of a DADA2-based pipeline for fungal internal transcribed spacer 1 (ITS1) amplicon data sets. *JCI Insight*, *7*(1). https://doi.org/10.1172/jci.insight.151663

Roux, S., Matthijnssens, J., & Dutilh, B. E. (2021). Metagenomics in virology. *Encyclopedia of Virology*, 133–140. https://doi.org/10.1016/B978-0-12-809633-8.20957-6

Sam, Q., Chang, M., & Chai, L. (2017). The fungal mycobiome and its interaction with gut bacteria in the host. *International Journal of Molecular Sciences*, 18(2), 330. https://doi.org/10.3390/ijms18020330

Shah, S. A., Deng, L., Thorsen, J., Pedersen, A. G., Dion, M. B., Castro-Mejía, J. L., Silins, R., Romme, F. O., Sausset, R., Jessen, L. E., Ndela, E. O., Hjelmsø, M., Rasmussen, M. A., Redgwell, T. A., Leal Rodríguez, C., Vestergaard, G., Zhang, Y., Chawes, B., Bønnelykke, K., … Nielsen, D. S. (2023). Expanding known viral diversity in the healthy infant gut. *Nature Microbiology*, 8(5), 986–998. https://doi.org/10.1038/s41564-023-01345-7

Shamash, M., & Maurice, C. F. (2022). Phages in the infant gut: A framework for virome development during early life. *The ISME Journal*, 16(2), 323–330. https://doi.org/10.1038/s41396-021-01090-x

Shang, J., Jiang, J., & Sun, Y. (2021). Bacteriophage classification for assembled contigs using graph convolutional network. *Bioinformatics*, 37(Supplement_1), i25–i33. https://doi.org/10.1093/bioinformatics/btab293

Shang, J., Tang, X., Guo, R., & Sun, Y. (2022). Accurate identification of bacteriophages from metagenomic data using Transformer. *Briefings in Bioinformatics*, 23(4), bbac258. https://doi.org/10.1093/bib/bbac258

Shkoporov, A. N., Khokhlova, E. V., Fitzgerald, C. B., Stockdale, S. R., Draper, L. A., Ross, R. P., & Hill, C. (2018). ΦCrAss001 represents the most abundant bacteriophage family in the human gut and infects *Bacteroides intestinalis*. *Nature Communications*, 9(1), 4781. https://doi.org/10.1038/s41467-018-07225-7

Shkoporov, A. N., Clooney, A. G., Sutton, T. D. S., Ryan, F. J., Daly, K. M., Nolan, J. A., McDonnell, S. A., Khokhlova, E. V., Draper, L. A., Forde, A., Guerin, E., Velayudhan, V., Ross, R. P., & Hill, C. (2019). The human gut virome is highly diverse, stable, and individual specific. *Cell Host & Microbe*, *26*(4), 527–541.e5. https://doi.org/10.1016/j.chom.2019.09.009

Smith, L., Goldobina, E., Govi, B., & Shkoporov, A. N. (2023). Bacteriophages of the order crassvirales: What do we currently know about this keystone component of the human gut virome? *Biomolecules*, 13(4), 584. https://doi.org/10.3390/biom13040584

Soverini, M., Turroni, S., Biagi, E., Brigidi, P., Candela, M., & Rampelli, S. (2019). HumanMycobiomeScan: A new bioinformatics tool for the characterization of the fungal fraction in metagenomic samples. *BMC Genomics*, *20*(1). https://doi.org/10.1186/s12864-019-5883-y

Staley, J. T., Castenholz, R. W., Colwell, R. R., Holt, J. G., Kane, M. D., Pace, N. R., Salyers, A. A., & Tiedje, J. M. (1997). The microbial world: Foundation of the biosphere. Published by American Society of Microbiology.

Stone, E., Campbell, K., Grant, I., & McAuliffe, O. (2019). Understanding and exploiting phage–host interactions. *Viruses*, 11(6), 567. https://doi.org/10.3390/v11060567

Sun, Y., Zuo, T., Cheung, C. P., Gu, W., Wan, Y., Zhang, F., Chen, N., Zhan, H., Yeoh, Y. K., Niu, J., Du, Y., Zhang, F., Wen, Y., Yu, J., Sung, J. J. Y., Chan, P. K. S., Chan, F. K. L., Wang, K., Ng, S. C., & Miao, Y. (2021b). Population-level configurations of gut mycobiome across 6 ethnicities in urban and rural China. *Gastroenterology*, 160(1), 272–286.e11. https://doi.org/10.1053/j.gastro.2020.09.014

Sun, Z., Huang, S., Zhang, M., Zhu, Q., Haiminen, N., Carrieri, A. P., Vázquez-Baeza, Y., Parida, L., Kim, H.-C., Knight, R., & Liu, Y.-Y. (2021a). Challenges in benchmarking metagenomic profilers. *Nature Methods*, 18(6), 618–626. https://doi.org/10.1038/s41592-021-01141-3

Tang, J., Iliev, I. D., Brown, J., Underhill, D. M., & Funari, V. A. (2015). Mycobiome: Approaches to analysis of intestinal fungi. *Journal of Immunological Methods*, 421, 112–121. https://doi.org/10.1016/j.jim.2015.04.004

Thielemann, N., Herz, M., Kurzai, O., & Martin, R. (2022). Analyzing the human gut mycobiome – A short guide for beginners. *Computational and Structural Biotechnology Journal*, 20, 608–614. https://doi.org/10.1016/j.csbj.2022.01.008

Tiew, P. Y., Mac Aogain, M., Ali, N. A. B. M., Thng, K. X., Goh, K., Lau, K. J. X., & Chotirmall, S. H. (2020). The mycobiome in health and disease: Emerging concepts, methodologies and challenges. *Mycopathologia*. https://doi.org/10.1007/s11046-019-00413-z

Townsend, E. M., Kelly, L., Muscatt, G., Box, J. D., Hargraves, N., Lilley, D., & Jameson, E. (2021). The human gut phageome: Origins and roles in the human gut microbiome. *Frontiers in Cellular and Infection Microbiology*, *11*. https://doi.org/10.3389/fcimb.2021.643214

Tun, H. M., Peng, Y., Massimino, L., Sin, Z. Y., Parigi, T. L., Facoetti, A., Rahman, S., Danese, S., & Ungaro, F. (2024). Gut virome in inflammatory bowel disease and beyond. *Gut*, 73(2), 350–360. https://doi.org/10.1136/gutjnl-2023-330001

Vavreckova, M., Galanova, N., Kostovcik, M., Krystynik, O., Ivanovova, E., Roubalova, R., Jiraskova Zakostelska, Z., Friedecky, D., Friedecka, J., Haluzik, M., Karasek, D., & Kostovcikova, K. (2022). Specific gut bacterial and fungal microbiota pattern in the first half of pregnancy is linked to the development of gestational diabetes mellitus in the cohort including obese women. *Frontiers in Endocrinology*, *13*. https://doi.org/10.3389/fendo.2022.970825

Venturini, C., Petrovic Fabijan, A., Fajardo Lubian, A., Barbirz, S., & Iredell, J. (2022). Biological foundations of successful bacteriophage therapy. *EMBO Molecular Medicine*, 14(7), e12435. https://doi.org/10.15252/emmm.202012435

Vieira, M. F., Duarte, J., Domingues, R., Oliveira, H., & Dias, O. (2025). *PhageDPO*: A machine-learning based computational framework for identifying phage depolymerases. *Computers in Biology and Medicine*, 188, 109836. https://doi.org/10.1016/j.compbiomed.2025.109836

Wang, W., Ren, J., Tang, K., Dart, E., Ignacio-Espinoza, J. C., Fuhrman, J. A., Braun, J., Sun, F., & Ahlgren, N. A. (2020). A network-based integrated framework for predicting virus–prokaryote interactions. *NAR Genomics and Bioinformatics*, 2(2), lqaa044. https://doi.org/10.1093/nargab/lqaa044

Wang, X., Dong, Q., Huang, P., Yang, S., Gao, M., Zhang, C., Zhang, C., Deng, Y., Huang, Z., Ma, B., Jiao, Y., Zhou, Y., Wu, T., Zou, H., Shi, J., Sheng, Y., Wang, Y., Consortium, C., Tang, L., … Chen, L. (2024a). The genetic diversity and populational specificity of the human gut virome at single nucleotide resolution. *bioRxiv*. https://doi.org/10.1101/2024.12.30.630844

Wang, X., Sun, J., Zhang, X., Chen, W., Cao, J., & Hu, H. (2024b). Metagenomics reveals unique gut mycobiome biomarkers in psoriasis. *Skin Research and Technology*, *30*(7). https://doi.org/10.1111/srt.13822

Wang, Y., Fan, H., & Tong, Y. (2023). Unveil the secret of the bacteria and phage arms race. *International Journal of Molecular Sciences*, 24(5), 4363. https://doi.org/10.3390/ijms24054363

Wei, T., Lu, C., Du, H., Yang, Q., Qi, X., Liu, Y., Zhang, Y., Chen, C., Li, Y., Tang, Y., Zhang, W.-H., Tao, X., & Jiang, N. (2024). DeepPBI-KG: A deep learning method for the prediction of phage-bacteria interactions based on key genes. *Briefings in Bioinformatics*, 25(6), bbae484. https://doi.org/10.1093/bib/bbae484

Wheatley, R. M., Holtappels, D., & Koskella, B. (2025). Evaluation of bacteriophages as a signature of microbiome health: A systematic review and meta-analysis. *The Lancet Microbe*, *6*(12), 101196. https://doi.org/10.1016/j.lanmic.2025.101196.

Wu, X., Xia, Y., He, F., Zhu, C., & Ren, W. (2021). Intestinal mycobiota in health and diseases: From a disrupted equilibrium to clinical opportunities. *Microbiome*, *9*(1). https://doi.org/10.1186/s40168-021-01024-x

Xie, Z., & Manichanh, C. (2022). FunOMIC: Pipeline with built-in fungal taxonomic and functional databases for human mycobiome profiling. *Computational and Structural Biotechnology Journal*, 20, 3685–3694. https://doi.org/10.1016/j.csbj.2022.07.010

Yadav, H., Jain, S., Nagpal, R., & Marotta, F. (2016). Increased fecal viral content associated with obesity in mice. *World Journal of Diabetes*, 7(15), 316–320. https://doi.org/10.4239/wjd.v7.i15.316

Yan, Q., Huang, L., Li, S., Zhang, Y., Guo, R., Zhang, P., Lei, Z., Lv, Q., Chen, F., Li, Z., Meng, J., Li, J., Wang, G., Chen, C., Ullah, H., Cheng, L., Fan, S., You, W., Zhang, Y., … Sun, W. (2025). The Chinese gut virus catalogue reveals gut virome diversity and disease-related viral signatures. *Genome Medicine*, 17(1), 30. https://doi.org/10.1186/s13073-025-01460-6

Yang, S.-Y., Han, S. M., Lee, J.-Y., Kim, K. S., Lee, J.-E., & Lee, D.-W. (2025). Advancing gut microbiome research: The shift from metagenomics to multi-omics and future perspectives. *Journal of Microbiology and Biotechnology*, 35. https://doi.org/10.4014/jmb.2412.12001

Yu, S., Ge, X., Xu, H., Tan, B., Tian, B., Shi, Y., Dai, Y., Li, Y., Hu, S., & Qian, J. (2023). Gut microbiome and mycobiome in inflammatory bowel disease patients with *Clostridioides difficile* infection. *Frontiers in Cellular and Infection Microbiology*, *13*. https://doi.org/10.3389/fcimb.2023.1129043

Zeng, S., Almeida, A., Li, S., Ying, J., Wang, H., Qu, Y., Paul Ross, R., Stanton, C., Zhou, Z., Niu, X., Mu, D., & Wang, S. (2024). A metagenomic catalog of the early-life human gut virome. *Nature Communications*, 15(1), 1864. https://doi.org/10.1038/s41467-024-45793-z

Zhan, H., Wan, Y., Sun, Y., Xu, Z., Zhang, F., Yang, K., Zhu, W., Cheung, C. P., Tang, W., Ng, E. K., Wong, S. K., Yeoh, Y. K., Kl Chan, F., Miao, Y., Zuo, T., Zeng, Z., & Ng, S. C. (2024). Gut mycobiome alterations in obesity in geographically different regions. *Gut Microbes*, *16*(1). https://doi.org/10.1080/19490976.2024.2367297

Zhang, F., Aschenbrenner, D., Yoo, J. Y., & Zuo, T. (2022a). The gut mycobiome in health, disease, and clinical applications in association with the gut bacterial microbiome assembly. *The Lancet Microbe*, 3(12), e969–e983. https://doi.org/10.1016/s2666-5247(22)00203-8

Zhang, M., Zhang, T., Yu, M., Chen, Y.-L., & Jin, M. (2022b). The life cycle transitions of temperate phages: Regulating factors and potential ecological implications. *Viruses*, *14*(9), 1904. https://doi.org/10.3390/v14091904

Zhang, P., Shi, H., Guo, R., Li, L., Guo, X., Yang, H., Chang, D., Cheng, Y., Zhao, G., Li, S., Zhong, Q., Zhang, H., Zhao, P., Fu, C., Song, Y., Yang, L., Wang, Y., Zhang, Y., Jiang, J., … Ma, S. (2024). Metagenomic analysis reveals altered gut virome and diagnostic potential in pancreatic cancer. *Journal of Medical Virology*, 96(7), e29809. https://doi.org/10.1002/jmv.29809

Zhang, Y., & Wang, R. (2023). The human gut phageome: Composition, development, and alterations in disease. *Frontiers in Microbiology*, 14, 1213625. https://doi.org/10.3389/fmicb.2023.1213625

Zhao, G., Wu, G., Lim, E. S., Droit, L., Krishnamurthy, S., Barouch, D. H., Virgin, H. W., & Wang, D. (2017). VirusSeeker, a computational pipeline for virus discovery and virome composition analysis. *Virology*, 503, 21–30. https://doi.org/10.1016/j.virol.2017.01.005

Zheng, T., Li, J., Ni, Y., Kang, K., Misiakou, M.-A., Imamovic, L., Chow, B. K. C., Rode, A. A., Bytzer, P., Sommer, M., & Panagiotou, G. (2019). Mining, analyzing, and integrating viral signals from metagenomic data. *Microbiome*, 7(1), 42. https://doi.org/10.1186/s40168-019-0657-y

Zuo, W., Michail, S., & Sun, F. (2022). Metagenomic analyses of multiple gut datasets revealed the association of phage signatures in colorectal cancer. *Frontiers in Cellular and Infection Microbiology*, *12*. https://doi.org/10.3389/fcimb.2022.918010

Index

Note: **Bold** page numbers refer to tables and *italic* page numbers refer to figures.

N

O

P

Q

R

U

V

W

Y

Z